KNOWING MANCHURIA

KNOWING MANCHURIA

Environments, the Senses,
and Natural Knowledge on
an Asian Borderland

RUTH ROGASKI

The University of Chicago Press
Chicago and London

The University of Chicago Press, Chicago 60637
The University of Chicago Press, Ltd., London
© 2022 by The University of Chicago
Published 2022
Printed in the United States of America

31 30 29 28 27 26 25 24 23 22   1 2 3 4 5

ISBN-13: 978-0-226-80965-6 (cloth)
ISBN-13: 978-0-226-81880-1 (e-book)
DOI: https://doi.org/10.7208/chicago/9780226818801.001.0001

Published with support of the Susan E. Abrams Fund

Library of Congress Cataloging-in-Publication Data
Names: Rogaski, Ruth, author.
Title: Knowing Manchuria : environments, the senses, and natural knowledge on an
    Asian borderland / Ruth Rogaski.
Description: Chicago : The University of Chicago Press, 2022. | Includes
    bibliographical references and index.
Identifiers: LCCN 2021050922 | ISBN 9780226809656 (cloth) | ISBN 9780226818801
    (e-book)
Subjects: LCSH: Natural history—China—Manchuria. | Borderlands—China—
    Manchuria. | Manchuria (China)—Description and travel.
Classification: LCC QH21.C6 R64 2022 | DDC 508.51/8—dc23/eng/20211022
LC record available at https://lccn.loc.gov/2021050922

CONTENTS

ILLUSTRATIONS

Color plates after page 216

THE FLYING VOLES OF GANNAN AND THE CHALLENGE OF KNOWING MANCHURIA'S NATURES

According to official reports, hundreds of rodents fell from the sky over Gannan county on the night of April 4, 1952. Villagers in this remote corner of northeastern China awoke the next morning to find sickly voles scattered in haystacks, piled on rooftops, and even squirming on beds next to slumbering women and children. With their practical knowledge of the natural flora and fauna of their environment, the villagers suspected that these rat-like animals were not native to the area. The government of the People's Republic of China had urged citizens to consider anything out of the ordinary in nature to be vectors of germ warfare launched by American planes from the Korean War front. Heeding government warnings, by noon the villagers had collected, killed, burned, and buried every vole they could find.[1]

The alleged vole drop was one of many bizarre manifestations of nature reported by the newly established government of the People's Republic of China in the early 1950s.[2] Oddly out-of-place natural objects had been discovered in multiple locations across a remarkably large distance: in addition to the airborne voles of Gannan, located in the Manchurian-Mongolian grasslands, there were masses of flies found on top of the snow at Shenyang in the Liao River basin, four hundred miles to the south, and strange leaves in Kuandian in the Changbai Mountains five hundred miles to the east. There were even swarms of spiders found in Dalian at the tip of the Liaodong Peninsula, a point seven hundred miles south of Gannan, a distance almost the same as that from New York to Atlanta. A map of these disparate points included in an investigative report (fig. 0.1) reveals the outline of a vast space sometimes known as Manchuria.

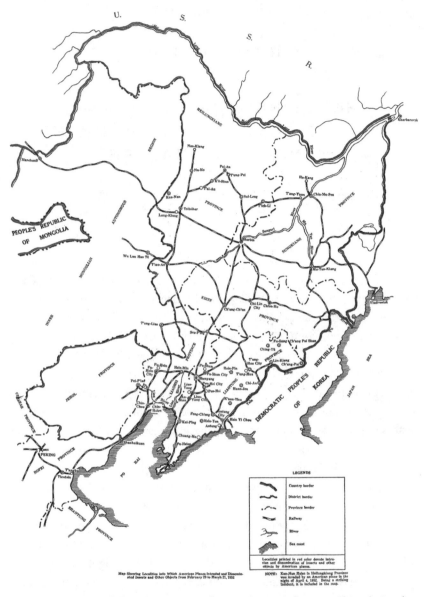

Fig. O.1 Locations of alleged US germ warfare attacks in northeastern China during the Korean War. *Report of the International Scientific Commission for the Investigation of the Facts Concerning Bacterial Warfare in China and Korea* (Peking, 1952), 165.

The Chinese government quickly mobilized the nation's top scientists to investigate the mysterious phenomena. Through painstaking study, the scientists confirmed what the Gannan farmers had already intuited: the voles, flies, leaves, spiders, rodents, and even bacteria found in 1952 did not belong to the region but were alien biological invaders dispatched by the forces of imperialism. The unknown voles that fell on Gannan resembled the naturally occurring local species but differed in the number of pads on their feet. The leaves discovered in Kuandian were from a type of tree found only in South Korea. Autopsies performed on humans in Shenyang who had died from mysterious diseases revealed bacteria never before encountered in that city.[3] With their confident marshaling of entomology, botany, zoology, bacteriology, and pathology, the scientists of New China demonstrated that they knew the nature of Manchuria. Through their knowledge, the People's Republic claimed the power to determine for itself which nature belonged within the borders of its territory and which was a foreign invader.

Located at the intersection of China, Mongolia, Russia, and Korea, known both as an "unknown frontier" and as a "cradle of conflict," Manchuria was a place of natural anomalies and violent contestation for centuries before the Korean War.[4] Indeed, the Manchu emperor Kangxi (r. 1666–1722), one of the most influential rulers of the former Qing empire (1636–1912), would not have been surprised to hear about the flying voles of Gannan. For Kangxi, the easternmost region of his domain (which extended hundreds of miles beyond the current PRC boundary) was a land where strange phenomena appeared as a matter of course. It was a place of fiery mountains and perpetual ice; a place of uncanny transformations where wood became stone, fish turned into rock, and sea creatures could morph into deer. Its dense forests hosted such a profuse diversity of plants and animals that they could not all be identified. In Manchuria, one could even find dragons coursing through the earth.[5]

Manchuria may have been a place of natural wonders, but at the same time it was a place of brutality and war. From the mid-seventeenth to the mid-twentieth century, the Qing empire, the Russian empire, the Japanese empire, Chosŏn Korea, the Republic of China, and the People's Republic of China all struggled for control of the territory. In addition to these polities, French, German, British, and American agents probed Manchuria's flora and fauna, plumbed its mines, and surveyed its fields in the hopes of exploiting its natural resources. By the twentieth century, battles over control of Manchuria helped spark World War II. Manchuria was a space where continent-spanning empires intersected and clashed; its natural resources fueled worldwide dreams of political expansion and economic exploitation. These projects of control frequently intersected with projects to comprehend Manchuria's perplexing environments. As the PRC scientists who investigated the phenomenon at Gannan could attest, knowing the nature of Manchuria was central to claiming possession of it.

Inspired by both the flying voles of Gannan and a Manchu emperor's curiosity, this study is a history of knowledge making about the environments of Manchuria from the seventeenth to the twentieth century. It explores the relationship between creating natural knowledge of a place and effecting political domination of a place, finding the seams where those endeavors overlapped and probing the gaps where they diverged. Because it was a highly contested borderland, Manchuria offers an opportunity to compare strategies of knowing nature that are associated with different polities. This study introduces a diverse cast of characters—including Chinese poets, Manchu shamans, Korean mathematicians, Russian botanists, Japanese bacteriologists, American paleontologists, and Nanai hunters—and considers how they made sense of Manchuria's terrain, flora, and fauna. Their work in Manchuria is not important because of any contributions to a single dominant stream of scientific knowledge at a distant metropole. Instead, Manchuria is a valuable site in and of itself, a place where we can gain new comparative perspectives on the creation of natural knowledge.

At its heart, this is a study of how humans came to know a place. As such, it tries not to privilege one form of knowledge over another. The chapters encompass bacteriological assays and dragon sightings, knowledge gained through wielding a telescope and that achieved by wielding a hoe. To encompass these different approaches, I present all forms of knowledge as the product of intimate, complex entanglements of individual humans and highly specific environments, and follow our actors as they engaged intensely with Manchuria's mountains, rivers, forests, and plains. This approach allows us to compare different ways of knowing without invoking a presumed universal standard of science. By recovering a multisited, multiactor history of knowledge creation, this study instead argues the importance of the local, the importance of examining the processes through which humans come to know the ground beneath our feet.

While this study is a history of ways of knowing a place, at the same time it is a history of the place itself. Manchuria has emerged over the past twenty years as a major focus of scholarly inquiry, and numerous works have explored its cultural, political, and social history.[6] This rich scholarship has marked

Manchuria as a place of importance in the history of East Asia, in the history of empire, and ultimately in the history of modernity. In spite of its crucial role in world history, the nature of Manchuria itself—its expanse, its terrain, its environments—remains elusive. It is difficult to grasp a sense of space and distance, a sense of scale, a sense of the environment that shaped (and was shaped by) human activities.[7] This work brings the nonhuman world back in as a crucial participant in the shaping of the human history of northeast Asia.

Probing the intersection of environment and human perception in Manchuria throws into stark relief how ways of knowing shifted in concert with political and economic change. As polities wrestled control from each other, sunk borders into the land, and above all, sought to intensify the extraction of wealth from the environment, what constituted the "natural wonders" of Manchuria shifted from furs, herbs, and *qi* energy to mass-produced cash crops and fossil fuels. Natural knowledge, and thus the nature of Manchuria itself, changed from a "land where the dragon arose" to a global center of strip-mining and contagious disease. This shift was accompanied, quite literally, by a different way of perceiving the environment, an ever-intensifying "regime of attention" that looked for minute differences in fossilized shells in order to locate mineral deposits or that obsessively traced the invisible movements of bacteria so they could be turned into agents of war.[8] Such approaches may have stripped enchantment from certain parts of the world, but they allowed for the rise of different sorts of fantasies, ones centered on the endless accumulation of national wealth and power. As a region where Asian actors predominate, what emerges when we look at Manchuria over the *longue durée* is not a straightforward tale of Western domination of indigenous groups, or an East-versus-West story. Through a focus on Manchuria, we can see how all empires deployed knowledge toward goals of universalism and supremacy while other ways of knowing were appropriated, repressed, or lost.

When farmers and scientists at an insignificant locale on a vast prairie tried to make sense of small rodents in 1952, they were part of a centuries-long process of knowing Manchuria: inheritors of a legacy of war and wonderment on a harsh but politically crucial northern frontier. To understand how they and other actors came to know Manchuria, I deploy three intertwined approaches: an attention to the specifics of space, an openness to different forms of knowledge, and a curiosity about the role of the senses in creating human understanding of the environment.

The Multiple Spaces of Manchuria

This book is about Manchuria, but the very concept of Manchuria itself poses multiple dilemmas. "Manchuria" is the name frequently used to describe a part of northeast Asia once associated with the Manchu people, scattered groups of non-Chinese farmers and hunters living in the region of the Changbai Mountains who united to form China's last imperial dynasty, the Qing (1636–1912). Today, Manchuria is often thought of as being the equivalent of the PRC's three northeasternmost provinces: Liaoning, Jilin, and Heilongjiang, collectively referred to in Chinese as "The Northeast" (Dongbei). This Northeast is shaped by significant borders. To the southwest, one man-made border—the Great Wall—has served to separate this region from the plains and mountains of "China proper" for centuries. While geographically indistinct, traditional Chinese phrases meaning "beyond the Wall" (*saiwai, guanwai*) reference this location as a frontier from the perspective of China. Elsewhere, other, more seemingly natural border objects sketch out its shape. To the southeast, the border with North Korea is formed by rivers that emerge from the slopes of Mount Paektu (Chinese, Changbaishan, or Long White Mountain), a massive stratovolcano that straddles the border between the PRC and North Korea. To the north and east it is separated from Russia by the main flow and tributaries of the Amur River (Chinese, Heilongjiang, or Black Dragon River). These two landmarks are sometimes used to define Manchuria: in Chinese, "White Mountain/Black Water" (*Baishan heishui*) is a frequently invoked metaphorical name that stands for the entire region.

While the "White Mountains and Black Waters" seem to form natural borders embedded in the very earth, the outline of something that grew to be called Manchuria has shifted across the centuries, making the space of this imagined territory difficult to establish with cartographic certainty. The history of this part of northeast Asia is complex in the extreme, and multiple polities have laid claim to parts of its terrain. To introduce Manchuria to contemporary readers, we might try to imagine it as a territory as environmentally diverse as the United States west of the Rockies, contested by political entities far more culturally and linguistically divergent than Western European empires.

Manchuria can be conceptualized as the homeland of the Manchus, but this homeland was never a stable entity. Once it moved the court to Beijing and established rule over a Han Chinese majority, the Manchu dynasty pushed the borders of its "homeland" hundreds of miles north of its original tribal territories, incorporated other indigenous peoples of the region into its ranks, and sought to preserve the territory from the Great Wall to Siberia as a specially administered preserve.[9] In the seventeenth and eighteenth centuries, the Qing court created distinct borders to the south, east, and north designed to prevent Han Chinese agriculturalists, Korean foragers, and the Russian military from accessing the region's valuable natural resources, which included ginseng, furs, forest products, and agricultural land.[10]

The rise of new modernizing empires in Asia and the decline of the Qing in the nineteenth and twentieth centuries radically altered the shape and fate of this region. By the mid-nineteenth century, imperial Russia formally annexed large swaths of northern and eastern Manchuria. As the Qing loosened control, increasing numbers of Koreans and Han Chinese settled in Manchuria's heartlands, and by the time of the Qing collapse in 1912, Han Chinese migrants from regions south of the Great Wall formed the vast majority of the population. After 1912, the central government of the Republic of China struggled to establish control over the region in the face of political, economic, and military involvement by other Eurasian polities and the establishment of semi-independent Chinese warlord regimes.[11] Attracted by the area's natural resources—which included large deposits of coal and oil—Japan made increasing inroads into Manchuria in the late nineteenth and early twentieth centuries, fighting against the Qing and Russian empires for influence in the region. In 1931, Japanese forces occupied all of Manchuria, an event that some see as a precursor to World War II.[12] Japan ruled the region from 1932 to 1945 via the puppet state of Manchukuo—the only time a distinct "Manchuria" was defined through "international" boundaries.[13] With the defeat of Japan at the end of WWII, competing Chinese factions claimed the territory, and in the late 1940s, the Chinese Communists used Manchuria as a power base to launch their successful war against Chiang Kai-shek's Nationalists.[14] By the mid-twentieth century, much of this long-contested frontier had become an inviolable part of the People's Republic of China, and using the very moniker "Manchuria" to designate a region with a separate identity and history became (and still is) a politically suspect enterprise.[15]

Many of those working on the history of Manchuria, including myself, are taking up questions that were originally established in the pioneering work of Mark Elliott. In his seminal essay "The Limits of Tartary," Elliott demonstrated how Manchuria had been mapped and imagined as a place with a singular identity beginning with the efforts of the Qing court in the seventeenth century. The creation of a Manchuria distinct from China south of the

Great Wall, sometimes depicted by Chinese nationalists as a separatist scheme devised by imperialist powers (particularly Japan), was in fact a nativist project conceived by the Manchu rulers of China itself. This project of making Manchuria required multiple strategies of ritual, literature, administration, and cartography. Elliott highlighted how Manchuria went from "undifferentiated frontier" to a "bounded geographical place" imbued with cultural significance, a process that he summarized, with a nod to Yi-Fu Tuan, as a transition from "space to place."[16]

This work starts from the premise that Manchuria is not one place but many. While Manchuria has been readily conceptualized as a bordered region in the cartographic imagination, the very vastness of Manchuria's space makes it difficult to conceptualize it as a singular entity. The map in figure 0.2 attempts to convey both the geographical expanse and the environmental complexity of this space. If we think of Manchuria defined as the northeasternmost territories of the Qing empire at its height, the region covered over five hundred thousand square miles, from the Great Wall outside of Beijing to the Stanovoy Mountains in Siberia, and from the Pacific Ocean to the Mongolian Plateau. To envision this expanse, we can see it as large as France, Germany, and Poland combined, or compare it to all the land in the United States east of the Mississippi (minus Florida), or perhaps think of it as almost twice the size of Texas. The distance from the southernmost point at the tip of Liaodong Peninsula to its northernmost regions above Sakhalin Island is over 1,400 miles, or about the same distance as Miami to Montreal; from east to west, from the Pacific coast at Vladivostok to the Central Asian origins of the Heilongjiang (Amur) River is an expanse of another thousand miles, about the distance from New York to Minneapolis.

Given these vast distances, it should be no surprise that the area of what can be thought of as Manchuria encompasses a diversity of environments. The region might be known through the shorthand of "White Mountain/ Black Water," but this phrase belies its environmental complexity: its expanse includes arctic taiga, prairie grasslands, coastal rainforests, semiarid deserts, and alluvial wetlands. Given this environmental complexity, we might well ask whether Manchuria holds together as a region at all. Indeed, if we follow Sue Naquin's suggestion that we define regions by the way natural resources (mineral, plant) shape "the material world of ordinary people," then Manchuria is not one place but many.[17] Perhaps Manchuria can be seen, following Kate Brown, as "no place," a borderland with no definite boundaries whose multiple environments "shaped the land into an enigma—untidy, formless,

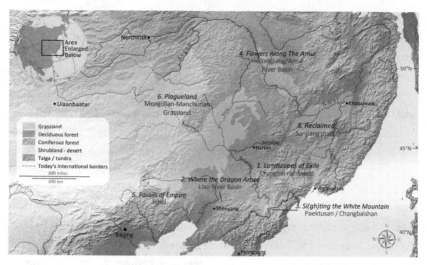

Fig. O.2 Manchuria's multiple natures: Chapter locations and environments.
Map by Jeff Blossom. See plate 13 for color version.

eluding definition."[18] The first challenge to knowing Manchuria, then, is imagining Manchuria as a composite portrait of multiple landscapes: not just an imagined abstraction of the "White Mountain/Black Water" but an amalgam of multiple local environments pressed into being by rival states, stitched together by military routes and rail lines, and formed into a singular dream by those who coveted its resources.

Knowing Manchuria disaggregates this imagined whole to show the great efforts needed to attempt its construction. Accordingly, each chapter is set in a different site that reflects Manchuria's environmental diversity. The map in fig. 0.2 provides a geographical guide to the locations of the chapters. Chapters 1 and 2 are centered on the rain forests of the Changbai mountain range and the valleys of the Liao River basin, environments that constituted the early power bases of the Manchus. Chapter 3 focuses on Manchuria's famous "White Mountain" (known in Korean as Paektusan and in Chinese as Changbaishan), a 9,000-foot-tall stratovolcano on the current border between North Korea and the PRC. Chapter 4 follows the botanical landscapes along both banks of the "Black Water"—the Black Dragon River (Chinese, Heilongjiang; Russian, Amur) that extends from Central Asia to the Pacific Ocean and serves as the current border between the PRC and Russia. Chapter 5 is centered on a borderland within a borderland: the coal-bearing, arid hills of Jehol, sandwiched between Beijing and Manchuria. Chapters 6 and 7 take us to the grasslands that link Manchuria with the Mongolian Plateau. Chapter 8 centers on the alluvial wetland delta formed by the intersection of the region's great Three Rivers (Sanjiang)—the Ussuri, Songhua, and Heilongjiang—as they wind their way to the Pacific. Together, these sites reflect the expanse of what might be considered Manchuria to the cardinal directions: north to Siberia, south along the Bohai Gulf, east toward the Pacific Ocean, and west to Central Asia. Significantly, many of these environments were not (and are still not) contained within the confines of one national boundary. Mountains and rivers may have been bifurcated by boundaries, but human vision and experience transcended those lines. Manchuria's many natures—its white-capped mountains, dense green forests, serpentine rivers, and even its un-seen and subterranean entities, from "dragon veins" to bacteria—traveled and expanded between and beyond political borders. This study, while primarily sited in terrain that is now China, considers how people on different sides of the "mountains and waters" attempted to create meaningful landscapes from the region's multiple natures.[19]

By deconstructing Manchuria as a monolith and focusing on specific border-transcending landscapes, this study argues for the importance of the experience of local environments. Manchuria—writ large—holds a central place in the historiography of East Asia. As early as the 1930s, Owen Latti-more argued for the centrality of Manchuria as a way to understand sweeping trends in Chinese history.[20] Manchuria was central to Korea's visions of an expansive past, Russia's faded dreams of a thriving Far East, and, most impor-tantly, to Japan's imperial debacle. This study turns our focus to the particulars of local terrain in shaping the process whereby the frontier was known and desired by East Asia's centralizing powers. At each local site, multiple actors—some sojourning from metropole, others traversing the ground of home—encountered and made sense of a specific place, their experience shaped by the subtle incline of a slope or the presence of wildflowers in a field.

By recovering the local, this study hopes to do for borderland terrain what other scholars have done for borderland identities. In her important work on identity in Qing imperial ideology, *Translucent Mirror*, Pamela Crossley insists that early modern local cultures in Asian borderlands were not just combinations of larger identities (Han, Manchu, Korean) as defined by cen-tralizing states. By drawing our attention to the shape of local societies, using terms that evoke local places (like "Liaodongese"), Crossley hopes to escape what she calls the "slough of hybridity," the tendency to see the local only as the condition of being positioned in between larger entities of greater value.

For Crossley, local cultures were not "hybrids"; they were in and of themselves "coherent . . . with a history and a discrete geographical contour."[21] By focusing strongly on the space of specific sites, this study sees places not as always-already native parts of larger polities, but as sites that become so through processes of "hierarchy, aggression, allegiance, and submission" used to impose order.[22] Manchuria was an assemblage created by empires, its many natures forced to cohere. By writing a disaggregated composite portrait of its multiple natures, I avoid seeing it as a "timeless and spaceless abstraction" and trace the process of how knowing its natures was used to bring Manchuria into being.[23] In this process, we see how intersections of empire and natural knowledge made Manchuria the site of tragic environmental and human disasters: Manchuria is, after all, the home of Asia's largest open-pit coal mine and the birthplace of modern germ warfare. At the same time, there are intimate, local spaces in Manchuria's many terrains that might open up hopeful glimpses into different ways of knowing the nonhuman world.

From "Science and Empire" to Natural Knowledges and Local Environments

Knowledge of nature plays a role in the creation of expansive polities. This basic premise has long been at the foundation of scholarship on "science and empire." In classic modes of this scholarship, agents from Europe journeyed to non-European shores; collected, categorized, and mapped the flora, fauna, and terrain; and used this knowledge to establish colonial governance over indigenes and to extract resources for use by the metropole. Since the 1980s, such science-and-empire narratives have frequently been used to critique the domination and exploitation of Western powers over non-Western places.[24] In recent years, the history of science has moved away from science-and-empire binaries of dominance—Europeans as active creators of knowledge and natives as passive recipients—to recognize the role of non-Western indigenous agents in the creation of what becomes thought of as "Western" science.[25] But how are we to think of the relationship between frontier environments and the polities that sought to control them when the agents doing the exploring, categorizing, and mapping were not European but Asian? In explorations of Manchuria, elites from Beijing, Seoul, and Tokyo wielded tools of scientific inquiry on sites they perceived as frontiers and in pursuit of projects of political consolidation independent of direct Western involvement.[26] This perspective of Asians both as creators of natural knowledge and as active makers of empire differs considerably from work that recovers non-European participation within European-led projects of knowledge creation.

Furthermore, when one takes a long-term perspective on knowledge making in Manchuria, the position of imperial and indigenous identities becomes blurred, leaving us to ponder who is a local and who is an outsider. Russian botanists arrived on the banks of the Amur River from St. Petersburg via a six-month ocean journey around South America and then across the Pacific, but even ostensibly "indigenous" Manchu or Korean elites from Beijing or Seoul embarked on arduous overland journeys of hundreds of miles to reach the same locations, all the while perceiving of Manchuria's fringes as distant and remote destinations. To explore these frontier environments, all comers, no matter where their point of origin, had to employ the help of local guides: hunters and foragers from indigenous groups who straddled borders in northeast Asia, alternately evading or assisting different regional authorities depending on the circumstances. As East Asian polities became incorporated into a Western-dominated global system of empires and nation-states in the nineteenth and twentieth centuries, what once counted as "imperial" knowledge within Asia could take the position of "the local" as Asian polities were reconfigured as indigenes in the eyes of Western imperial powers. The result was a series of "nested" imperialisms that encompassed the landscape in layered and interpenetrating hierarchies. The creation of natural knowledge in Manchuria was a complex process in which the identity of who was "imperial" and who was "native" was not stable across time.[27]

Finally, with a few important exceptions, typical studies of knowledge creation focus on agents from one empire: Spanish, Dutch, British.[28] Manchuria was a borderland so complex that over the centuries we must take into account over half a dozen different polities: the Qing, Russian, and Japanese empires; the kingdom of Chosŏn Korea; the Chinese and Soviet republics; in addition to the involvement variously of Great Britain, France, Germany, and the United States. Manchuria joins places such as the Caribbean and the "middle grounds" of North America as a meeting place of multiple competing polities, but those involved in contesting Manchuria were more divergent—culturally, linguistically, politically—than any other example we find in the current scholarship. Ultimately, Manchuria is a zone that obliterates the possibility of binary approaches and blurs distinctions between the indigenous and the imperialist. Positioned between multiple competing polities, Manchuria is a particularly fecund example of what Fa-ti Fan has called a "cultural borderland," a place where knowledge was created through complex transactions "between metropole and colony, between colonies, and among Europeans, creoles and autochthons"—but without Europeans playing a predominant role.[29]

If we are trying to understand how different polities came to know Manchuria, how, then, are we to accommodate knowledge of nature that is created by such diverse entities? How can we think about human-environment interaction from the early modern period to the twentieth century, across multiple polities, multiple actors?

One promising technique is to shift the location of our attention. Instead of focusing on actors and activities emanating from one metropole, this study maintains a focus on specific sites in the frontier itself—Manchuria's mountains, rivers, wetlands, and plains—and considers actors from different locations as they walk on, across, and off these singular stages. This focus on sites follows the spirit of recent history of science that seeks to "put knowledge in its place."[30] But while this scholarship uses space to illuminate the nature of scientific knowledge, this study uses space to illuminate both the nature of knowledge and the nature of a place: to see how environments shaped human knowledge and how human knowledge in turn shaped specific environments. Like many recent works on the United States and Europe, *Knowing Manchuria* navigates the "shared ground" between environmental history and the history of science.[31]

Exploring this shared ground requires a more expansive way of thinking about what it means to know nature. Many of the actors considered in this study engaged in activities clearly seen as branches of science, including cartography, botany, and bacteriology. Yet instrumental knowledge about nature was not strictly confined to what we recognize as science: early modern Han Chinese literati applied exacting language to create natural histories of Manchuria's forests through poetry; Qing officials categorized flora and fauna of the Heilongjiang River through the principles of Chinese medicine; while borderland hunters, foragers, and shamans all produced different kinds of mental maps of the White Mountain. To encompass these and other ways of knowing Manchuria, I frequently use the term "natural knowledge." There is nothing fancy about this term: the historian Cameron Strang straightforwardly defines "natural knowledge" as "knowledge that humans develop about nature" and calmly notes that it includes "practices other than those that we would comfortably call scientific."[32] A shift from science to knowledge allows the historian to include the work of multiple actors, a greater diversity of perspectives, and a more expansive range of time periods. It opens us up to consider what the historian of science James Delbourgo has called "the Knowing World," a more global history that intentionally "destabilizes any particular brand of knowledge's supposed domination."[33]

A consideration of natural knowledge allows a chipping away at ready-made binaries that structure the world into easily defined terms of East and West, rational versus spiritual, colonizer and colonized, but it does not mean an abandonment of considerations of power. It allows us to examine both the physical techniques of a particular way of knowing and the stories and dreams that inspire these techniques: a Qing emperor's dream of tapping into the cosmological power of the earth by supplicating the gods, or a Japanese paleontologist's dream of extracting endless energy from the earth by examining fossils. The reason I do this—put different types of knowledge on a level playing field—is exhaustion from the crises of climate and earth harm that science has put us in, an exhaustion perhaps felt more intensely by those who work on the history of Asia. This exhaustion has in turn kindled, for Asianists as for those working on other environments on other continents, a renewed sense of the value within other ways of knowing the world.[34] The turn to natural knowledge is done with an understanding that some forms of knowing have been (but did not *have* to be) more closely linked to extreme political domination and environmental destruction than others. Attention to natural knowledge also entails a hope that the past might provide us new, less destructive possibilities for engaging with the world in the present.

Using natural knowledge requires that all forms of knowledge creation be viewed in the same way; evaluated, to the extent possible, on the same scale. If we focus on the dynamic spaces where the human body and the environment meet, it is possible to see all forms of knowledge as stemming from different "regimes of attention"—different modes of deploying the human senses in engagement with the environment.[35]

Knowledge and the Senses

As David Turnbull observed in his description of the "traveling turn" in the history of science, "all knowledge is constructed at specific sites through embodied engagements."[36] This study frames the construction of knowledge about Manchuria's natures as a kinesthetic project, emphasizing the knowledge produced by the senses as humans moved through particular terrain. In this regard, I adhere to Richard White's simple but insightful assertion: "If space is the question then movement is the answer."[37] I chose this approach because I intuitively feel that this is how history is best done: by focusing in on a particular scene and placing ourselves, to the best of our abilities, within the shoes of the historical agents we study, to see what they see, pay attention to what they feel.

Ultimately this close attention to the intersection of human senses and terrain illuminates the role of the physical environment in shaping knowledge. Walking through a field of thousands of wildflowers along the Amur River brought botanists face-to-face with the challenge of diversity. The ability of *Yersinia pestis*—the microorganism that causes the plague—to thrive within a wide range of mammals in the Manchurian-Mongolian grasslands drove scientists to disembowel tens of thousands of bodies in order to peer inside them. The arrangement of dense forests and the remnants of past volcanic eruptions rendered the White Mountain invisible to the surveyors who tried to locate it. While some debate whether nonhuman entities have agency or whether human perception is independent of historical context, my goal is simply to view humans as intimately embedded in and inseparable from the nonhuman landscape.[38]

To imagine this embeddedness, I used a variety of approaches and sources. First and foremost are the words of the observers themselves as they narrated their experiences in essays, field notes, scientific publications, or poetry. The chapters scrutinize Chinese poetry, Korean travelogues, Japanese scientific publications, and Manchu shaman songs to see what each says about human engagement with the nonhuman world. I also made use of the digital humanities tool kit, frequently referencing topographical maps, historical digital data, and approaches such as viewshed analysis, which attempts to re-create lines of vision of historical actors in their settings.[39] To supplement these approaches, I used my own body. I traveled to the sites under consideration and placed myself as closely as I could to the environments traversed by historical actors, recognizing that both environments and styles of human experience had inevitably changed but holding the hope that some connection to the past still remained. Approaching Mount Paektu from different directions (except, unfortunately, from the southeastern, or North Korean, side) helped me imagine what it might be like as an early explorer trying to find the mountain in the midst of the wilderness. Walking the endless rice fields of the Sanjiang Plain and boating through the small patches that are left of its original wetlands (once as large as the state of Tennessee) helped me recognize the enormous human effort required in the agricultural transformation of Heilongjiang. Hiking up dusty roads to the exposed-shale outcroppings of Yi County in what was once known as Jehol revealed the rocky remnants of a once-lush Cretaceous-era rain forest. Splitting open a piece of shale in my hands to uncover a tiny, exquisitely detailed fish fossil delighted me in much the same way described by the Kangxi emperor in his jottings three hundred years ago.

Throughout the book, I try to adhere as closely as I can to the seam between the human senses and nature: following closely as a paleontologist counts the tiny whorls in an ancient snail shell or as an explorer rafts down the rapids of a rain-swollen river. The sense of sight is foremost, for it captures a wide array of experience, from a surveyor squinting into a telescope to the ritual vision-flights of shamans. Vision is also a critical topic for understanding the significance of changes in knowledge over time. In particular, it allows us to appreciate the radical disjuncture that is the anatomical and atomized vision of modern science, whether that of a botanist scrutinizing the minute parts of a dissected flower or a physician searching for bacteria in the bloody recesses of a dissected corpse.[40]

Although vision is privileged, other senses are considered, including those that go beyond the standard five. Human movement and work involving multiple senses can generate important ways of knowing. Inspired by the anthropologist Tim Ingold's concept of dwelling perspective, this study frequently pauses to consider human skills and perceptions that rise out of skilled everyday living within a specific environment. Such sensibilities are sited throughout the body, the product of a combination of senses or something that is felt "in the flesh." They might be thought of (or dismissed) as "intuition," but they frame the knowledge used in both artisanal skills and in crucial activities of survival.[41] Examples here include the embodied knowledge generated by hunters who gauged distance by the number of days it took to carry a heavy deer carcass back to camp, foragers who could "read" an environment as they walked, or laborers who understood the nature of Manchuria by dragging heavy iron plows through its intractable mud. While it can be difficult to tease these forms of "kinesthetic knowing" from texts that privilege the visual, the chapters try to pause to consider forms of knowing based on bodily movement, knowledge that incorporates sensations at the very intersection where the human and nonhuman meet.

Mapping This Book

The chapters weave together these three threads—specific sites, multiple polities, and the embodied experiences of individual actors—to contemplate how

Manchuria's many natures became known. Chapters 1 and 2, "Landscape of Exile" and "Where the Dragon Arose," examine two dramatically different descriptions of similar journeys from China south of the Great Wall into Manchuria: the first undertaken by a southern Chinese poet exiled to a frontier outpost in 1659, and the second an expedition of the Manchu emperor's imperial cortege almost twenty-five years later. Together the chapters examine how Chinese and Manchus (along with an occasional European) used different forms of poetry, history, and objectivity to detect two different landscapes, one a tragic home of sorrowful ghosts, the other triumphant and brimming with sacred energies.

The next two chapters together consider how humans made sense of Manchuria's tallest mountain, Paektusan (chapter 3, "Si(gh)ting the White Mountain"), and its longest river, the Amur (chapter 4, "Flowers along the Amur") in the eighteenth and nineteenth centuries. We find elites from Seoul, Beijing, and St. Petersburg traveling to these sites with scientific instruments and a concern for border making, but it was the expert and embodied knowledge of "unbordered" local inhabitants—ginseng foragers, hunters, and shamans—that guided elites to produce knowledge that combined exact measurements with the power of gods.

Chapters 5 and 6 explore knowledge making in the grim pursuit of other forms of power in the twentieth century. This section ponders the roles of two different sciences, paleontology and bacteriology, in the context of intense imperial competition for Manchuria's modern natural resources. In "The Fossils of Jehol" we see how the meaning of the unique stone creatures of southwestern Manchuria (the Jehol biota) shifted with the arrival of the modern desire to consume fossil fuels. "Plagueland" probes the often horrifying ways that Japanese and Chinese scientists linked the bacterial, rodent, and human environments of the Mongolian-Manchurian grasslands in their pursuit of the bacillus that causes the plague: scientific legacies that played a role in the medical atrocities perpetrated by Unit 731, the biological warfare organization developed by the Japanese Imperial Army during World War II.

A brief interlude ("Scientific Redemption," chapter 7) revisits the flying voles of Gannan that began the introduction, and interprets the PRC's germ warfare investigations as a way of redeeming Manchuria's nature from its association with conquest, degradation, and disease—a strange but effective way to stake the flag of a modern, scientific "New China" into the bordered soil of Manchuria. Finally, chapter 8, "Reclaimed," considers how a generation of Chinese youth sent to the border with Russia from the 1950s to the 1970s transformed China's largest freshwater wetland into China's most productive farmland using shovels, hoes, and tractors. By positioning this chapter last, I suggest that for today's China, knowing the nature of Manchuria was done through a combination of technology and the muscular, intensely physical sacrifices of its youth.

The chronology of the study, ranging from the seventeenth to the twentieth century, suggests an arc of change over time. One of the distinguishing characteristics of natural knowledge about Manchuria before the twentieth century was the explicit, joint pursuit of the goals of enchantment and precision. Natural history and cartography combined with mythological histories, a sense of energy in the earth, and ritual to create an enchanted and bounded place. Telescopes and quadrants were employed in projects that included the siting of geomantic dragon veins. This form of knowledge was clearly used toward projects of empire making and the violent enforcement of borders, but something sacred lingered within. Modernity may have stripped a certain kind of enchantment from this part of the world, but it was predicated on the rise of a different sort of fantasy, one centered on endless growth and the prosecution of total, globe-spanning war. This shift was accompanied, quite literally, by a different way of sensing nature, one that looked for minute differences in a shell in order to locate oil or opened thousands of bodies in order to track disease.

As convincingly detailed in pioneering environmental histories by David Bello and Jonathan Schlesinger, Manchuria had long been a site of organized natural resource extraction and was clearly being drained of its abundance in fur-bearing animals, pearl-bearing mollusks, and wild medicinal plants as early as the seventeenth century.[42] But by the late nineteenth century, there was a shift to a different approach to extraction, from commodities to meet the demand of court elites and luxury markets to fossil fuels and agriculture to meet the demand of industry and mass armies. As they did everywhere to frontiers around the world, nation building, capitalism, and global markets led to an intensification of the plunder of Manchuria's natures. There is something distinctive about Manchuria, however. As the historian of technology Victor Seow has observed in his pioneering study of the Fushun mine in Manchuria, with the advent of the twentieth century—or perhaps more importantly, the advent of capitalist penetration and the extraction of fossil fuels—state-backed actors ripped open Manchuria's dragon veins in order to unearth coal seams.[43] Perhaps, however, if we look closely enough, we might find that dragon veins still exist.

This history of knowing Manchuria does not establish an inevitable (and continuing) arc of destruction. Indeed, this multisited history may open up spaces for thinking about other ways of being and knowing. The linking of the disparate moments and distant spaces considered here results in a portrait of Manchuria that is admittedly a collage. Paul Carter, in his pathbreaking spatial history *Road to Botany Bay,* acknowledged the nonlinear, assembled nature of spatialized writing, likening it to the production of "unfinished maps."[44] For Carter, this nonlinear approach, as it "makes no claim to authoritative completeness," is better able to shine light on the possible alternative roads embedded within the past. Spatial history avoids "organizing its subject matter into a nationalist enterprise" and thus is capable of "suggesting the plurality of historical directions."[45] It is my hope that through this spatial assemblage I might convey to readers a strong sense of the stunning expanse, abundance, and diversity of Manchuria's environments while also suggesting the possibilities of different ways of human engagement with the environment that are embedded in Manchuria's past.

Ultimately, by following those who engaged with the natures of Manchuria, I hope to animate a different mode of thinking about space in East Asia, one that is not limited by the nation-state. This mode acknowledges that the tragedies of the recent past are embedded in the very land itself. At the same time, it calls into question and challenges the insistence on territorial domination that resulted in these tragedies. Through attention to locales and to those who transcended borders, I see a multiplicity of roads. This study does not flinch from histories of imperialism and conquest, but in response to those histories, it strives to reclaim the full measure of human ways of connecting with the environment. My goal is to encourage a sense of wonder about embodied engagement with the world, an engagement possible when we are cognizant of political borders but not bound by them.

LANDSCAPES OF EXILE

Nostalgia and Natural History on the Journey to Ningguta

> To know these landscapes beyond the realm—what the ears and eyes
> cannot fathom, what the mind cannot measure, what cannot be named
> and categorized—is it not the case that Heaven has left it all to await
> the exiled official?
> —ZHANG JINYAN, *Ningguta shanshui ji*
> (A Record of the Ningguta Landscape, 1668)

On a chilly March morning in 1659, a small group of prisoners exited Beijing's massive city walls under armed guard and began a long journey of exile. Their destination was Ningguta, a frontier outpost deep in the forests of Manchuria, eight hundred miles to the northeast. The prisoners included the young Chinese poet Wu Zhaoqian (1631–1684), one of the most brilliant minds of his generation, along with several other high-ranking officials and prominent literati who had been sentenced to death (commuted to exile) for corruption in the imperial examinations. Torn from the familiar climes of their ancestral homes in southern China, after an arduous four-month journey Wu and his fellow exiles found themselves in a strange place, a land where the environment was extreme and unknown. It was also, in their eyes, a land without history—a vast region whose mountains and rivers had not been cataloged in the annals of past dynasties.

This chapter follows Wu and his fellow exiles as they tried to make sense of the environments they encountered in what they called the *jue yu*: the "extreme region" or "the territory on the edge." Once they arrived at their final destination of Ningguta, the exiles perceived themselves to be in a place beyond the boundaries of what was knowable. They struggled to describe their unfamiliar surroundings using familiar templates, and when those failed, they searched for new ways to describe and categorize what they saw. As they created knowledge about the nonhuman world, the exiles attempted to fit Manchuria within the borders of the human world—to determine, through their perception of nature, exactly where this place belonged. Was it part of China, a foreign land, or the extreme edges of the world? The exiles' writings reveal the problem of creating natural knowledge at the fringes of the empire.

We will begin by following their long journey of exile to the northeast, across mountains, rivers, settlements, and the distances that separated them. By placing the exiles within specific sites and moving with them, we can gain a kinesthetic sense of the multiple spaces of what was to become known as "Manchuria." The path taken by the exiles was a well-defined route used for centuries by emissaries and armies traveling between China proper and the northeastern borderlands. Many of today's superhighways and high-speed rail systems follow the same path, and those of us who have lived and worked in China have unwittingly replicated the exiles' journey, hurtling through the terrain at hundreds of kilometers per hour. But we cannot take this space for granted. As we see Wu's progress unfold at the leisurely pace of an oxcart or simply at the pace of human perambulation, the terrain takes on different meanings. Slower speed produces more detailed observations of the environment and intensifies the activity of place making. As observed by Tim Ingold and Jo Lee, the simple act of walking "fosters a realization of emotional and environmental conditions . . . situated somewhere between an external outward-looking vision and an internal self-reflective vision."[1] By walking northeast with the exiles, we see Manchuria unfold as a process of discovery: the meeting of human perception with the physical presence of an alien territory.

Indeed, we can see the journey of Wu and his colleagues as an unsung part of a larger global age of discovery. As Peter Perdue has suggested, exiled Qing scholars can be seen as an Asian equivalent of European early modern explorers, literate adventurers who journeyed far from home and made sense of distant, exotic landscapes for audiences in the metropole. Instead of embarking for strange places aboard oceangoing ships, however, Chinese exiles for the most part journeyed overland (by foot, on horseback, on river barges, and perhaps by ignoble oxcart), and their explorations were continental: the deserts of Central Asia, the tropics of southeast Asia, and the boreal forests of northeast Asia.[2]

The men that departed Beijing's gates in 1659 made for strange explorers: Wu Zhaoqian was a twenty-eight-year-old literary celebrity, a young man accustom to a silk-clad life in the heart of southern China's most refined cities. His older companions, also sophisticated southerners, were erudite scholars and powerful court officials. They traveled to new, exotic lands, but their fate as exiles was an old, familiar role. Indeed, for almost two millennia, scholars who ran afoul of emperors had been banished to the frontier regions of China's ever-shifting imperial expanse. Whether they were sent to the foothills of the Himalayas or the tropics of Vietnam, such men had anchored Chinese understandings of nature at remote sites across multiple dynasties.[3] The irony in the early Qing was that the newly ascendant Manchu emperors, foreign rulers of a conquest dynasty, punished Chinese by banishing them to the dynasty's homeland. This "new world" the exiles encountered thus had a unique and confusing character: simultaneously a remote "foreign" frontier, yet at the same time part of "their own" country.

Like their early modern European counterparts, early Qing exiles had to make sense of new natures using old traditions. In the sixteen and seventeenth centuries, Europeans experienced a "gusher of novelty" in objects as empires expanded to encompass the marvels of the New World.[4] To make sense of the strange things they encountered—from trees that cured fever to lizards that swam like fish—European scholars triangulated between received texts and direct observation.[5] As scholars sought to make sense of these "things without names," they turned to their books for guidance. They combed the natural histories of the ancient Greeks, biblical treatises, materia medica, travel literature, bestiaries, and compendia of wonders, seeking to identify the new through references to familiar objects from home. Ultimately, what Tony Grafton has called the "shock of discovery" in Europe helped to move the needle away from a reliance on texts in the direction of direct empirical observation and description.[6] But while observers began to feel more comfortable taking nature and not books as the focus of study, classical texts continued to influence knowledge about nature well into the seventeenth century, the result of "accommodation with texts rather than wholesale innovation."[7]

Seventeenth-century Chinese exiles also experienced a "shock of discovery" as their empire expanded—encountering phenomena such as strange plants, astonishing snowfalls, and skies that stayed light until midnight. As they tried to make sense of their new environment, exiles exercised new empirical skills of observation and even strove for what we can see as a certain kind of objectivity. At the same time, their perceptions were also informed by a number of received templates. Foremost among these templates was an aesthetic of nostalgia, a sensibility that inspired the sojourner to look out over an ostensibly natural landscape and see historical ruins—poignant traces of the rise and fall of previous dynasties.[8] As they traveled northeast beyond the Great Wall across the frontier toward Manchuria, Wu Zhaoqian and his companions saw not just water and land; they also saw the officials, kings, and soldiers who had lived, fought, and died on the land in centuries past. In this vision, the temporal human dimension of landscape appeared as a constant ghostly presence that suffused geography with sadness. History entailed human emotion, and both were intimately entangled with ways of knowing the nonhuman world.

The ghosts of history may have haunted their vision of landscape on the road, but once the exiles' journey was finished and they arrived at their destination—the "extreme region" (*jue yu*) of Ningguta—the ghosts disappeared. The dense temperate rain forests of the Changbaishan Mountains struck the exiles as a profoundly "other" landscape, but even more disorienting was the lack of history to guide their perceptions of their new environment. Search as they might, the rusticated scholars found few, if any, specific references to this region in their classical Chinese histories and poems. Some of the exiles found the lack of textual references profoundly disorienting—a land without what to them counted as human history was by definition unknowable. Some wondered whether it was even possible to use knowledge derived from the "Central Plain" of Han China to analyze a land perceived as perched on the extreme edges of the earth. The experience of Manchuria inspired a new direction of knowledge making: a tense and sometimes contradictory triangulation of empirical observation, received history, and human emotion.

Qing elites later became avid (and nostalgic) consumers of the products of Manchuria's wilderness: luxury items such as furs and freshwater pearls; food products such as fish, meat, mushrooms, and honey; medicines such as ginseng; even tail feathers from Manchurian cranes and eagles. Manchuria's nature was understood through its consumable items: indeed, as Jonathan Schlessinger has observed, "to know the frontier, or any place, was to know its products."[9] For as long as the Qing was in existence, government bureaucrats compiled statistics enumerating the natural products extracted from Manchuria, items tabulated and organized according to color, quality, and value.

These early Han exiles contemplated other problems in their writing: they wondered whether the nature of Manchuria was knowable at all, and if it was, how they could make sense of it using familiar modes of thought. They subjected their environment to the universalizing standards of what they believed to be the cultural metropole, and were mostly preoccupied by how they—the Han literati, writing in Chinese—could make sense of the environment. In the words of the exiled official Zhang Jinyan that form the epigraph to this chapter, the exiles assumed that to "measure," "categorize," and "fathom" nature on the empire's frontiers was a task reserved for them alone.[10]

Ultimately, the creation of natural knowledge was inextricably tied to the process of defining which nature belonged in this newly emerging empire. Here David Bello's concept of Hanspace is useful. Chinese thinkers for centuries sensed a certain landscape as distinctively Han: a mostly agrarian landscape, "bounded by ecological links forged by distance and distinct types of flora and fauna," and defined by its particular *qi* energy as being different from non-Han or barbarian spaces.[11] Wu Zhaoqian's impulse was to perceive some territory as "Han" and some as "other." As Wu slowly progressed beyond the Great Wall, he perceived himself moving between these different worlds, crossing political and cultural borders between Han and non-Han that were seemingly embedded within nature itself.[12] But as he slowly moved northeast, the borders between the two spaces seemed to move and shift along with him. He attempted to make sense of the environment through familiar references from dynastic histories and ancient poetry, but somewhere in the forest, this strategy eventually gave out. Disoriented by the perceived shifting natural boundaries between the land of the Manchus and the land of the Han, the exiles began to entertain the idea that what they understood to be the underlying principles of nature were universally valid; applicable even in a non-Han space, even in the absence of a Han history. These "universal principles" would allow them to create knowledge that could encompass all corners of the new empire, if not the entire globe. By creating natural knowledge, Chinese exiles helped bring Manchuria into China.

On an early spring morning in 1659, Wu Zhaoqian and his fellow exiles departed Beijing's city wall to begin their long journey north, carrying with them cultural strategies for seeing the world. Before we can access the significance of the landscape that Wu describes on the road to Ningguta, we first need to understand the way he had been trained to think and write about nature, and see how that writing intersected with matters of power and territory. We begin, then, by placing Wu in the significant landscape of his birth—the rich southlands of the Yangtze River delta—at a time of political crisis and uncertainty.

The Southern Literati of the Ming-Qing Transition and Their Inscribed Landscapes

Wu Zhaoqian was exiled to Manchuria at the age of twenty-eight because he failed an exam.[13] It was an important examination and Wu had failed spectacularly, in a way that gave offense to the imperial throne. This was not the outcome that Wu was expecting when he took the civil service examinations in the populous and sophisticated southern Chinese city of Nanjing in 1657. For centuries before that moment, young, elite Chinese men had dedicated their lives to studying the Confucian classics, poetry, and history in order to sit for the imperially sponsored examinations. Passing the exams would qualify these men to launch a career of wealth and power within the imperial bureaucracy.[14] Under normal circumstances it would be quite natural for Wu to take multiple levels of the exams and pass them—he was, after all, one of the brightest young stars of learned society, from the wealthiest and most sophisticated center of Chinese culture: the Jiangnan region in the Yangtze River delta.

But circumstances were far from normal in the 1650s when Wu took the exam. These were the early years of the new Qing dynasty, founded by Manchu invaders from regions in the far northeast, beyond the Great Wall that had traditionally divided Han civilization from northern non-Han "barbarians." The Manchu army of conquest (with its large contingent of Chinese collaborators) had descended from the Korean border regions to take over the Ming capital of Beijing in 1644. The juggernaut continued marching over a thousand miles south to the Yangtze, where it vanquished Chinese resistance, sometimes quite brutally, in the mid-1640s.[15] In spite of the cataclysmic violence that accompanied the conquest, the new dynasty quickly resumed the examination system: after all, the military-minded Manchus needed to recruit Chinese administrators to help them run densely populated Chinese regions of the empire. By doing so through examinations based on the Confucian classics, these former barbarians hoped to position themselves as upholders of a sacred Chinese tradition and thus gain the support of the society they had conquered. But the Jiangnan literati remained a suspect lot, with their vast wealth, powerful knowledge, and deep connection to the environment of the southlands.

The young exile Wu Zhaoqian was enmeshed in this learned, powerful, and beautiful world of southern literati. Wu was born in 1630 in Wujiang, a city on the eastern shore of Lake Tai in the heart of the Yangtze River delta.

Today, the population of Wujiang is approximately 1.5 million, and the entire Lake Tai basin, a metropolitan continuum that includes Shanghai, Wuxi, Suzhou, Zhenjiang, and Hangzhou, is home to some fifty million people (in an area a bit smaller than the Netherlands).[16] In spite of its current industrialized profile, the region was traditionally known as one of China's most scenic locales, with a lush environment fitting of a place on the same latitude as New Orleans. Even today, in spots that have not been occupied by manufacturing plants and corporate campuses, this is a watery environment where placid canals, languid rivers, and flooded rice paddies coat the land with a shimmer of water and melt the line where the earth meets the sky. Tourists flock to the area's preserved "water towns" to witness a classically beautiful Chinese scenery that matches our vision of an unchanging China—whitewashed villas with dark, delicately curving tile roofs, secluded verdant gardens, and graceful stone bridges arching over peaceful canals.[17]

During the Ming empire (1368–1644), this area was not only known for its elegant landscapes; it also boasted the richest families, the most extensive libraries, and the most sophisticated lifestyles of the realm. Wu Zhaoqian had the fortune to be born into one of Wujiang's most illustrious families. Wu Zhaoqian's father, Wu Puxi, had achieved the highest examination degree and had served for decades as an official to the Ming dynasty.[18] The young Zhaoqian was groomed at an early age to follow in the ancestors' footsteps: to climb the examination ladder to power and status through literary greatness.

Significantly, the verdant landscape of Jiangnan figured prominently in Wu Zhaoqian's earliest literary works, and his ability to describe the delta environment was the foundation of his claim to fame. At a stage when most students were just beginning to write basic prose, the eight-year-old lad reportedly impressed local literati with examples of the rhapsody (*fu*) genre, a notoriously difficult and archaic form of poetry that was traditionally used to describe both natural and built environments.[19] These long prose poems could run thousands of characters long and were designed to display the author's knowledge of specialized, often obscure vocabulary. Whether narrating palaces or imperial hunting parks, *fu* typically included exhaustive listings of names of things in the world: varied species of animals and plants, technical jargon for architectural detail, and multiple terms for designating mountains, rivers, and forests, all adorned with vivid, even extravagant adjectives, written in complex Chinese characters that only the most exquisitely literate people could read. For over a thousand years, *fu* had provided both a vehicle for displaying erudition and a template for describing nature. Mastery of *fu* could cement the author's identity as an authority on objects, an adept with catalog-like knowledge who had the audacity and power to name all the things of the world.[20]

As we can see by one of Wu Zhaoqian's early undated works, the "Spring Rhapsody" (*Chun fu*), the Jiangnan landscape was a world where the human and the nonhuman were deeply entangled. The poem begins with a lyrical description of his homeland's scenery:

> Now in its time, the cold gives way to the cusp of spring's flourish,
> I gaze afar upon the fair mountains and rivers, the brilliant colors on the
>   pavilion-topped bank—
> At the dike's edge, gentle waves turn to mist; jade fragrances rise from the
>   luxuriously greening fields,
> The branches are draped in silk-like beauty, flower-sweet breezes perfume
>   the scene.
> Tung-tree blossoms cover the hills in rich brocade,
> And orchid leaves coat the riverbanks in glittering silver.[21]

With a few lines, the young poet conveyed the gentle beauty of a Jiangnan early spring, when life emerges from a cold winter to decorate the landscape in color and fragrance. This sensory-laden landscape is neither a rustic pastoral nor a vision of unspoiled nature. Instead, in the poet's mind, the beauty of Jiangnan's nature mirrored the sensual material delights of late-Ming urban society. Wu uses luxurious domestic imagery of silver, perfume, and silk brocade to portray the external environment, suggesting a nature that mirrors the man-made environments enjoyed by sophisticated southern literati.[22]

This interweaving of the human and nonhuman worlds extended beyond domestic interiors to the interior of the human heart. *Fu* also expressed the inner world of the writer, and within the next few lines of "Spring Rhapsody," the narrative turns from exposition of the external environment to intimations of human emotion. The incipient beauty of spring emerging from winter's landscape serves as a painful reminder of the passage of time and the cruelty of separation, and the mood of the poet shifts from delight to melancholy:

> But I regret the impermanence of the scenery,
> ponder the vagaries of sorrow and joy.
> While gazing afar, I cherish thoughts of my youth;
> Recalling the thousand leagues that separate us only deepens my pain.[23]

The subject of the poem then shifts from nature to the tragedies of the human past. After painting Jiangnan's beauty, Wu's *fu* conjures a world of dark castles populated by anxious lords who face the dangers of war and political intrigue. The lords must travel far away from their beloved wives who remain at home, bemoaning their tragic separation from their men. Wu's spring landscape was shot through with worry and despair.

We may be tempted to read Wu's poem as providing direct insight into the poet's own life. As the child of an official in the Ming imperial government, Wu had certainly been subjected to long, lonely journeys and painful family separations. The "law of avoidance," a fundamental regulation of the Ming (and also the Qing) bureaucracy, dictated that officials could not serve in their home provinces. Following this principle, Zhaoqian's father, Wu Puxi, had been charged with overseeing administration in Yongzhou prefecture, a site over seven hundred miles from his home in the rugged mountains of semitropical Hunan province.[24] The irony of the posting would not have been lost on the erudite Wu clan: Yongzhou was best known as the site of banishment for one of the most famous exiles of Chinese history, Liu Zongyuan (773–819), and it was clear that the elder Wu had been assigned to a place synonymous with remote wilderness.[25] As a child, Wu Zhaoqian had accompanied his father on one epic journey to Yongzhou and back, but he mostly remained behind in Wujiang, living in a state of sad separation from his father.

Wu's "Spring Rhapsody" may also be alluding to the most tragic event from the poet's youth: when Wu was fifteen, his family life was shattered by the Qing invasion. In 1645, armies of the new Manchu dynasty swept through Jiangnan, laying siege to cities of millions, perpetrating massacres while at the same time forging politically fraught bonds of collaboration with the region's Han Chinese elites. Wu's father, still serving the Ming in distant Hunan, was first caught up in the chaos brought by the peasant armies of the Chinese rebel Li Zicheng. Even after the Qing had wrestled the capital back from the hands of the rebels and replaced the Ming, Wu Puxi remained loyal to the fallen dynasty and briefly served the Southern Ming court in exile.[26] Wu Zhaoqian's poem clearly links the Jiangnan environment to his emotions at this tumultuous time of dislocation and rootlessness.

To convey this link between emotions and nature, Wu's poem uses numerous phrasings from works penned by famous literati who lived in unsettled times during China's imperial past. Some phrasings derive from "On Spring Sentiments," by Wang Bo (650–676), who wrote his famous *fu* while journey-

ing to the mountains of Sichuan, in the western frontier of the Tang empire
(618–907). Wu's poem also contains references to the work of Qu Yuan (343–
278 BCE), an exiled minister-poet who was (and still is) the literary archetype
for the cashiered official. Ironically, Qu Yuan had been banished to the lower
Yangtze River area at a time when even that region was considered a frontier.
As described by the literary scholar Nicholas Morrow Williams, Qu Yuan's
poetry provided imperial China's literati with the locus classicus for "thinking
of the beautiful landscapes of Jiangnan, at their lushest and most vital in the
springtime, as themselves the very motive of melancholy."[27]

Wu's youthful poem thus reveals the environment of Jiangnan as a prime
example of what Richard Strassberg has called an "inscribed landscape," a
condition in which nature was, in the eyes of literati observers, "inextricably
linked with language and history."[28] Poems like Wu Zhaoqian's early *fu* present the author's perspective on the nonhuman world as a human-authored
text. If the human and nonhuman worlds are co-constitutive, for Chinese of
the Ming-Qing transition, history was the shuttle that helped weave the two
together.

Wu's ability to conjure that inscribed landscape allowed the youthful poet
to establish his reputation at the center of China's learned elite. Indeed, writing masterfully about nature was one of the skills that would allow men like
Wu to access elite social connections, which could open the door to the halls
of political power. But the path to achieving that power—through the imperial examinations and service to the emperor—was fraught with risks. Political missteps could result in being forever separated from the cherished landscapes that had nurtured this power in the first place.

The Qing conquest of the 1640s had devastated parts of the Yangtze delta
region and traumatized its population, but by the 1650s, daily life for elites
seemed to continue much as before, with time passed in poetry writing and
appreciation of the landscape—activities that dovetailed well with preparation for the imperial examinations. By his midtwenties, Wu Zhaoqian had
become known as one of the "Three Phoenixes of the [Yangtze] Left Bank."
He socialized with the era's most celebrated poets and complex political figures, including Qian Qianyi (1582–1664) and Wu Weiye (1609–1671), older
scholars who had served as officials in both the Ming and the Qing dynasties.
Wu Zhaoqian did not wear his fame lightly, and the poet's contemporaries
described him as an aloof eccentric who carried himself with a smug sense of
pride.[29] Wu had reason to be confident. The Manchus had revived the imperial examinations, and young Han men like Wu who were talented in poetry

could rise into the power-holding class of scholar-officials through literary works, the careful cultivation of connections, and success on the exams.

The year Wu Zhaoqian took his exams—1657—turned out to be a year of devastating scandal. The Chinese officials in charge of evaluating the tests administered in the old southern capital of Nanjing were accused of taking bribes and favoring their friends and relatives. Those students who had not passed protested bitterly, no doubt fueled by personal grudges and factional rifts.[30] Imperial censors called for investigation. Dozens of men who had passed the exam in Nanjing, including Wu Zhaoqian, were summoned in the dead of winter by imperial decree to journey seven hundred miles north to Beijing. There these men who had been raised in the southern heart of Han Chinese civilization were suddenly thrust into the center of a northern Manchu-dominated world. They were forced to retake the examination surrounded by scimitar-wielding Manchu guards and in the presence of the Shunzhi emperor himself. In the end, Wu submitted nothing but a blank paper—when forced to write at the command of the Manchu emperor, he wrote nothing at all.[31] Scholars have offered little insight into why Wu failed to demonstrate his brilliance under these conditions: was it an intentional act of political resistance against the Qing, or simply the result of paralyzing fear?[32] Whatever the reason, it was the imperial examinations—the very vehicle through which Wu had hoped to translate literary genius into political power at the center of the empire—that led to Wu's banishment to the empire's far northeastern fringe. For his remarkable act of defiance, Wu Zhaoqian was sentenced to lifelong exile in Ningguta, a frontier outpost 1,500 miles from his home, in the land of the Manchu's origins.

In the eyes of the Jiangnan literati elite, the environment that awaited the exiles in the northeast was dark, dangerous, and inherently evil. On the eve of Wu Zhaoqian's banishment in 1659, the famous Jiangnan poet-official Wu Weiye, himself haunted by his decision to collaborate with the Manchu regime, anticipated Wu Zhaoqian's journey in a poem, "A Song of Sorrow for Wu Jizi."[33] In this poem, the elder scholar contrasted the refined life of the south with the foreboding northern environment that awaited the young poet:

In life's thousand miles and ten thousand miles
There's nothing but the overwhelming despair of parting,
But why is it that you alone have to suffer like this?
O, mountains that are not mountains!
Water that is not really water,

A life that is not really life;
A death that is not really death.

You were a student of the Classics, conversant with the Histories,
Born into Jiangnan's silken air, penning elegant poems without compare;
But the innocent man has been slandered.
One day you were taken away in fetters, your protestations useless;
Banished to the remote frontier, cut off by a thousand mountains.
We send you off, crying without ceasing,
What has the exile left to rely on?
We are filled with dread that you'll never return—
You know your journey can never be altered.

In the eighth month, desert snowflakes begin to fall,
Snow falls deep as the camel's waist, so high it covers the horses' ears.
White bones of battles past piled in gleaming heaps;
No ferries cross the Black River, but a few passengers wait on the banks,
Dreading the tigers lurking ahead and the hydras lurking behind.
Here there are crawling vermin that slither forth from their lairs,
And fishlike things large as mountains, endlessly huge—
A flap of their fins stirs the wind, the foam from their mouths like rain.
Here the sun and the moon reverse their courses and dive to the bottom of
     the sea,
In broad daylight you can encounter men who are half-human, half ghost.
Alas, how horribly tragic!

If you have a brilliant son, be cautious, not joyful;
There's a reason that Cangjie cried through the night,
Misfortune all begins with the reading of books–
Behold the tragic case of Wu Jizi![34]

Here Wu Weiye wrote an intensely personal, emotionally moving original
poem that linked Manchuria's nature to the fate of scholars who ran afoul of
political power. Literary talent went hand in hand with political persecution:
according to the poet, giving birth to an intelligent son is a harbinger of ill
fortune, and even Cangjie, the mythical inventor of writing, experienced great
sorrow as a result of his literary brilliance. In this new chaotic world dominated by the Manchus, the good are rewarded with nothing but suffering, and

even the course of nature is inverted. The exile is condemned to a life that isn't life, something like death that isn't quite death. Even the supposedly stable touchstones of the nonhuman world melt away and become unrecognizable in the remote land of the north: "Mountains are not mountains, waters no longer waters."

Wu's poem describes the bleak environment that awaited the young poet by employing tropes about the barren northern landscape that Chinese poets had used for centuries. While these zones were hundreds, if not thousands, of miles away from the main centers of Han Chinese culture, literary templates were readily available to describe them. The frontiers between Chinese and northern nomadic peoples had been the site of war and dislocation since the earliest years of the Han empire (206 BCE–220 CE). The frontier theme was particularly prominent in the poetry of the Tang dynasty (618–906), when renowned poets (including Li Bo, Du Fu, Bai Juyi, and many others) wrote poems describing bleak journeys to the frontier beyond the Great Wall as a way to bemoan the fate of common people under an oppressive, war-hungry state. These frontier poems are full of exhausted soldiers, conscripted laborers, and exiled officials who are reluctant observers of a foreboding environment. Crossing rivers and mountain passes, they leave behind the verdant "central plain" and enter the desert frontier. Through their tearful eyes, they see bones of dead soldiers piled in white heaps, sterile brown-and-gray vistas or landscapes covered in snow and ice, with the ruins of fallen cities standing mute in the distance. Wu Weiye's poem, for all its power, relies on many of these familiar tropes.[35]

In the years after the Manchu conquest, some Chinese literati retooled the general metaphors of the frontier poem tradition and used them to denigrate the homeland of their Manchu conquerors. The northeast frontiers for centuries had been referenced in Chinese texts as the land of the "White Mountains and Black Waters" (*baishan heishui*): the black water being the Heilongjiang River, the white mountains the Changbaishan range. As a literary shorthand, the terms had vaguely conjured a faraway barbarian land characterized by imposing mountains and rivers. Now that the tribes from the land of the Black Waters and White Mountains had violently conquered China, the phrase took on a more foreboding meaning in the hands of poets who survived from the Ming. Wu Weiye's contemporary, the famous poet Qian Qianyi, in his *Toubi ji* (Collection of poems on renouncing the pen [to take up the sword]), expressed his resistance to the new dynasty by casting the mountains and rivers of the Manchu homeland as haunted places. In his poetry, Qian vilified the northeast with lines such as "Wandering goblins from the Black Water wail on the grasslands / War demons from the White Mountain howl to the sound of the Tartar flute."[36] Wu Weiye's lamentation for Wu Zhaoqian similarly paints the Manchurian landscape as a hellish place dominated by the mysterious Black River, its depths infested with vermin and monsters. The exile awaits on its banks, terrified to cross over into oblivion. Even before he exited from Beijing's city gate to begin his journey of exile, Wu Zhaoqian knew what awaited him was a land replete with ghosts.

Wu captured his journey to this haunted frontier in dozens of poems that he wrote on the way to Ningguta in the spring of 1659. His writings used many of the same tropes that appeared in Wu Weiye's parting poem: black water and white bones, monsters and gloom. But Wu's poems also reflect something more: a kinesthetic sense of an encounter between human senses and the nonhuman environment. Wu Zhaoqian's poems were written while the young man traveled through the landscape, viewing the environment as he moved. The poems note specific landmarks—roads, cities, mountains, and rivers—that allow us to follow clearly the route of his journey north. While his poetry was undoubtedly suffused with standard tropes, at the same time this poetry can also allow insights into the observer's immediate experience of the environment. As the scholar and translator Jonathan Chaves suggests, the poetry of the Ming-Qing transition period often served as a form of real-time travelogue, one "concerned with capturing the experience of journeying through an actual landscape."[37]

This "actual landscape" was a product of a certain type of vision created through dynamic encounters among cultural desires, the embodied sensory perception of human actors, and the physical terrain. Back in China proper, Wu Zhaoqian had confidently navigated this intersection through his poetry. The Jiangnan landscape had allowed him to turn the naming of nature into power. But as a convicted criminal moving beyond the Great Wall, Wu's templates for interpreting nature slipped against the changing terrain. As he moved forward in this landscape, Wu struggled to make sense of what he saw. What was familiar ground, and what was alien land? How far could his Chinese cultural templates be stretched into the frontier? Where did China's nature end, and the nature of the "other" begin? Would this environment always be suffused with the ghosts of Chinese tragedy, or might there be other ways to make sense of the landscape beyond the Great Wall?

Ghosts of the Past on the Road to Ningguta

Wu began his journey from his Beijing prison cell in March 1659, traveling eastward by oxcart out of the massive city gates toward where the Great Wall meets the Bohai Gulf. Although he traveled slowly as a dejected prisoner, he was traveling along an important route taken by legions of imperial armies. Moving first toward the coast, then turning northward up the coastal corridor, Wu's road to exile replicated (though in reverse order) the path followed by the Manchu forces in the first decades of the seventeenth century as they traveled from the northeast to the Great Wall pass at Shanhaiguan (Mountain-Ocean Pass), and then inland to capture Beijing. Wu was exquisitely aware that he was moving through a scarred terrain: ruined villages and mass graves could easily be seen from the road. While we cannot be exactly certain of the extent to which Wu was familiar with the specific history of the Manchu conquest, it is clear from his writing that he remained intentionally silent about the most recent battles the landscape had witnessed. Instead, Wu saw ruins from wars of centuries past, where ancient dynasties had endlessly fought over contested lands. The following text and accompanying map (fig. 1.1) lay out the route of Wu's journey and allow us to reflect on the way that Wu saw history embedded in the landscape.

The place where Wu was going was not yet called "Manchuria." As soon

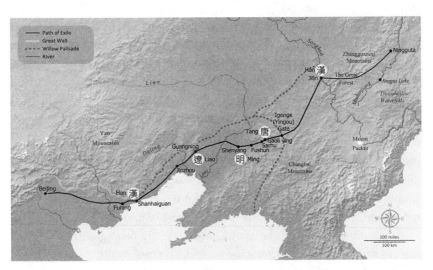

Fig. 1.1 Wu Zhaoqian's 1659 journey of exile from Beijing to Ningguta, in dynasties. Map by Jeff Blossom. See plate 14 for color version.

as he moved east of Beijing, Wu's poems were brimming with an old term used to name this region: *Liao* (辽 / 遼). The character itself literally means "far away, distant," or, by extension, "the Pale," and had been used in Chinese records for centuries to describe the territory lying to the north of the Bohai Gulf, particularly the land encompassing its central peninsula and its interior reaches. Since the conquests of the Han dynasty (221 BCE–220 CE), this region was positioned at the contested edge between an expanding imperial China and multiple competing northeast Asian peoples, today characterized as groups including Mongol, Korean, and Tungistic tribes.

Today, a map of northeast Asia shows this territory divided firmly within the clear boundaries of nation-states. The peninsula itself is part of Liaoning province of the People's Republic of China. To the east lie North and South Korea; to the northwest, the Inner Mongolian Autonomous Region of the PRC; and farther inland, the Republic of Mongolia. This bordered clarity masks a remarkable ethnic, social, and political complexity that marked the region for millennia. While marked as "the Pale" for Han Chinese of the Central Plain, the region was the stage for political and military struggles of multiple non-Han groups.

Since the very first empires of China, the Qin and Han, "Liao" was also the name given to a major river that begins on the Mongolian Plateau and drains into the Bohai Gulf. The land of the river's basin was divided between what lay to the west of the river (Liaoxi) and what lay to the east (Liaodong). The Liao River region was a site of conflict between different dynasties associated with China and polities associated with Korea (Koguryŏ) from the beginning of the common era until the defeat of Koguryŏ in 668 CE.[38] "Liao" was also the name given to the dynasty formed by the Khitan people who dominated the region at the beginning of the last millennium until their dynasty's demise at the hands of the Jurchen Jin dynasty in 1125. With the return of Han Chinese rule after the overthrow of the Mongol Yuan dynasty in 1368, the Ming dynasty positioned fortifications from Beijing up through the Liao valley to defend the approach to the capital from non-Han "barbarians," primarily groups of Mongols and Jurchens. With the establishment of the Manchu military confederation and the launch of the Qing conquest against the Ming, the Liao territory became a war zone, a status that lasted for decades, from the late sixteenth to the mid-seventeenth century.[39] As Wu Zhaoqian looked out over this terrain from his oxcart, he was highly cognizant that he was traveling through a battleground, one that had known warfare not just within his lifetime but across the centuries.

Wu's first poem "from the exile saddle" was written in Funing, a city some 150 miles due east of Beijing. Today, the town of Funing is a part of Qinhuangdao, a port city of three million people on the Bohai Gulf. The beach resort of Beidaihe (once famous as a popular retreat for Communist Party leaders) lies just to the north. In the seventeenth century, Funing was a walled city with a population of several tens of thousands, located strategically along the main road from the Bohai coast to the capital.[40] While Funing was still located within the flat plain shared by Tianjin, Beijing, and other major North China cities, it was a transitional place, where the rugged peaks of the Yan Mountains met the North China Plain. Today, these mountains are a familiar site to anyone who has landed at the Beijing airport: sharply angled peaks ranging between two thousand and six thousand feet high that dominate the landscape north of Beijing. To this day, the Great Wall, the fortification designed to keep invaders out of China proper, snakes along the top ridges of these mountains. For Wu Zhaoqian, the Yan Mountains' abrupt rise at Funing signaled the division between China proper and the beginning of an alien terrain:

Departing from Funing Early in the Morning, Written on the
Wall of the Inn

Through the long night, the sound of drums and horns crossed the bitter
   winds,
The traveler awakes before dawn, startling the tethered horses.
My journey follows the path of the geese on their March flight home;
On the dusty road, Lulong lies before my eyes,
I glance back to the South: there's nothing but the "Eastern Pale."[41]

At Funing, Wu senses that he is on the verge of entering a different, foreboding world. He experiences a sort of disorientation—spending sleepless nights surrounded by a militaristic soundscape of war drums and cavalry. He notes with irony that it is spring, and the geese are happily heading northward toward their home, but the exile traveling in the same direction is leaving his home behind. Ahead of him are the rugged mountains leading to the Great Wall. He turns his head to the south, in the direction of his home, only to find that what lies to his south is more of the land of Liao—the true south, China proper, is no longer visible.

From Funing, Wu traveled thirty miles northeast to Shanhaiguan, the great pass where the Great Wall begins, a site that had marked the beginning of the frontier for more than a millennium:[42]

At Shanhaiguan

Surrounded by a thousand peaks, here begins the Great Wall,
The Han Nation once marked this as the limits of the Central Plain.
To the south the Wall borders the imposing Liao Sea,
Resting upon the Yan Mountains, it guards the northern pass.
. . . On the high tower, who still remembers the world of the Central
     Mountains?
Peering at the far horizon, the gray grasses merge with the gloomy sky.[43]

Exiting the Pass

From the frontier tower I turn my head—steep mountain ridges rise one
     after the other
Noise and confusion—the imperial messengers galloping in the dust . . .
Perpetual dark clouds and snow in the Land of the Yellow Dragon,
Willows only starting to bud—it is spring on the Purple Frontier.
Before Lady Meng Jiang's Stone, a myriad horses are stationed,
Yet those who guard the Pass are people of the Han[44]

These poems contain many of the elements used for centuries to describe the northern frontier around the Great Wall: lone towers, gloomy clouds, perpetual snows. The landscape serves to confirm the poet's dread at leaving behind the "Central Mountains," the world of China proper. At the Great Wall, Wu enters a liminal world where physical boundaries become indistinct. He is surrounded by gloom, a landscape where both land and sky are colored gray, white, and brown, with scattered snows still visible in the hills. At the same time, Wu distinctly sensed that he has reached a place where cultural boundaries *should* be very clear: this was the place that had separated Han from non-Han peoples since antiquity. In these poems written around Shanhaiguan, where the Great Wall meets the "Sea of Liao" (Liaohai, or Bohai Gulf), the word "Han" frequently appears, although Wu interestingly refrains from naming or describing the "other," the Manchus who then ruled all of China and were his jailers. But even this border is ambiguous: near the temple erected in honor of Lady Meng Jiang, the virtuous wife of antiquity who traveled to the newly built Great Wall in search of her laborer husband, Wu notes that this monument to loyalty is guarded by Han bannermen who have become loyal to the Manchu dynasty.[45]

From Shanhaiguan, Wu continued northeast. To his right, along the narrow, flat plain, he frequently caught glimpses of the Bohai Gulf, while the jagged peaks of the Yan Mountains rose to his left. This region between the ocean and the mountains was known as Liaoxi, "west of the Liao River" or the "Western Pale." The Liaoxi corridor had been a passageway into the northeastern frontiers recognized even before the first dynasty of China, a strategic opening into lands on the fluid and shifting borders between Han settlement, Mongol societies to the north and west, and Tungistic and Korean peoples to the east.[46]

Wu passed through the walled cities of Jinzhou and Guangning near the top of the oval of the Bohai Gulf and then continued inland, away from the ocean waters and into the broad plain formed by the Liao drainage basin. Beyond Jinzhou, Wu passed through the valley between two north-to-south-flowing rivers: the Daling, flowing southeast from the Mongolian Plateau, and then the Liao, flowing southwest from the northern highlands. The valley between them had formed the crossroads between multiple peoples in northeast Asia for centuries. In the eyes of the poet, the contemporary landscape gave way to a ghostly historical landscape—scenes of past battles that allowed Wu to comment on the trauma of the conquest while avoiding mention of his new rulers.

As recently as the 1620s, just thirty years before Wu's journey, the region between the Liao and Daling Rivers had been fiercely contested during the decades-long struggle leading up to the Manchu conquest. Many seventeenth-century observers noted that the area was full of ruined and abandoned villages.[47] Wu sees the wars of the past physically etched onto the landscape of the present, but instead of directly referencing the most recent conflict, Wu allows his vision to range across the centuries, blurring the temporal boundaries between contemporary and ancient events. Outside of Jinzhou, Wu observes the ruins of old watchtowers that had survived "a hundred battles," suggesting a landscape weary of war, where the most recent battles were but a small part in a long continuum of violence. On the road to the city of Guangning, Wu's thoughts shifted to five-hundred-year-old events, noting that the area had once been the domain of a renegade Khitan prince from the Liao dynasty (907–1125).[48] Wu does not mention that Guangning was also the site of a crucial battle in the Manchu conquest, captured by Nurhaci's forces from its Ming commanders in 1622.[49] As he approached the Liao River, the sight of ruined city walls, empty encampments, and forgotten graves from the recent conquest moved Wu to tears: "The general's headquarters has withered away/ the mountains and rivers along the Gulf have all grown wild / Ten years of attacks have demolished the defensive walls."[50] In his vague phrasing, it is difficult to know which era's wars Wu laments.

After crossing the Liao River, Wu Zhaoqian formally entered "Liaodong" (the Eastern Pale, east of the Liao River) and entered into the southern heartland of the Manchurian Plain. Formed by the drainage basins of the Liao River to the south and the Songhua River to the north, the Manchurian Plain (also called the Song-Liao Plain) forms a vast flat expanse in northeast Asia between the Greater Khingan Mountains to the west, the Lesser Khingan to the north, and the Changbai Mountains to the east. The southern part of the Manchurian Plain in the drainage basin of the Liao River had long been a site of Han Chinese settlement, and Han farmers had tilled the land there for centuries. It was also the site of a unique regional identity that Pamela Crossley designates as "Liaodongese."[51] The Ming court exerted control over this politically sensitive territory by establishing a network of twenty-five military garrisons and fostering increased Han Chinese settlement throughout the region.[52] But just to the east lay the power base of the Jianzhou Jurchens, the group that gave rise to the dynasty that defeated the Ming and conquered China.

One month and four hundred miles after leaving Beijing, Wu arrived in Shenyang, a major city in the heart of the Manchurian Plain on the banks of the Hun (or Shen) River. Today, Shenyang is a massive metropolis with a population of over eight million. In the seventeenth century, Shenyang was a large walled city that had served as the interim capital of the emerging Manchu Qing dynasty before the conquest of Beijing. The city continued to be an important site for Qing civil and military administration in the northeast even after the establishment of Beijing as the capital of the empire in 1644. In his poetry, Wu avoided using the official Chinese name given to the city by the Qing dynasty, "Shengjing" ("Flourishing Capital," or in Manchu, "Mukden"), a name that reflected the city's political status as a regional Manchu capital. Instead, Wu called the city by its Ming-dynasty Chinese name, "Shenyang" (literally "The Upper Bank of the Shen River"). Wu stayed at Shenyang for several weeks: the relatively relaxed carceral regime for literati exiles could include time for dining, drinking, and poetry writing with other Han exiles. In spite of the relative freedom, the forced march had to continue, and in late May, Wu departed the massive city gates of Shenyang and journeyed again toward the northeast, following the Hun River toward Ningguta. As one poet sending him off from the relative comfort of Shenyang put it, "*saiwai* [beyond the Great Wall] is unbearable, but now you must go even more *saiwai*."[53] Tellingly, the exiles sensed multiple demarcations within the shifting terrain of Manchuria: what was considered a frontier could become a center when compared with the distant lands that lay beyond.

To the east of Shenyang, the landscape changed, and the broad Manchurian Plain started to give way to forested low mountains crisscrossed by narrow river valleys. Wu had entered the heartland of the territories dominated by the Qing dynasty's ruling house, the Aisin Gioro, and was now in the southern foothills of the Changbai (Long White) Mountain range. The Long White Mountains are a distinguishing feature of the Manchurian landscape, extending for hundreds of miles along a southwest-to-northeast axis roughly straddling the present-day border between North Korea and the People's Republic of China. The tallest peak in the Long White range, known eponymously in Chinese as the Long White Mountain (Changbaishan), is a massive semiactive volcano that stands nine thousand feet above sea level.[54] While the fame of this mountain makes it a symbol for the entire region, physically it is an outlier, since most of the mountains in the Changbai range are far shorter and much less spectacular. Unlike the white-capped Mount Changbai, the Changbai Mountains on the whole are green and rounded, ranging from two thousand to five thousand feet high, etched by rapid rivers and cloaked in mixed coniferous and deciduous forests. While the mountains ranging into present-day North Korea have an alpine height and microclimate, the Changbai Mountains that rise to the east of Shenyang resemble more the Appalachians than the Alps. Their gentle slopes and fertile river valleys formed the original power base for the Jianzhou Jurchens, the confederation of Jurchen tribes that would go on to consolidate their power under the ruler Nurhaci (1559–1626) and become the core of the Manchus.

A two-day journey east of Shenyang along the Hun River valley brought Wu through a landscape at the heart of Manchu power. Wu stopped first at the walled fortress of Fushun, the first major Ming fortification conquered by the newly ascendant khan Nurhaci in 1618.[55] Just thirty miles to the southeast of Fushun, along the Suzi River, a windy tributary of the Hun River, lay Nurhaci's original power base, the outpost of Hetu Ala, where Nurhaci established headquarters for his Eight Banners military force in the late sixteenth century. Between Fushun and Hetu Ala lay the plains of Sarhu, where in 1619 Nurhaci's armies defeated tens of thousands of Ming troops (along with their Korean and Mongol allies) that had been amassed in an attempt to root the khan from his power base.[56] Today the Dahuofang Reservoir covers much of the plain where the battle was fought, but in the seventeenth century, Wu's eastward path would have taken him within sight of the hills that encircled the fateful plain.

Outside of Fushun, instead of seeing a landscape where the Ming had been defeated, Wu instead envisioned a place where Chinese armies had fought and prevailed one thousand years in the past. In his poem "Gaoli ying" (The Korean Encampment), Wu mused about the remains of a fortress of the Koguryŏ kingdom (37 BCE–668 CE) in the region. Koguryŏ was an expansive proto-Korean state that controlled much of southern Manchuria from its capital in Pyongyang. In the seventh century, armies of the Chinese Tang empire (with its capital situated a thousand miles to the southwest in the city of Xi'an) launched a series of expeditions against Koguryŏ, eventually capturing its outposts in Liaodong and ultimately causing the downfall of the kingdom.[57] Wu's poem imagined the "bloody battles" launched by the expansionist Tang emperor Taizong and marveled that the sad ruins of the fortress of the enemy had survived from the distant past down to the present day. Wu mentions a campaign where the Chinese were victors, but reflecting the tradition of the great Tang poets such as Du Fu, the poem evinces no joy in victory and recalls instead the horrors of war.[58] On the question of Nurhaci's victories, however, Wu was silent.

About sixty miles northeast of Fushun, Wu arrived at a significant manmade border: the Willow Palisade. Built by the Manchus beginning as early as the 1630s, the Willow Palisade was a sort of organic—and far less militarized—Great Wall. Constructed with a combination of ditch works, plantings of willow trees, and wooden fences, the Willow Palisade formed an arboreal line demarcating the area of traditional Han settlement in southern Liaodong from the territory of the Manchus. In 1659, when Wu passed through it, it was still a work in progress. When its arrangement was finished in the 1680s, this section of the palisade stretched from the area of the Great Wall to the border with Korea, making a wide loop around the Liaodong Peninsula.[59] A total of sixteen gates were placed at various strategic spots around the palisade, each manned by a small contingent of banner soldiers who were charged with inspecting the passes of anyone who tried to travel beyond the barrier. In theory, only bannermen and others in the service of the emperor were authorized to proceed—although this line of trees produced "a flimsy barrier" that was only partially successful in enforcing a segregation of the Manchu homeland.[60]

Wu marked his arrival at the Willow Palisade's gate with another poem. Wu approximated the Manchu name of the gate—"Igenge," meaning "origin"—with the transliteration "Yin'gou," Chinese characters that sounded a bit like "Igenge" but meant "Shady Ditch." Wu uses Chinese transliterations of Manchu place-names at this point in his journey, mimicking the sound of the language and ignoring the more noble meaning of Manchu words.[61] This shift in language was significant: for Wu, the Willow Palisade seemed as though it were both a natural and a cultural border, a place where he was about to enter yet another realm of removal from China proper into uncharted territory:

> At Yin'gou Gate
>
> The myriad mountains pile into the brilliant sky for a thousand fathoms,
> The palisade cliffs arrayed like a heroic strategic pass . . .
> The smoke of homes rises into a yellow cloud at dusk, cattle pasture on the
>   windswept white grass—
> The reckless and fierce battles of the past are a distant thought;
> Today, the northern Liao is already "Within the Pass."[62]

Wu creates a shift in terrain at the Willow Palisade, as if nature itself transformed at this line. Although the mountains at this juncture are low, Wu exaggerates the drama of the scenery, portraying the place as a "heroic strategic pass." But instead of a frontier military outpost with pawing horses and howling winds, Wu detects a more pastoral landscape, noting the presence of a small village, with herds of cattle grazing on the still-dormant grass. While the area had been a site of violent frontier battles, Wu was struck by a new sense of geography: with the existence of this Manchu-built boundary marker over three hundred miles to the northeast of the Great Wall, the part of Liaodong he had just traversed, a territory once decidedly a barbarian realm "beyond the pass," had become part of the lands "within the pass"—a term used to characterize China proper.

Once he crossed the "Shady Ditch" Gate, however, Wu entered a forested world where frontier-poetry allusions no longer held. Instead of the standard non-Han frontier landscape of gray-brown deserts and barren mountains, Wu encountered the deep-green temperate rain forest of what is now Jilin province. Signs of Han settlement grew rare, and Wu saw little evidence of ancient civilizations in the landscape. To Wu, the landscape appeared practically uninhabited, with only courier stations for changing horses in lieu of towns. As Wu's fellow exile Fang Gongqian had put it, in the eyes of the exiles, the road from the Willow Palisade at Yinggou Gate to Ningguta was "full of nothing but mountains, rivers, and swamps."[63]

This swampy road to Ningguta took Wu across the intersection of the Liao and Songhua River drainage basins. Following the road northeast meant ferrying across a tangle of winding tributaries of these riverine systems, including the Huifa, Lianhe, Nianma, and Meihe Rivers. Spring snowmelts and early summer rains had swelled the flows and turned the banks into muddy morasses. Wu's poem titles, such as "Stuck for Five Days at the Nianma River" and "Written . . . on the Occasion of Once Again Not Being Able to Cross the Deep River Waters," describe the difficulty of the passage. Crossing a rushing river on the day of the Duanwu festival (also known as the Dragon Boat Festival, which celebrates the life and death of the great poet-exile Qu Yuan), Wu sardonically observed: "At least the Miluo River [where Qu Yuan drowned] was in Jiangnan [South China]—I'm beginning to think that Qu Yuan wasn't so pitiable after all!"[64]

By the beginning of August, Wu reached the great Songhua River at Girin Ula (Jilin), and crossed the river on Li Qiu, the first day of the autumn season in the lunar calendar. In his poem "Crossing the Huntong [Songhua] River," Wu captures the wide expanse of the river at the Jilin crossing: gazing from his boat toward the horizon, he was unable to see the opposite shore. The Songhuajiang formed yet another significant zone of passage for Wu, signaling his arrival into the last leg of his journey into an unknown realm. Known in Manchu as the Sungari Ula, in Chinese as the Songhuajiang, the Songhua flows through the heart of Manchuria. Its main stream emerges from the caldera lake on top of the volcanic Mount Changbai (Paektu), and as it flows in a northwesterly direction for five hundred miles toward the lowlands of the central Manchurian Plain, it gathers multiple rivulets that descend from the Changbai range. At the city of Songyuan (Boduna in Manchu), the Songhua merges with the Nen River, makes a dogleg turn to the northeast, and meanders across the Sanjiang Plain for another six hundred miles until it empties into the great Amur River (Sahaliyan Ula in Manchu; in Chinese, Heilongjiang). The terrain surrounding the Songhua at the center of Manchuria is remarkably flat, and the river flows languidly in twists and turns in channels several kilometers wide, forming oxbow lakes and broad floodplains.[65] The Songhua River was an important riverine pathway for the Qing empire: its connection to the Heilongjiang (Amur) River to the far north meant that the Songhua was a potential zone of conflict with Russia. When Wu crossed it in 1659, the Qing empire was beginning to bolster its riverine defenses in this crucial region.

Ironically, even as he crossed the Songhua River surrounded by Manchu bannermen, Wu chose to describe this area as Xuantu, one of the "Four Han Commanderies" established by a great Chinese warrior-emperor of the Han dynasty, Han Wudi (156–87 BCE) as "far eastern" outposts in the second century BC.[66] The region, Wu admitted, was also the place where the Jurchen Jin dynasty (1115–1234)—considered the predecessor to the current Manchu dynasty—had begun its march to hegemony in the northeast five hundred years before. Even as he tried to hold onto this complex vision of dynasties past, Wu lamented that the evidence for history was fading from sight. In a poem about the environs around Girin Ula, he stated a desire to "read the ancient inscription on the broken monument" but found that "the carefully carved words have faded away, and the stone is covered in weeds."[67] For Wu, here on the banks of the Songhua River, Han-dominated history was beginning to disappear.

After crossing the Songhua, Wu entered the final leg of his journey to Ningguta, heading northeast along the Jiaohe River through the Zhangguancai Mountains—a frontier where, according to Wu, they "found shelter amongst the trees and ate game hunted with bow and knife."[68] The Zhangguancai Mountains (Manchu, Julgen sain alin, or "Good Luck Mountains")[69] are a series of gently sloped hills ranging from two thousand to three thousand feet, and are considered a northern section of the Changbaishan range. Rising up from the eastern edge of the Manchurian Plain, the Zhuangguancai Mountains trap the northern progress of the continental monsoon, thus ensuring that the area receives an abundant rainfall. As a result, the area was covered in dense, primeval forests.[70] For Wu Zhaoqian, who had already crossed marshes, grasslands, mountains, and plains in the 1,500-mile trek from his home in Jiangnan, the forests were unlike anything he had ever seen. The poem "The Great Wuji" (*wuji* being a Chinese transliteration of the Manchu word *weji*, or "dense forest") describes Wu's fear and fascination at entering this unfamiliar environment:

> In the morning, we depart the rocky shelter, clouds chaotic on the
>   mountaintops,
> By evening, we dwell in the dark forest ten thousand cubits deep.
> A flood of trees rings the sky for more than a hundred miles—
> Dense pines that have covered the land for more than a thousand years.
> The frightened cries of the ice-dwelling sable ring through the forest;
> Among the dormant trees, bears hibernate, suspended in solitude.
> I've heard that anywhere you go, you can observe the work of Yu;
> But given the rugged road, I doubt he ever made it to this remote place.[71]

"The work of Yu" refers to Yu the Great, a mythical god-engineer from antiquity who marked out the "nine provinces" of China as he ranged the earth harnessing rivers and controlling floods. To suggest that Yu hadn't made it as far as this region was to imply that it was truly beyond the pale, unknown even to the one who had set the very boundaries of the civilized world.

After a four-month journey across hundreds of miles of mountains, rivers, and forests—and across thousands of years of Chinese history—Wu Zhao-qian arrived in Ningguta in the late summer of 1659. It was August, and the weather had already grown cold.

Creating Knowledge in a Land without History

Exiled from Jiangnan's filigreed whitewashed cities, Wu now dwelt in Ning-guta, a tiny, rugged frontier town at the northeast fringes of the empire. Located originally near the present-day town of Hailin in Heilongjiang province, Ningguta in the mid-seventeenth century was a military outpost of about three hundred families, nestled in a narrow valley along the Hailang River, a tributary of the Mudan River (from the Manchu *mudan*, meaning "curved"). Dwellings were humble cabins constructed of logs and bark. A stacked stone wall protected the inner buildings populated by Manchu officers and soldiers, and a crude wooden stockade surrounded the "outer" settlement of Han exiles and merchants. The mountains of the Zhangguangcai and Laoyeling branches of the Changbai Mountains rose to the east, north, and south. The land on the approach to the stockade was flat and under cultivation, but outside of the valley the landscape was forest and scrub.[72] Summers were short and pleasantly warm, but winters were long, cold, and remarkably full of snow. Today, the local economy tries to capitalize on winter tourism—the area is home to China's "Snow Town" (Xuexiang), a "snow village" extravaganza designed to rival the Spring Festival ice sculptures of Harbin. Snowfall accumulations can reach over six feet, while the twenty-four-hour average temperature in January hovers around 1 degree Fahrenheit.[73]

Wu was horrified by the extremes of nature in Ningguta. In letters to his parents written in the first few years of his exile, Wu describes his surroundings:

The bitter cold of Ningguta is like nowhere else on earth. From the beginning of spring to the middle of the fourth month, a great wind blows night and day, like the sound of thunder and the peal of lightning, so you can't see a foot in front of you. From the fifth month to the seventh month, the weather is constantly overcast and rainy. In the middle of the eighth month, big snow starts to fall. Rivers are frozen solid by the beginning of the ninth month. Snow covers the whole land and freezes solid; even if the sun shines warmly on it during the day, it won't melt. Everywhere you look for hundreds of miles, it's infinite white snow. By the middle of the third month, the snow starts to melt, but the grasses still don't sprout. The locals here say that since Han officials started arriving this year, the weather has grown unusually warm, but we southerners still find it unbearably frigid. Luckily, I have plenty of fur robes, and so I haven't frozen to death!

Uncle Fang (Fang Gongqian) often says, "People say there's a road to hell—but if you've been to Ningguta, even ten hells won't frighten you." He also says, "Those who get exiled to Shenyang are as lucky as if they got into heaven!" All of what I am saying I have actually experienced: it's not an exaggeration. Up until now, I haven't dared to tell you the extent of the hardships of this place—I was afraid of hurting you."[74]

Although Wu was trapped in a cold and alien environment, he found warm and familiar companionship with other Jiangnan natives who had been exiled to Ningguta. The "Uncle Fang" Wu referenced was Fang Gongqian (1596–1667), a prominent scholar from a wealthy Tongcheng family and a fellow exile. Considered one of the most talented minds of the late Ming, Fang had risen to great heights in the imperial bureaucracy and served the court as imperial tutor. After the fall of the Ming, Fang once again achieved positions of influence under the Manchu regime. As fate would have it, Fang was the chief examiner in the scandal-ridden Nanjing examination of 1657. Accused of corruption, Fang fell from the pinnacle of Qing intellectual life and was exiled to Ningguta along with his five sons. The sixty-three-year-old Fang became fast friends with the twenty-eight-year-old Wu, and the young poet frequently turned to Fang for lodging, support, and literary commiseration.[75] Fang shared Wu's horror of conditions at Ningguta and poured his observations into a torrent of poetry and essays composed during the two years he spent in exile.

It is clear from even a cursory glance at Fang's work that the scholar viewed Manchuria's environment with disdain. Fang gave his collection of nine hundred exile poems the wry (and self-aggrandizing) title *Writings from the "What Uncouthness?" Abode* (*Helouju wenji*). This title was an allusion to a quote from the Confucian classic *Analects*. Confucius, when asked by his disciples how he could stand to live among the uncouth barbarians of the east, replied by saying, "If the gentleman lives among them, what uncouthness could there be?"[76] This quote obviously had great significance for the sophisticated southern Chinese scholar living on the eastern fringes of the

Manchu's world. Even if he was in a barbarian land, like the great Confucius had two thousand years before, Fang would carry himself with dignity and bring Chinese civilization to the frontier.

In spite of his attachment to Confucian tradition, Fang found that ancient texts could not help him make sense of this new environment. He scoured his knowledge of centuries of Chinese history but could find no information about the ground he walked upon. Since he could not find reference to Ningguta within his books, the scholar's task of naming and classifying the things of nature proved exceedingly difficult. Frustrated by his inability to read his place, Fang complained that the northeastern frontier was, in essence, a "land without history." Without this history, it was impossible to know where to begin to write: "As for Ningguta, it is not in the geographies, so we do not know what territory it belongs to. It is not in the histories, so we do not know what [state] it belonged to. For a thousand miles there is no stone marker to consult, no antiquities to inquire of."[77]

It is not surprising that Fang's perspective on the knowability of his location derived first and foremost from his reading of the (Chinese) written word. From their comfortable studies, scholars in late imperial China could access a wide variety of books that described the boundaries and terrains of the empire from the most ancient times. The earliest "mythogeography" of China and surrounding territories, the *Shanhaijing* (Classic of Mountains and Seas) dated to the third century BCE. The *Yugong* (Tribute of Yu), attributed to Yu the Great, was a section from one of the original Confucian classics that described China's "nine provinces." Standard histories of the imperial dynasties from the Han (206 BCE–220 CE) on all contained geographical treatises, and later dynasties completed comprehensive atlases of their respective realms. Most detailed and comprehensive were gazetteers (*difang zhi*) compiled by local elites in cities and counties throughout China. While they focused primarily on local society, these gazetteers also included information on local mountains, rivers, flora, and fauna. Finally, travelers' writings were a genre that had become particularly abundant during the late Ming, when scholars like Xu Xiake (1587–1641) roamed the remote corners of the empire, climbing mountains, observing the qualities of stone and earth, and probing the sources of rivers.[78] Fang lamented that "Ningguta" was neither mentioned in the Chinese dynastic histories nor appeared in ancient geographies. Han scholars had not visited and recorded the flora and fauna. As the name did not appear in Chinese books, the place could not appear clearly in the mind's eye—the landscape, quite literally, was difficult to read.

＊

A landscape without history required a great deal of work to know, and Fang Gongqian, for one, did not seem to be particularly interested in investing energy in the task. Without textual referents for inspiration, Fang found the terrain of Ningguta pedestrian at best. His *Ningguta zhi* (Record of Ningguta), also known as *Jue yu jilue* (A brief record of a place at the edge), bristles with the author's grim view of the Manchurian landscape. Indeed, his euphemism for Ningguta, *Jue yu*, could also be read as "A Place without Hope": "What kind of place is Ningguta? There's no reason to go there and no reason to come back. This old man actually went and came back: how could it not be the doing of Heaven?"[79] Fang even used his exposition of the Manchu place-name to mock the landscape. The name Ningguta referred to the place of the Six (*ninggu*) Headmen (*da*). The Manchu word was transliterated into Chinese using characters that mimicked the sound but changed the meaning: common transliterations were "Ning-gu-tai," meaning "peaceful ancient escarpment," or "Ning-gu-ta," "peaceful ancient pagoda." As a classically trained scholar, Fang was compelled to probe the philology of place-names, but he did so in a manner that not only disregarded the Manchu meaning of Ningguta but also denigrated the locale. Fang opined that Ningguta, in spite of its name in Chinese, had "neither escarpment nor pagoda . . . only an uneven mound not even worth climbing."[80]

Fang's frustrations were a familiar theme in exile nature writing, and similar sentiments accompanied the observations of exiles from past centuries. Indeed, the Ningguta exiles saw themselves as part of this long tradition of frustrated frontier naturalists, and their list of perceived predecessors stretched back to the Han dynasty, fifteen hundred years before. In their writings, they frequently invoked the example of Liu Zongyuan (773–819), the Tang-dynasty scholar whose "Eight Records of Excursions in Yongzhou," with its bitter essays on "Prisoner's Mountain" and "Fool's Creek," was a classic example of how exiled literati projected their anger onto the landscape.[81] Exiles held that perception of the environment in an unfamiliar land was intimately intertwined with human emotion, and some found those emotions so powerful that they overwhelmed their vision of the nonhuman world.

Other exiles considered themselves more emotionally open to the northeastern landscape and thus, in their estimation, more capable of making accurate observations of nature. One such exile was Zhang Jinyan (1599–1670), another displaced southern literatus who wrote extensively about the Ningguta environment. Sixty-two years old when he was banished to the frontier, Zhang had already survived a long and tumultuous career that spanned both the Ming and Qing regimes. Sentenced to exile for allegedly composing seditious writings against the Manchus, Zhang arrived in Ningguta in 1661, where he joined his fellow Jiangnan compatriots in puzzling through Manchuria's environment.[82] Zhang's *Record of Ningguta's Mountains and Waters* (*Ningguta shanshui ji*) contains twenty-two short essays describing the features of the area. Only three passages are dedicated to the built environment, including the settlements of Ningguta, while the rest of the work names and describes the region's mountains, cliffs, rivers, and streams. Here and there, Zhang's *Record* also contains remarkable commentary on the way that human history, emotions, and perception created the exiles' understanding of the nonhuman world.

In Zhang's perspective, exiled Han Chinese literati in Ningguta had a very important role to fill: they were in essence responsible for creating valid knowledge about the environment. In the eyes of these exiles, local knowledge was insufficient. As evidence, Zhang offered that on the "ten thousand mile" road to his exile in Ningguta, he asked locals the names of the many mountains and rivers he passed, but none could produce an answer.[83] Zhang did not specify who these "locals" were or the language they spoke, but Zhang lay the responsibility for this mutual inability to communicate at their feet. Locals, Zhang concluded, did not understand their surroundings because they were both habituated to them and at the same time frightened by them. In their eyes, the landscape was "filled with mountains everywhere," but they were dangerous places where tigers and wolves lurked, and thus best avoided.[84] As new arrivals from the southlands, Chinese exiles noticed the unfamiliar and the new; thus, they could view nature with attentive eyes. In addition, their classical Chinese education equipped them with an understanding of the ways that nature should be parsed. As far as Zhang was concerned, literati Han Chinese exiles from the south were the only ones who had the capacity to observe and record: the only ones whose "senses had the ability to reach, minds had the ability to measure, and words had the ability to categorize" the environment. Thus, Zhang declared (in a quote used as the epigraph for this chapter) that natural knowledge of the frontier had to "wait for the exiled official to achieve."[85] Through their efforts at naming and categorizing, the exiles would bring clarity and meaning to the lands "beyond the realm."

While exiles felt that they alone were qualified, the task was not easy. Prefaces to Zhang's *Record* problematize the exiles' emotions—resentment, indignation, and anger—seeing them as an obstacle to the creation of natural knowledge. According to Zhang and his colleagues, writing about the frontier required a purging of intense nostalgia for home and a diminishing of the negative emotions that could cloud the eyes and intellectual faculties: a state that could be reached only through a certain spiritual self-discipline acquired over the course of many years. The inevitable hardship of place could blind men to their surroundings and make them take their personal misfortune out on the innocent mountains and rivers around them. The result was writing that reflected the inner life of the observer but failed to capture the external environment. Taking the Han literati tradition to task, Zhang's work claims that much of China's frontier nature-writing tradition was suffused with this problem.

Zhang's critique is remarkably similar to the historian Edward Schafer's analysis of Tang-dynasty exiles to the mountain frontiers of Guangdong and Vietnam. Schaffer notes that Han Chinese banished to the Vietnam border during the Tang dynasty (618–907) were "plagued" by nostalgia, an intense longing for a pre-exile life that kindled a bitterness toward their surroundings. Schaffer argues that this sentiment burdened exiles with a "purblind insensitivity" to their new environment. Hobbled by nostalgia, exiles were unable to "open their eyes and ears" to what was around them, and as a result, they relied on "established apparitions"—tropes from received knowledge—to represent nature. Exiles were thus unable to generate novel empirical observation of their surroundings, and so remained "prisoners of their ecological lexicons."[86]

A preface to the *Record of Ningguta's Mountains and Waters* suggests that Zhang Jinyan had overcome the problem of emotion. Because of his unique personality and his ability to be detached from the concerns of the world, the sixty-five-year-old exile "found joy in the mountains and waters" and had become "settled (or at peace) with his locale." Having come to terms with his situation, he could approach the environment without anger and was thus qualified to "convey and narrate" his surroundings.[87] Commentary on Zhang's writing in the *Record* praises the closeness of his descriptions to the reality of nature: "In the space between the ink and paper, the reader's eyes can directly engage with the landscape." In other words, for his contemporaries, Zhang's words were such a faithful copy of nature that there was no distinction between reality and writing.[88] The exiles claimed that their circumstances gave them the ability to leave behind human emotion and grapple directly with the new. Of course, the knowledge they produced, like all human knowledge, remained mediated by their human desires and perceptions. Nevertheless, in their pursuit of a closer connection between the eye, the word, and the world, the exiles of Ningguta began to see their own human desires and perceptions as a *problem*, an obstacle to the apprehension of nature.

Zhang Jinyan claimed to be able to view his surroundings without the veil of emotion that plagued his colleagues such as Fang Gongqian, but he did agree with Fang on one point: Ningguta was a place without names, without categories, without history: "Alas! Since ancient times, how could we have known of Ningguta's landscape? If we look at maps or check [textual] descriptions, all refer to it simply as a 'remote and barren area.' You never hear that Ningguta's forests, springs, streams and hills actually have much that should merit the 'cultured hermit's' appreciation."[89] Zhang criticized Chinese scholars' tendency to rely solely on the authority of the ancients to understand the frontier, and was particularly adamant that Chinese scholars needed to explore, analyze, and write about the far-flung territories that lay beyond the pale: "If one travels to the Kaiming Gate, explores the caves of Danxue, climbs the peaks of Mount Buzhou, or ascends the cliffs of Kongtong [Kongdong] in search of marvels and wonders, [one will find mountains] that are not recorded in the traditional books. Yet we in the modern era still want to use evidence from a couple of lines in ancient texts or [comparisons with] the Five Directions in order to verify [their existence]: such a task is quite difficult."[90] Evoking exotic locales from ancient geographies and legends—including the Kaiming Gate atop the heavenly Mount Kunlun and Mount Buzhou, the mythical peaks said to hold up the sky—Zhang argued that in his current day, scholars could travel to the far corners of the earth and discover unknown geographies. Furthermore, one could not study these places using ancient concepts such as the five phases, which grouped all the phenomena of the world, from bodily organs to directions, into five different categories. Instead, the scholar's own faculties equipped him to measure and parse the natural world. Indeed, Zhang questioned the ability of texts to establish knowledge of the environment, and he vigorously advocated instead for the direct observation of nature: "One can understand the principle of mountains by observing them with one's own eyes from a close range, or one can stand at a distance and be ignorant about them. One can be casually familiar with something and find it strange, or one can [actually] survey it."[91]

Zhang's writing about the need for using "one's own eyes" and "surveying" phenomenon directly places him within a major intellectual trend of the Ming-Qing transition era—that of *shixue,* or practical learning (also called substantive or concrete learning or studies). Stemming from the ideas of earlier influential thinkers such as the twelfth-century philosopher Zhu Xi, who encouraged "the investigation of things" (*ge zhi*), *shixue* emphasized the importance of producing knowledge geared toward the "real world" as opposed to meditative or purely literary goals: knowledge that was "of use" rather than for the purposes of empty speculation. It has been associated most closely with scholars of the Ming-Qing transition, towering figures such as Gu Yanwu (1613–1682), Huang Zongxi (1610–1695), and especially Fang Yizhi (1611–1671), men who lamented the passing of the Ming and feared that superfluous, abstract learning was partially to blame for its collapse.[92] *Shixue* can be seen as a form of empiricism that emerged in the seventeenth century and continued to influence major intellectual trends through the end of the Qing. But as Ben Elman has shown, while *shixue* certainly emphasized "descriptive knowledge of the natural world," most of its practitioners were also deeply concerned with analyzing texts as well.[93] The exiles to Manchuria maintained a similar approach to understanding the borderland: informed by textual tradition yet dedicated to the goal of "observing with one's own eyes from close range."[94]

This recognition of the limits of textual evidence helped fuel an enthusiasm for firsthand observation of Ningguta's terrain. The exiles' writings reveal frequent and systematic excursions to nearby mountains and rivers. In this activity, they were in part replicating the lives they would have had had they remained in Jiangnan: day trips to enjoy the beauty of local scenic spots were common leisure activities for literati in the south. But while they mimicked the spirit of Jiangnan jaunts, these trips to the environs around Ningguta were different: somewhere between aesthetic excursion and geographical exploration.

Records of their roamings, such as this poem by Wu Zhaoqian describing what was in essence a whitewater-rafting journey with his colleagues, give insight into the spirit of embodied adventure and discovery the exiles felt on these potentially dangerous trips into the wilderness:

An Ode, with Ten Rhymes, on the Occasion of a Boat Trip to
White Cliff Pass with Friends, Taken at Spring's End When the
Ice on the River Had Melted

The crystalline ice has disappeared, and we cautiously lower our solitary rafts
   in the flow.
At first we are terrified as the boundless landscape zooms past our eyes.
Ducks and geese float in the morning light; fishlike dragons slither in the
   mist.

Azure spring grows at the water's edge, while the crimson sun splits the
corners of the sky.
Suddenly we hear the shallow rapids thunder; then in a blink of an eye, the
waves settle again into silence.
An unknown wind drives the skiffs, we are stunned at how fast the snowy
peaks go by . . .
The vast and mighty waters fill our minds, leaving our bodies in peril on this
remote frontier.[95]

Exiles' writings bear evidence of these dramatic exploratory journeys and
contain firsthand descriptions of landmarks around Ningguta that are still
recognizable today. Zhang's *Record of Ningguta's Mountains and Waters* begins
with a description of Ninggu-tai, or "the Ninggu Escarpment." This mountain
is raised and flat on top, like a platform, hence the name *tai* (platform) in Chinese. Zhang vividly outlines the strange stone formations of the escarpment,
the difficulty of accessing it, the rivers that encircle it, and the way the rivers
cascade over its cliffs. From this description, it is clear that Zhang is naming
what is today known as Diaoshuilou Falls, a sort of mini–Niagara Falls at the
northern part of Jingpo Lake, approximately fifty miles south of Ningguta.
Zhang provides vivid descriptions of the slopes, caves, outcroppings, flora,
and fauna of several more mountains: Shephard's Mountain, Cliff Mountain,
Tiger Mountain, Officer's Road Mountain, Sleeping Buddha Mountain. Like
his discussion of the terrain around Jingpo Lake, Zhang's detailed writing suggests empirical observation, a series of in-person and physically challenging
explorations through the forests, mountains, and rivers of the region.

Zhang's descriptions did not come easily. Without relying on history,
without recourse to received geographies, Zhang struggled with how he was
to name the features of this foreign place—especially the region's most distinctive feature, its mountains. After all, Zhang observed, mountains could
not just name themselves:

It is said that mountains and rivers are named after their natures [or follow
their essences]: Mount Hua is magnificent, and is therefore named "Hua"
[magnificent]. Song Mountain is tall (*song*), and so it's designated as Mount
"Song." . . . Mount Tai is called "Tai" [Great] because it is uniquely large
among its surrounding mountains. But it's not as if this was the Creator's
intent from the moment the mountains were carved out of the vast waste.
Even if Cangjie hadn't invented writing, or Yu the Great hadn't surveyed the
mountains and rivers, if the Yellow Emperor hadn't performed sacrifices

[to sacred mountains], the world's mountains still arose towering out of the primordial murk at the beginning of creation. How could it be, then, that their names as mountains or as sacred peaks could arise spontaneously from their "natures" (*xing*)?[96]

In this sophisticated musing, Zhang allows that there is a nonhuman world separate from human perception and meaning making. Scholars today debate whether or not there was a concept of nature in premodern China—a sense of a unified nonhuman world separate from the world of humans. While the term *ziran* used to designate "nature" in modern Chinese is undoubtedly a neologism, it is clear from this passage that Zhang had a concept of the environment around him being separate from the human world. Things like mountains and rivers (but especially mountains) emerged from a primordial waste, brought into being by "the creator" (*zaowuzhe*, literally "that which made things").[97] The categories, names, and classifications bestowed upon these things did not emerge directly from nature per se but were a result of outside observation—the application of human perception and intelligence.

Zhang Jinyan no doubt felt himself particularly well qualified to apply his intelligence to Ningguta's mountainous landscape. Back in China proper, he had cultivated a reputation as an expert on mountains, collaborating on the publication of essays about famous mountains and a revised, massive seven-volume history of Mount Tai.[98] Mount Tai (Taishan), the Eastern Sacred Mountain, was the most storied mountain in China, worshipped by generations of emperors and common folk throughout the empire. The mountain was possibly one of the most "inscribed" of all landscapes in China, existing in texts dating back to the hoary imperial past. At the same time, the mountain itself was literally inscribed—poetic names of cliffs, grottoes, and passes were carved like labels directly onto its physical features.[99] In Zhang's perception, the landscape of Ningguta was blank, so the scholar had to inscribe it with meaning himself.

Possessed of immense erudition about the natural world, Zhang nevertheless wrestled with the problem of extending knowledge from China proper to interpret the nature of the "extreme territory." Was it possible to use principles derived from life south of the Great Wall to make sense of life at the edges of the world? The exiles were surrounded by many novel phenomena; for example, on their excursions, they encountered myriad plants that they could not identify (*bu neng ming zhe*; literally, "those which cannot be named"). Novel flora made Zhang recognize the limits of received knowledge: he marveled that there could be things in nature that were entirely unknown even to Shennong, the god of agriculture who had identified the hundred herbs of Chinese medicine.[100]

At the same time, Zhang noted that the environment of this remote, inhospitable place still bore a puzzling resemblance to the environment of Jiangnan. There were trees clearly identifiable as pine, linden, birch, willow; plants that included peonies, lilies, ferns. There were also plants that *seemed* familiar but on closer inspection were quite unknown: plants that seemed like jasmine, others like sweet osmanthus or like geraniums, but were actually something else. Similarly, Zhang noticed that mountains in this region seemed similar to and yet departed from the principles that ordered nature in China proper. As with mountains within the Great Wall, the mountains around Ningguta arranged themselves into undulating folds that stretched out to create uninterrupted ranges. Why, then, was the Ningguta escarpment cut off, a singular elevated plane surrounded by water and strangely devoid of vegetation? It was clear that a toehold of familiarity remained, but the nature of the frontier manifested so much diversity that it was hard to grasp.[101]

Nevertheless, this "shock of discovery" did not completely dislodge traditional templates for making sense of nature. In particular, the cosmologies of *qi* from below the Great Wall could still apply in the northeastern frontier. Zhang postulated that the Ningguta escarpment's remarkable features were caused by the active *yang* action of the region's rivers. The swift movement of the waters had cut the escarpment off from other mountains around it, thus interrupting the *qi* that formed the rest of the range. Zhang also noted that the top of the escarpment was devoid of trees or plants. He attributed this phenomenon to the clash of the waterfall's active *yang* with the obscure or dark *yin* of the mountain's *qi*, which thus cooled and withered the plant life above the waterfalls.[102] Zhang also used *qi*-based thinking to explain the living things of Ningguta. The presence of different plant life in each region, according to Zhang, was a reflection of the *qi* of the earth for each region. As the quality of *qi* of the earth in each locality was different, it gave rise to different life-forms. The *qi* of the "Land at the Edge" was different from that of the lands below the Great Wall but not radically so: hence many of the plants looked familiar but differed in particular ways.[103]

As Zhang parsed the Ningguta environment according to *qi*, yin, and yang, he also attempted to place its phenomena within hierarchies dictated by other traditions of China proper. Mountains, in particular, were objects to be ranked, and Chinese history was replete with mountains assigned special significance above all others. Unlike our contemporary world, the hierarchy for mountains in China was historically not based on height alone. The most exalted mountains in China were the sacred mountains (*yue*) of the five cardinal directions: Hengshan in Shanxi was the sacred mountain of the north, another Hengshan in Hunan was the sacred mountain of the south. Songshan was the sacred mountain of the center, Huashan in Shaanxi was the sacred mountain of the west, and Taishan in Shandong was the sacred mountain of the east.[104] These mountains had been the focus of admiration, worship, and sacrifice since the beginnings of imperial China, and as a mountain expert, Zhang was exquisitely aware of this important tradition.

While maintaining reverence for China's significant mountains, Zhang criticized the ancients for belittling the mountains of remote regions. The classics, he lamented, considered even the finest towering mountains on the frontier to be nothing but dirt hills simply because they were outside of China proper. As a result, frontier mountains that were equally as lofty as the sacred mountains of China were not considered worthy of placing within the hierarchy of the five directions. It seemed particularly unfair to Zhang that the northeast's greatest mountain, Changbaishan, the Long White Mountain, had not been more lauded in China's annals of mountain lore: "Since crossing the Liao River and entering the frontier, I have gazed in the distance at Mount Changbai, stretching unbroken for 1,000 *li*. This origin of the two rivers, so great and yet quite close, today is still not ranked among the sacred mountains. Does it have to be larger by 1,000 or 10,000 *li* [before it is honored]?"[105] Zhang considered the massive Mount Changbai to be more than worthy of both veneration and exploration. Not only was it just as large, if not larger, than many of China's sacred mountains, but Zhang had discovered another important fact: the underlying principle that distinguished the sacred mountains of China proper also operated for Changbaishan: "It is said that the mountains of Shanxi [a province in north-central China] all face the north out of respect for the Sacred Mountain of the North [Heng Shan, located in the northern part of Shanxi Province]. I've observed that the majority of this region's mountains face west, and there are also those that face south. Since Ningguta is located in the northeast [of Mount Changbai], is this not because these mountains are facing their lord?"

Here Zhang applied principles at the intersection of Chinese geography and geomancy to understand patterns exhibited by Manchuria's mountains. Mountain ranges were conceptualized as visible manifestations of the earth's *qi* energy: the pattern of their ridges was visible evidence of the movement of this *qi*. Through what the art historian Stephen Whiteman has called a "sacred chorography," mountain ranges could be seen as flowing from the energy of a significant origin.[106] Here Zhang stated as fact that the ranges of Shanxi were arrayed so their slopes faced the direction of the region's sacred mountain. Through analogy, he postulated that the mountains of the Ningguta region were arrayed in a way that followed the patterns of nature evinced in China below the Great Wall: the mountains of Manchuria all faced the most significant mountain in the region, the slopes of the mighty Long White Mountain, Changbaishan.

With this meditation on mountains, Zhang put forth a basic claim: this place might be beyond the borders of civilization, but it was entirely within the boundaries of knowable nature. The nonhuman world in the far northeast held many unusual phenomena: the exiles found themselves in an extreme environment, a place they called the "remote region," the "place at the edge," even "the land without hope." Through a process of emotional reckoning, textual analysis, and empirical observation, exiles came to an important conclusion about their ability to make sense of this land without history. Although the environment was foreign, it was possible that the principles that created, structured, and animated nature here were the same as in China proper. Because nature itself adhered to the same "universal" principles everywhere, even in the far northeast, the standards, approaches, and language for naming things in China proper would hold here as well. The exiles could extend their universal principles for creating knowledge to encompass, order, and embrace the environments of Manchuria.

Conclusion

In early modern China, as in early modern Europe, ways of thinking about nature were in flux, poised between textual knowledge and the knowledge of the senses. Like their European contemporaries in the age of discovery, Chinese scholars during the transition to the Qing were confronted with an unfamiliar terrain and grappled with how they might write an environment that had mostly gone unrecorded in the geographies of their native culture. Early Qing exiles did not cross oceans to reach new shores. Nevertheless, literati from south China such as Wu Zhaoqian, Fang Gongqian, and Zhang Jinyan underwent an arduous journey of over fifteen hundred miles (roughly the distance from Miami to Minneapolis), crossing multiple ecosystems, climates, and terrains, and finally entering a place they perceived as radically foreign to the land of their birth. Along the way, the ghosts of the past haunted their journey. Once they arrived at their destination, the ghosts subsided, but the exiles were still terrified by what they experienced as a bleak—and blank—environment.

In Ningguta, the Han exiles saw their task as calming their emotions and training their eyes, ears, and mind to name a nature that could not name itself. This impulse to suppress the self can be fruitfully interpreted as reflecting an impulse toward something akin to objectivity. As Lorraine Daston and Peter Galison have argued, objectivity has a history. It is not a singular concept but has encompassed and emphasized different things at different times. These historical emphases have included not only practices associated with science, such as quantification and the use of instruments to register data, but also more abstract ideas, such as a "belief in a bedrock reality independent of human observers" and a "desire for emotional detachment."[107] To write the passage to their new environment, exiles initially relied on the exiles of the past whose poetry had painted northern borderlands as a tragic, battle-scarred frontier. As such, nature was a cultural and historical landscape that served primarily to reflect the exiles' personal sorrows. But bodily experience with the novel environs of Ningguta—through movement involving mountain hikes, plant collecting, and whitewater rafting—inspired some to think of themselves as slipping the confines of history and engaging more directly with their environment. If objectivity is "first and foremost . . . the suppression of some aspect of the self," then the natural knowledge of the Ningguta exiles, even if it was expressed in poetry, should be included in the history of this concept.[108]

This natural knowledge had implications for the definition of the empire. As exiles slowly moved through Manchuria at the measured, deliberate speed of human or animal perambulation, the boundaries between familiar and unfamiliar, alien and native, were literally shifting under their feet. As Wu Zhaoqian journeyed in a northeasterly direction from Beijing, he described the rivers and mountains as being inscribed either within the history of the Han or beyond the boundaries of Han reach. And yet the definition of what constituted Han or non-Han territory was unstable. Exile poets mourned the loss of their Jiangnan landscape, but it did not necessarily constitute a Han *landschaft*, like the German forest or the British lowlands.[109] The empire was simply too large to be represented by any one terrain, and which landscape counted as "native" relied on a combination of political, historical, and environmental cues. By the time Wu entered the vicinity of the Great Wall, he perceived the territory behind him as already within "the Pale," a liminal borderland between what was Han and what was foreign. The wars that ancient

Chinese empires had fought with northeastern peoples in the region rendered possession uncertain: the lands of the Liao River were simultaneously part of the empire and part of the other. The Willow Palisade marked a significant passageway into an outer world, but Wu read its presence to signify that a great deal of land north of the Great Wall (or "outside the pass") was now in fact "within the pass," or part of China proper. Only the dramatic environmental shift near Ningguta, with its impenetrable forests and endless snows, was experienced as a bona fide entry into an alien land. But even here, after a period of transition, the exiles concluded that nature could be known through the same set of universals that allowed for the knowing of nature within China proper. As they observed, they consciously wrote the nature of Manchuria into the empire.

*

Even as they created new knowledge, this first generation of exiles could never entirely escape the melancholy of their condition. As the years passed, the once vibrant community dissipated. Zhang Jinyan died in exile in 1670, at the age of seventy-one, just two years after writing his *Record of the Ningguta Landscape*. Fang Gongqian, finally ransomed by wealthy friends back in Nanjing, was able to return to the south years before. Wu Zhaoqian was not so fortunate: he was fated to stay on the frontier for a total of twenty-four years, long after his friends had died or escaped. During this time, Wu created certain bonds with his Manchu captors: he developed particularly close relationships with Ningguta's top officials, including the military governor Bahai (d. 1696) and the vice governor Sabsu (d. 1701).[110] In 1673, Bahai hired Wu Zhaoqian to tutor his sons in the Chinese classics, and Wu's own son (whose childhood name was "Suhuan," literally "Return to Suzhou") grew up together with Bahai's sons.[111]

Wu Zhaoqian's exile in Manchuria coincided with one of the great existential threats of the early Qing: the rise of a Russian military presence along the Amur (Heilongjiang) River and its tributaries. As early as the 1640s, Cossacks and adventurers seeking inroads into the rich resources of "southern Siberia" probed the Amur and pressed its local peoples to offer up tribute of furs and food. Resistance by indigenous groups and Qing forces failed to deter the Russians, who by the 1650s had established permanent settlements on the Amur and were venturing southward toward Ningguta. By the 1670s, the Kangxi emperor (r. 1661–1722) had devised a long-game strategy of building Qing forces in the northeast with the goal of eventually neutralizing the Russian threat. Other enemies to the far south also demanded Kangxi's attention: Chinese generals who had once served the Qing had turned against the empire in what would become the Revolt of the Three Feudatories (1673–1681). But protecting the lands of his ancestor's origins remained a Qing priority. Kangxi, the son of the emperor who banished Wu, made Manchuria a permanent fulcrum for the empire's security. As a result, the Ningguta of Wu's long exile was a place that hummed with martial energies.[112]

Wu's later poems reflect this sense of threat and resistance. Wu derided the Russians as evil "foxes" and "vultures," animal-like beasts of the northlands who were so wild that they had to be met with force. He glorified the exploits of the commanders Bahai and Sabsu in their battles against the Russians, portraying them as strong men who wield large swords and "let arrows fly from the bow with ease." In self-deprecating comparison, Wu states that when he tries his hand at a Manchu bow, he "cannot even bend the string." The poet even equated the Manchu generals of Ningguta with the great generals of the ancient Han dynasty who had battled against the "barbarian" Xiongnu tribes to the north (Chinese equivalents of the Huns) fifteen hundred years before.[113] It is perhaps strange to think that Wu, who was cast into exile by a Manchu emperor and whose father was a Ming loyalist, would laud Manchu warriors as valiant Chinese heroes of old. But beneath these lines in praise of Manchu victory there lies a hint of bitterness. This first generation of exiles had helped write Manchuria's nature into the empire—but for them, Manchuria still represented a landscape of defeat. It would take another journey to reveal Manchuria as a place of energy and promise: a land that harbored dragons.

WHERE THE DRAGON AROSE

Discovering the Dragon through Number and Blood

The land marked by such a dragon is indeed the most
excellent of all lands.
—GAO QIZHUO, Qing official and *fengshui* expert,
in a memorial to the emperor, 1730

On a chilly March morning in 1682, a massive procession rumbled out of Beijing's Dongzhi Gate.[1] The Kangxi emperor, with his consorts, his favorite son, his courtiers, guards, and a military train of thousands of soldiers embarked on a seven-hundred-mile journey northeast to visit the tombs of his ancestors in Manchuria. Kangxi (1654–1722, r. 1661–1722) was the first Manchu emperor to be born south of the Great Wall. He had traveled to Manchuria once before, in 1671, but on that occasion, he was only seventeen years old and his tour had not ventured much beyond the old Manchu capital of Mukden.[2] This time the emperor was twenty-eight years old, a vigorous, victorious ruler who had successfully suppressed—at great cost—a massive Chinese uprising against his dynasty in the southern provinces. With the Revolt of the Three Feudatories vanquished and China proper all but unified under Qing rule, Kangxi sensed it was time to explore the northeastern portion of his realm and see for himself the "land where the dragon arose" (*long xing zhi di*): the land that had given sustenance to the earliest endeavors of his revered ancestors, the founders of the Qing dynasty.

The entourage, called the "eastern tour," was so large that some estimated it stretched over six miles long. Clouds of dust kicked up by the passage of thousands of horses meant that the individual participants could see neither the beginning nor the end of the mass they moved within.[3] The tour traveled with its own food: thousands of pounds of rice and flour, huge vats of sesame oil, dozens of carts of vegetables, and thousands of live animals—ducks, chickens, pigs, cows, and sheep on the hoof—along with its own shelters, kitchens, medicines, and clothing.[4] The tour resembled the exodus of a nation or a cattle drive, albeit one that was led by the Son of Heaven.

Amid the hundreds of carts laden with provisions and tents, several carried more esoteric objects: a reference library of dynastic histories and geographical treatises, and a set of astrolabes, star charts, and telescopes: technologies that would be used by two important members of the entourage to create knowledge of the nature of Manchuria.[5] The retinue of thousands included two brilliant scholars who were close to the emperor: one, a thirty-seven-year-old southern Chinese literatus named Gao Shiqi, the other, a sixty-one-year-old Flemish Jesuit priest named Ferdinand Verbiest. Each had served as personal tutor to Kangxi and resident court expert in his respective field: Gao in Chinese history and literature, Verbiest in European astronomy and mathematics. The Kangxi emperor had both men serve him on the imperial expedition to the northeast. Commanded to observe and record the Manchurian landscape, Gao and Verbiest produced two very different narratives, each one deploying the scholar's highest skills for knowing the natural world: Gao, through history, and Verbiest, through number.

This chapter employs a close reading of the narratives created by Gao and Verbiest to scrutinize the relationship between the eastern tour of 1682 and the creation of knowledge about the nature of Manchuria. Toggling between historical texts, period maps, and GIS-assisted visualizations of the terrain (together with my own travels in contemporary northeastern China), I adhere closely to the route of the tour, following the experience of individuals as they traversed the contours of the land.[6] The tour passed through multiple environments: coastal marshlands, rugged mountains, broad valleys, expansive grasslands, and dense forests punctuated by networks of whitewater rapids. Each environment, each position, was not just a place that was passed through on the way to important destinations but was imbued with historical, scientific, and spiritual significance. Like Wu Zhaoqian on the road into Manchurian exile in 1659, the members of the imperial tour who entered Manchuria in 1682 created knowledge about the environments they encountered through an embodied process that entangled terrain, perception, and echoes of history. But unlike the Han exiles who painted a landscape of defeat, the members of the eastern tour sensed, celebrated, and even measured an energy of victorious power within the land of Manchuria, manifest in the metaphor and reality of the dragon.

Several of the most prominent scholars in Qing history have touched upon the eastern tour in their work. Descriptions from Kangxi's eastern tour are used to delightful effect in one of Jonathan Spence's earliest works, his *Emperor of China: Self Portrait of Kang-hsi*, first published in 1974—to date still one of our most insightful studies of the Qing dynasty's most influential ruler.[7] In his landmark work on imperial touring, *A Court on Horseback*, Michael Chang shows that such tours, while part of a centuries-long history of imperial rulership in China, were used to particular effect by the Manchu rulers of the Qing dynasty. Qing emperors left the confines of the Forbidden City in grand processions so frequently that they were more often on the road than in Beijing. As they resembled military expeditions, tours were clearly part of the Qing rulers' strategy for maintaining a martial spirit among court retainers and capital bannermen, but they were far more than that. Chang presents the tours as "political theater" designed to enact dominance through "multi-valent ritual practice," a form of power Chang dubs "ethno-dynastic rule": government effected through the personal influence of the emperor as the head of a dominant Manchu ethnic group.[8] Mark Elliott has perceptively linked the eastern tours to the creation of Manchuria as a "ritualized homeland," one aspect of a much larger project of the Qing court's creation of a distinctive Manchu territory and identity. Elliott briefly discusses Kangxi's second eastern tour, emphasizing the rituals Kangxi conducted at his family's tombs as well as mentioning the veneration of Mount Changbai that Kangxi conducted while at Girin Ula. Overall, the important work of the 1682 tour appears to take place primarily in urban centers, and the eastern tours in general are characterized as "excursions on a modest scale."[9]

My reading of Kangxi's 1682 journey echoes these overall conclusions but differs in its texture and emphasis. I focus in on the eastern tour as a remarkable and sometimes brutal eighty-day exercise in inscribing power and victory over wide swaths of the Manchurian environment. The rituals for ancestral tombs and the sacred mountain of Changbaishan were important goals, but the significance of the terrain was written daily in blood, poetry, and number all along the 1,500-mile long route. The tour was an effortful, mindful mission of collecting information and shaping meaning, enacted through bodily movement across intentional sites. This movement was no cakewalk. Through the eastern tour, the "ritualized homeland" was experienced as a daunting, dangerous frontier. The entourage did stop at the city of Mukden and enjoyed an urban setting for several days, but much of the journey was through rough, sparsely inhabited terrain, and included segments that intentionally diverged from the relative comforts of the imperial road. As they crossed whitewater rapids and navigated mountain precipices, members of the entourage—the bookish poet Gao, the aged Jesuit astronomer Verbiest, and even the emperor himself—all had close brushes with death. The tour was pushed forward in an almost maniacal way by the will of the Kangxi emperor, who sought to commune with the past not only through stately ritual but also by exposing his court to the same physical extremes experienced by his long-deceased ancestors. This chapter follows the tour's participants as bodies moving through space and probes how natural knowledge was created at the contact points between human experience and the environment.

The journey of 1682 also gives us a unique opportunity to explore the nature of Qing knowledge about nature. The tour was a kinetic manifestation of what the art historian Pat Berger, in her analysis of Qing architectural forms, has called a "collaged vision" of Qing rule.[10] The tour was a tour de force of landscape inscription that drew upon multiple techniques from multiple sources: poetry, the whole-scale slaughter of wildlife, history writing, mathematically precise astronomy, *fengshui* readings, and funerary ritual. Using the talents of Han Chinese, Europeans, and Manchus, the emperor's tour was an immense multilayered project designed to create a Manchurian environment that was both known and numinous, surveyed and sacred. Indeed, it is most useful to see the eastern tour as both a pilgrimage, a movement through space that imbued the terrain with spiritual significance, and as a surveying expedition, a movement through space that gave each point along the way significance because of its relationship to number.

Approaching the eastern tour as both pilgrimage and scientific expedition encapsulates Pamela Crossley's influential idea of "simultaneous rulership," a concept used to describe a Central Asian style of ruling in which the emperor manifested multiple personae that spoke to multiple constituencies through a cosmological and political "bundling of functions." The utility of this model, Crossley points out, lies not only in illuminating how the Qing maintained rule in an ethnically diverse empire; it also helps the modern historian avoid constructing "false dichotomies" between the "sacral" and the "rational," between "numinal" and "phenomenal" in the knowledge that the Qing created.[11] This perspective helps us make sense of a phenomenon that seems full of dichotomies: a voyage of discovery that traveled to a familiar place; a staged historical reenactment that was simultaneously a real military exercise; a tour of an exclusive Manchu homeland that relied on European and Chinese technologies; a journey that generated both spiritual resonances and geographical coordinates. We might be tempted to separate these categories of history, science, and spirit, but those on the eastern tour saw no contradictions in this combination.

The multivalent nature of this knowledge is perhaps best embodied in the form of the dragon, an entity that was conceptualized simultaneously as a physical being and as a symbolic metaphor. The eastern tour could be characterized as an expedition that was in search of dragons: an object of discovery that was simultaneously hidden in the Manchurian environment and everywhere evident. The dragon was both the person of the emperor and a symbol of the emperor, abundantly manifest in the language, architecture, and material culture of the Qing court. The dragon was at the same time a real (but elusive) biological entity and a manifestation of the energy of nature, visible foremost in formations of terrain, detectable in the stars, and sensed in the atmosphere of a place. Ultimately, dragons needed both to be discovered (already present in the land but found only through careful observation) and created (produced through human ritual, artistic, and literary effort). The tour inscribed Manchuria as a place where the potentially terrifying energy of the dragon rolled through a wild land and could be channeled and imbibed by the person of the emperor. The eastern tour would allow the Kangxi emperor, and by extension, the empire itself, to recharge and reconnect with the cosmic energies embedded in the environment where his ancestors had lived, conquered, and were buried. At the same time, the dragon (and its manifestation, the dragon vein) was not something that was just there—it had to be discovered, traced, and mapped into the reality of the empire. Indeed, following Crossley's simultaneous rulership warnings against dichotomies, this search for dragons forces us to transcend the literal and metaphorical, the visible and the invisible in Qing knowledge formations. The eastern tour was an embodied process of creating knowledge about the nature of Manchuria and imbuing it with new meanings: marking the land while moving through it, experiencing its history and energies while embossing—sometimes through sheer brutality—an imperial imprint upon the land.

Landscapes of History, Number, and Blood

In the shadow of the Long White Mountain, the myriad mountains tower,
Guards clear the road for the Son of Heaven as he travels to the east.
The ten thousand riders, assembled like clouds, all gaze at the distant Penghai Gulf,
In the melting ice of the third month, they cross the Liao River . . .

The flowing Black Water and the Pine Flower River
Swell with grandeur as the Six-dragon chariot passes.
The *qi* of Kings connects with faraway Muyeshan,
And the thousand cities along the way prepare the sacrificial offerings . . .

In the first passages describing his journey with the emperor, *Hucong dongxun rilu* (Diary of serving with the imperial eastern tour), Gao Shiqi presents poems written by the Qing court's top literati celebrating the departure of the imperial cortege.[12] These courtiers outdid themselves in their send-off, depicting the heroic landscape that awaited the passing of the emperor. Their poems resounded with images of the Long White Mountain (Changbaishan) and the Black Water (Heilongjiang), symbols of the land of the Manchus.[13] At the same time, the lines were resplendent with allusions used to laud Chinese emperors since the Han dynasty. The mountains tower unceasingly into the distance, their azure clouds thick with royal *qi*, while rivers swell in anticipation of the emperor's chariot, pulled by "six dragons" that soar into the heavens.

With the rivers, seas, valleys, and mountains of the northeast humming with imperial energies, these poems could not be farther in spirit from the stark, terrifying landscape of Manchuria portrayed twenty-three years earlier in Wu Weiye's send-off poem for Wu Zhaoqian, "A Song of Sorrow for Wu Jizi," discussed in the previous chapter. In that poem, Manchuria is a place where monsters lurk in the Black River, where the ghosts of defeated soldiers walk the earth, and where the landscape is unknowable: "Mountains are not mountains, rivers are not rivers." But the Han Chinese in the employ of the Manchu emperor were in the business of using Chinese culture to celebrate the ruling house. Through their literary skill, the environment of Manchuria was expressed as the apogee of imperial grandeur.

None was a more expert practitioner of this technique than the literatus and companion of the Kangxi emperor Gao Shiqi (1645–1703). Gao Shiqi was an interesting anomaly at the imperial court. Born into a poor family but possessed of a remarkable literary genius, Gao moved to Beijing and managed to obtain at the youthful age of twenty-six a position in the empire's top academic institution, the Hanlin Academy, even though he had no advanced degree. The young emperor was so impressed by Gao's literary skills that he invited the upstart scholar to serve as his private tutor. Through the 1670s and 1680s, Gao was a frequent companion of the emperor, composing poems, explicating the canon of dynastic histories, and performing various erudite literary tasks for the emperor's pleasure.[14] He also joined the emperor's closest entourage on several of Kangxi's imperial tours. Gao's narrative of the imperial expedition to Manchuria is a tour de force of the skills for which Gao was employed: poetry, history, and the literary adulation of the ruling dynasty.[15] Gao's daily entries commonly begin with brief observations: the date, the weather, the name of the destination locale, and a notation of any extraordinary occurrence—including conversation with the emperor—that took place during the day. Gao could provide vivid, detailed portraits of the terrain the tour passed though, describing the environments of forests, fields, and mountains along with any sightings of wildlife as they appeared before his eyes. The entries then turn immediately to the historical geography of the places the tour passed through. Gao would record the names of each locale through previous dynasties; the political unit to which the place had belonged, and note if any famous occurrences had taken place nearby. A description of the environment of a mountain might be accompanied by a half dozen quotes from classical poems, geographical treatises, and ancient histories that also mentioned the mountain. For Gao, both natural and human history were simultaneously visible in the environment, with human history given top billing.

Kangxi's inner circle on the tour also included a European, the Flemish priest Ferdinand Verbiest (1623–1688), who had been serving in the Qing court for even longer than the Chinese scholar Gao Shiqi. Born in Pittem, a small town fifteen miles south of Bruges, Verbiest arrived in China in 1654 during the early years of the Manchu dynasty and served as an astronomer and mathematician to the court of Kangxi's father, the Shunzhi emperor.[16] Through the 1660s, Verbiest weathered court intrigues, persecution, and imprisonment as Chinese officials tried to purge European Jesuits from the imperial Bureau of Astronomy. He survived because the young Kangxi emperor, recognizing Verbiest's skills as an astronomer, squelched the persecution and had the priest reinstated as head of the bureau.[17] In addition to supervising the bureau, designing new technical instruments for the court, and creating the empire's calendar (one of the most sensitive and important tasks in the realm), Verbiest served the youthful Kangxi as his personal tutor, giving him private lessons in astronomy and mathematics in the Manchu language.[18] Kangxi may have asked Verbiest to come along on the 1682 tour in order to show his old teacher the homeland of his ancestors, but he also put Verbiest to work as a technical expert: using his skills to record atmospheric phenomena, determine the latitude and magnetic declination of locations beyond the Great Wall, measure the heights of mountains, and answer any questions the emperor might have about celestial objects. Verbiest was well equipped for his task: several horses in the imperial train transported a selection of what the priest called his "*instrumenta mathematica*."[19] Verbiest had considerable enthusiasm for his role as the tour's expert on the heavens, but for the most part he endured the rigors of the journey not for the sake of scientific exploration but for the opportunity it afforded him to be near to the court nobles and speak to them of God.

Verbiest's narrative of the 1682 eastern tour is contained in a report filed at the Vatican and soon published in Paris as *Voyages de l'Empereur de la Chine dans la Tartarie*.[20] Verbiest's account emphasized his closeness to the person of the emperor and the special kindness the emperor bestowed upon him. It also contains frequent praise of the abilities of the Qing government: the smoothness of the roads it built, the discipline of its soldiers, the vision and intelligence of its leaders. In capturing some of the details of the Manchu customs, such as the imperial hunts he observed from the saddle, Verbiest was writing as a sort of journalist of the exotic or an ethnographer of the habits of a foreign people. But Verbiest viewed the Manchurian environment in terms of number. While Gao's record begins with numerous "sending-off" poems from court officials, lauding in florid language the momentous journey into the "soaring shadow of the White Mountains," Verbiest's account begins with a brief statement of the purpose of the journey and summarizes the distance of the route: "One thousand eighty miles (each of which contains one thousand paces)."[21] For Gao and other court literati, the mind's eye was full of geographical features that provoked emotions and awe even before the journey began. For Verbiest, the expedition marched not toward any beautiful or august destination but for a certain specific number of paces in an east-northeasterly direction.[22]

✳

For the first few days of the journey, the imperial retinue traveled through familiar territory due east from the walls of the capital, heading toward the Bohai coast along a route more or less the same as that taken by Wu Zhaoqian and his exiled compatriots. The map of the tour's route in figure 2.1 shows a familiar geography, from the Great Wall northeast across the Liao River plain, into the foothills of the Changbai Mountain range, and then on to the great Songhua River. Kangxi's courtiers provided even more detailed observations than Wu's poems, and their daily recordings provide vivid insight into the intersections of terrain, perception, and movement. The imperial tour moved

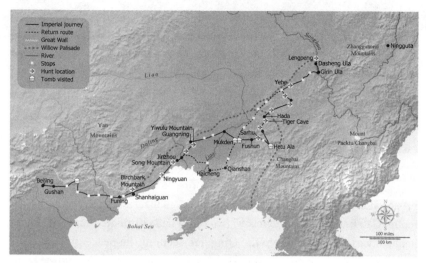

Fig. 2.1 Kangxi's second eastern tour. Map by Jeff Blossom. See plate 15 for color version.

with remarkable speed, progressing at a rate of twenty-five to thirty miles a day across the flat plains between Beijing and the modern-day cities of Tangshan and Qinhuangdao. It is hard for us today to imagine the landscape that spread before the entourage. This area is now intensively farmed and highly industrialized, a checkerboard of dense population centers, highway flyovers, cash crops, and exhausted coal mines. The many winding rivers that once crisscrossed the plain have since been straightened and channeled for irrigation. The first feature of the landscape noted by Gao Shiqi, "Orphan Mountain" (so called because it was the only elevated spot on the otherwise flat plain) no longer exists today, perhaps a victim of local mining. Gao's springtime sketch gives us a good sense of the wetland nature of the region as it appeared in the late seventeenth century. Across the flat terrain, still-dormant fields alternated with semifrozen marsh, rocky shoals, and low scrub. The passing of the massive entourage startled flocks of ducks and wetland birds.[23]

A week's journey from Beijing, the tour stopped outside the walled city Funing, 150 miles due east of Beijing near the Bohai Gulf, a place where the Yan Mountains come within ten miles of the ocean waters. With their stark, barren peaks, the Yan Mountains form the quintessential Great Wall landscape. At Funing, the Han Chinese exile Wu Zhaoqian had sensed a desolate landscape where bitter winds carried the sound of an invading barbarian army's drums. Twenty-three years later, the Han courtier Gao Shiqi also heard a sound like drums at Funing, but it was just the pounding of the nearby ocean tides. Instead of a desolate landscape, Gao saw an energizing environment, home of a thriving, successful empire: "I think back on how the ancient [scholars who traveled with the emperor] had to clutch their books in one hand and wield the sword in the other, following the army for ten thousand miles—but today the world is at peace and settled, and the lands through which the dragon chariot passes are all prosperous and happy."[24]

Beyond Funing, the tour pointed their horses' heads toward the northeast, and the entourage arrived the next day at Shanhaiguan, the spot where the Great Wall meets the sea and the crucial military gateway where Qing armies had entered China forty years before. Where Wu Zhaoqian had approached Shanhaiguan as the line separating civilization from the barbarian realm, Gao held that "our dynasty" (*wo chao*) had eliminated that line: Shanhaiguan was the place where the Qing had secured the peace of the empire and united "inner and outer into one family." Where Wu's springtime landscape at Shanhaiguan was brown and dreary in the extreme, Wu recorded a gorgeous morning scene with the sun rising from the sea in the east, bathing the sky in rosy mists.[25]

Over the next few days, the tour rode northeast along the Western Liao (Liaoxi) corridor, the strip of land that edges the Bohai Gulf and then opens up to approach the plains of the Liao River. Here the expedition passed crucial battle sites from the Qing conquest, including the cities of Ningyuan, Jinzhou, and Guangping.[26] Where Wu Zhaoqian had seen a landscape tragically strewn with the white bones of forgotten soldiers of past dynasties, Gao Shiqi saw proof of the courage of what he repeatedly referred to as "our dynasty" and "our army." Gao Shiqi's record included long, dramatic retellings of the military history of Kangxi's grandfather Hong Taiji (1592–1643) and his victories against the Chinese Ming dynasty in this area. Where Wu Zhaoqian had only seen Chinese victories and suppressed any mention of the Manchu conquest, Gao lavished his historical attentions on Manchu victories but entirely ignored their defeats. Liaoxi had been the site of fierce campaigns between the Ming and the Manchus for more than two decades, and several of these campaigns had been Manchu military failures. The great founder khan Nurhaci had even been mortally wounded while trying to take the city of Ningyuan in 1626. Nurhaci's son Hong Taiji similarly met defeat against these Ming fortresses and was able to achieve victory only in decisive battles at Jinzhou in 1642, a full sixteen years after his father's death.[27]

The details Gao included in celebrating Manchu victories could be quite dramatic, portraying Manchu military might like wild forces of nature. Traveling in the shadow of Pine Mountain (Songshan) outside Jinzhou on April 4, 1682, Gao recalled Hong Taiji's great rout of Ming forces at the site forty years before, a battle that left tens of thousands of soldiers lying dead in the waters of the Bohai Gulf, their corpses "bobbing up and down like geese on the waves."[28] On the evening of April 5, 1682, while crossing the Daling River in the midst of a cold spring rain, Gao was reminded of the surprise attack against Ming troops personally led by Kangxi's grandfather Hong Taiji that took place at this site in 1631. In Gao's detailed retelling, Hong Taiji's troops crossed the river and "directly rushed the enemy camp, leaving them no escape . . . the two armies met, cannons pealing like thunder, arrows falling like rain."[29] After successfully cutting off their route of escape with sophisticated flanking moves breathlessly detailed by Gao, Hong Taiji's forces chased the fleeing Chinese, cutting them down for miles. The effect of a Han Chinese scholar lovingly describing the slaughter of his fellow Han Chinese is rather jarring, but it illustrates the intensive work needed to transform this landscape of ruin into a place of military glory.

As if emulating the exploits of his Manchu forebearers, as soon as the eastern tour entourage crossed the Great Wall into this bloody frontier, the Kangxi emperor and his troops (accompanied by the emperor's nine-year-old son) left the flatland road and moved west into the Yan Mountains to hunt in massive "circle battue" formations described by Elliot as a central part of the "Manchu Way."[30] In tight formation, as many as five thousand of the emperor's bannermen created a vast circle that they closed step by step until a particular peak's wildlife was caught in the center of a human snare. Once the game was concentrated in one spot, the killing began. Where his great-grandfather and grandfather had slaughtered Chinese soldiers forty years before, now Kangxi and his men slaughtered deer, hares, and foxes "too numerous to count."[31] The Kangxi emperor certainly did not invent the imperial hunt: Michael Chang reminds us that the long history of hunting on imperial tours goes back at least to the Han dynasty (206 BCE–220 CE). Poets and courtiers throughout the imperial period regaled the bravery and boldness displayed on large-scale hunts but also critiqued the imperial hunt as a dangerous and wasteful extravagance: for example, Sima Xiangru (179–117 BCE), in his classic *Rhapsody on the Imperial Park*, portrayed a Son of Heaven who regrets his wantonness: "We threw away the days: following the celestial cycle, we killed and slaughtered."[32] Kangxi showed no such remorse.

As Mark Elliott and others have perceptively observed, the many hunts of the second eastern tour were exercises in military skill, but plotting the coor-

dinates of the hunts onto a historically informed map reveals that significant
hunts often coincided with battle sites from the conquest. Of all the animals
killed in the imperial hunts, the conquest of tigers was the only one that mer-
ited tabulation in the official imperial chronicle. The map in figure 2.1 indicates the locations of large-scale hunts where significant numbers of tigers
perished.[33] Just north of Shanhaiguan, where Qing armies first crossed the
Great Wall, the emperor shot and killed three tigers. Outside of Ningyuan,
where Nurhaci had been mortally wounded, Kangxi visited revenge on two
tigers. In the mountains around Guangning, the site of an important Manchu victory in 1622, two more tigers perished under imperial fire. In the area
around the Daling River where Hong Taiji had slaughtered Ming troops in
1631, another two tigers were killed. As Robert Marks has shown, the presence of tigers in the historical record is a reliable indication of a forested environment,[34] but Kangxi's hunts also marked Liaoxi as a landscape of conquest,
a site where Manchu dominance was impressed not only upon man but also
upon the beasts of nature. The Ming had been conquered decades before, the
Three Feudatories Revolt suppressed. There were, for the time being, no more
enemies of the dynasty within the empire's borders, but the army continued to
kill in the sites where their forebearers had defeated their enemies.

The geographical significance of this violence seems to have been lost on
the one European observer in Kangxi's train. Ferdinand Verbiest attended
the hunts at the emperor's invitation and was impressed by the skill and discipline that the Manchu bannermen (and Manchu hunting dogs) displayed
when faced with wild tigers and other prey. But for Verbiest, the hunts were
conducted not on landscapes where thousands of men had died in war, but
on a blank canvas of hills and scrub, a place without history. Unlike Gao, who
saw administrative boundary lines of previous dynasties, ancient fortresses,
and the sites of great battles wherever he looked, Verbiest saw only nameless
nature: mountains, endless rivers, and the "caves of wild beasts." He did note
that the landscape was strewn with the rubble of what had been many towns
and villages: proof the area had once been thriving and populous but was
now uniformly in ruins.[35] He could not have been entirely unaware of the
history of the earth-shattering conquest that had transpired upon the ground
he traversed, but he was not compelled to envision it. Gao looked out on the
environment and saw history that Verbiest could not see. But when Verbiest
looked out on the land, he saw things inscribed upon it that Gao could not
see: lines indicating the rise of the land above the equator, and numbers indicating the position of the land vis-à-vis the stars.

*

Twenty days after leaving Beijing, the eastern tour entered Mukden, the for-mer capital and administrative gateway for Manchuria. Using its Chinese des-ignation, Verbiest called the city "Xin-yam" (Chinese Shenyang) and he most likely appreciated the opportunity to sleep within the relatively new imperial palace complex instead of in riverside tents. It was here that the scientist-priest was able to set up his astronomical instruments and truly begin his observations. He was rather proud of his exacting use of the quadrant to pin-point the latitude of Shenyang. He placed it, "through frequent observations," at N 41°56′, a calculation far more accurate, he insisted, than previous read-ings done by "both Europeans and Chinese" who determined it to be only 41°. Through "many repeated observations," he also determined that Shenyang had no magnetic declination.[36] Verbiest did not specify what equipment he used, but we can assume that he used a mobile quadrant, the most common and widely used surveying instrument to determine earthly position in the field in the seventeenth century. A mobile quadrant would have been much more portable than the massive set of European-style astronomical instru-ments Verbiest had designed for the Kangxi emperor in 1674: the imposing bronze armillary spheres, altazimuth, celestial globe, and quadrants embel-lished with fierce dragons that are still visible today in the center of Beijing, perched atop a remnant of the old city wall.[37] But no matter the size or com-position of his instruments, Verbiest would have used the same technique to determine the latitude of Shenyang: employing a quadrant either to site the position of the sun at noon, or to measure the angle of the pole star in night-time observations.

While Verbiest busied himself with astronomical calculations during the four-day stay in Shengjing, an interesting natural history challenge appeared in the form of a living "sea bull" (Latin *vitulum marinum*, glossed in Elles-mere as "sea-calf") presented to the emperor by a visiting tribute embassy from Korea.[38] Kangxi asked Verbiest if European books had any mention of the strange "fish" they beheld. Verbiest recalled that the Jesuit library in Bei-jing possessed a book that contained an image of the animal "drawn from life." Curious to see if the Western pictures matched the animal, the emperor dispatched a courier to retrieve the books from Beijing. The couriers cov-ered the eight-hundred-mile round-trip within the space of a few days and delivered the two leather-bound Latin tomes to Shenyang. While Verbiest's narrative does not specify the titles retrieved from Beijing, catalogs show that the Jesuit libraries there included complete collections of Europe's most authoritative illustrated natural histories of the time, including those by Ulysses Aldrovandi, Conrad Gessner, and Jan Jonston.[39] One hopes that the delivered volumes included Jonston's *Historiae naturalis de piscibus et cetis*, published in Amsterdam in 1649. The illustration depicting the seal ("Seehundt," or *Vitus marinus*) in Jonston's work is far superior to the illustration in Aldrovandi (which looks something like a bulbous snake with human hands) or Gessner, whose cartoonlike illustration of a plump *phoca* captures the spirit but little of the detail of the seal.[40] Whichever version he viewed, Verbiest reported that the emperor was "much delighted" to find the "delineation in these volumes and the description were found to agree exactly with the specimen brought from Corea."[41]

Verbiest writes as if European natural history was able to solve a mystery for the emperor about the identity of the sea creature. He does not seem to realize that such an animal, while perhaps not frequently seen in its natural environment by elite members of the court, would be common knowledge for Manchus, Chinese, and Koreans alike. Known as *huwethi* in Manchu, *haibao* in Chinese, and *mulgae* in Korean, the seal was greatly appreciated not only for its fur, which appeared frequently in the clothing of wintertime Beijing, but also for its penis, which was used in materia medica as a remedy for cold disorders and flagging virility as early as the thirteenth century. Indeed, given the degree to which the seal was voraciously consumed for sartorial and medicinal purposes in northeast Asia, it may be the case that the gift from the Korean embassy represented an animal whose once-common presence on the rocky coasts of the northwestern Pacific was rapidly fading, even as early as the seventeenth century.[42]

These observations about cartography and natural history were in essence the only statements Verbiest made about his experience in Shenyang, other than calling the place "*urbs est satis integra et pulchra*" (which Ellesmere translated as "a tolerably handsome and complete city").[43] While Verbiest busied himself with his work, the other members of the tour experienced the city in an entirely different way. Tellingly, while the arrival of the "sea bull" in Shenyang made a great impression on the European Verbiest, neither Gao Shiqi nor the imperial chronicle record its presence at all. Instead, Gao and others in the entourage would have been impressed by visions of an even rarer beast that could be found in great abundance in the city of Shenyang: the dragon.

Sensing Dragons

As they entered into the city of Shenyang, the Manchu emperor and his re-
tainers were entering a city that was far more than "tolerably handsome": this
was the city shaped by the will of Kangxi's great ancestors Nurhaci and Hong
Taiji, the rulers who "rode the dragon" to imperial power. Shenyang was in
the heart of the Liao River basin, strategically located on the banks of the Hun
River, the gateway into the Changbai mountain region. The Ming had fortified
Shenyang's walls as a bulwark against expanding Jurchen power in the 1500s.
Kangxi's great-grandfather Nurhaci captured Shenyang in 1621, and the khan
made the city the capital of his "Later Jin" dynasty in 1625, just a year before
his death. Construction of an imperial palace complex in the city center began
under Nurhaci and accelerated under the rule of Nurhaci's son, Hong Taiji.
Just ten years after his father's death, Hong Taiji proclaimed the establishment of the Qing dynasty from Shenyang in 1636.[44] With the establishment of
Shenyang as the Manchu capital, the city became known in Manchu as "Mukden," or "the (place of) rising," from the Manchu verb *mukdembi*, "to rise,
to soar." The name, particularly when rendered in Chinese as Shengjing, or
"flourishing capital," suggests a place of prosperity, but it also intimates a literal rising upward: a suitable place for the launching of an imperial dragon.[45]

*

Unlike the malevolent beast that breathes fire, guards ill-gotten hoards, and
is slain by saints in the European imagination, the dragon in northeast Asian
culture is a powerful but primarily auspicious creature. The dragon is a deity
that brings life-giving rain, a generative energy in the cosmos and the earth,
and a close companion of the Son of Heaven. It is typically pictured as a reptil-
ian four-legged beast with claws and fearful teeth, but it is also a shape-shifter
whose mysterious appearances and disappearances held even the great sage
Confucius in awe.[46] Dragons were honored in popular Chinese religion as the
"lord of all the waters," and Dragon King temples were the sites of rain prayers
and rituals conducted by peasant communities and great officials alike.[47] Daoist adepts visualized dragons as manifestations of internal energies in their
esoteric bodily practices and were spied flying through the clouds on dragons
on their way to immortality.[48] In more mundane occurrences, dragons were
expressed within concepts as ubiquitous as times of the day, months of the
year, colors, directions, and the stars, associations that map the dragon onto
everyday dimensions of time and space.

While dragons permeated imaginaries at all levels of Chinese society, the beast was deeply intertwined with society's most powerful human, the emperor. Ancient kings were said to have possessed actual dragons, trained by dragon-rearing experts.[49] Sightings of dragons in the sky heralded the coming of a new emperor and a new dynasty as far back as the third century BCE, when the first emperor of China, Qin Shihuang, rose to power. Fifteen hundred years after the first emperor, the rise to power of the founder of the last Han Chinese dynasty, Zhu Yuanzhang, was preceded by multiple dragon sightings portending the doom of the Mongol Yuan dynasty and the rise of the Ming.[50] Once a new ruler had secured the throne, he became an embodiment of the dragon. Language related to the emperor was marked by references to dragons. The emperor wore "dragon robes" and was seated on a "dragon throne." His conveyance, a "dragon chariot," would be carried over a "dragon path." Imperial dishes, writing implements, saddles, quilts, vases, and rugs were all embellished with dragons. Almost everything the emperor touched and everything that touched him could be described using the word "dragon."[51]

For those members of the Qing court who lived in the confines of imperial palaces at Beijing, such word associations were made into visual metaphors in the architecture that surrounded the emperor. In the Forbidden City palace complex, built in the fifteenth century by the preceding Ming dynasty, dragons were an omnipresent sight. Huge colorful tile screens depicted dragons cavorting with pearls. The marble pathway into the central palaces over which the emperor's sedan chair was carried features what is said to be the longest dragon carving in the world: fifty-six feet long, nine feet wide, with nine dragons emerging in myriad twists and turns from a baroque tangle of marble waves. Dragons writhed on the emperors' multiple golden thrones, some rendered in relief within intricate carvings embedded in wood, others soaring on screens behind the throne, still others were liberated from their carving and depicted as three-dimensional beings that wrapped sinuously around armrests and chair backs. On the outside of palace buildings, carved dragon heads formed the ends of gutter and drainage spouts, dragons were stamped on the end of each roof tile, and dragon heads made of yellow tile guarded the ridged roof finials. On the interior of the Beijing palaces, one simply had to look up to find thousands (if not tens of thousands) of dragons. Paintings and carvings of golden dragons covered every square ceiling tile, and multiple tiny paintings of paired dragons graced every ceiling beam. Above the throne in the Hall of Supreme Harmony, a dragon's head emerges from the ceiling, as if poised to hurtle down upon the viewer, chasing a pearl suspended above the head of the emperor.[52]

As the first emperor of the Qing from Beijing, Kangxi was literally born into this world of dragons: he himself was an embodiment of the dragon, and surrounded by dragons at every turn. But there was a problem: many of the dragons that surrounded him in Beijing were dragons left over from the previous dynasty. Moreover, in the imperial legacy inherited from the immediate past, the "land where the dragon arose," or the geographical origin of the dynasty, was in central China: the county of Fengyang in Anhui province. Fengyang was the birthplace of Zhu Yuanzhang (1328–1398), the beggar-peasant who went on to overthrow the Mongols and become the first emperor of the Ming dynasty (1368–1644). Fengyang was celebrated as the "land where the dragon arose" in geographies and histories from the Ming, as well as in the architecture of the "imperial tomb" built by Zhu Yuanzhang for the remains of his once-impoverished parents at Fengyang.[53] Now, the Manchu Qing had taken on the mantle of the dragon: the language of the dragon became embedded in Qing imperial documents; the image of the dragon was worn on the person of the Qing emperor.[54] The question would now become, How could Manchuria, the land of the ancestors who had established the Qing, be made into the true land where the dragon arose? To realize this project, dragons had to be both found and created.

On the possibility of finding real dragons, Kangxi kept an open mind. As recorded by Gao Shiqi, Kangxi once almost saw a dragon in the lands north of the Great Wall.[55] Traveling in the Yan Mountains north of Beijing, the emperor rode by a small pool famed for the presence of an elusive dragon. As the emperor's horse passed, shadowy ripples appeared on the surface of the water. Pleased that a dragon—said to signal the presence of an emperor—seemed to appear as he rode by, the emperor asked Gao Shiqi his expert opinion: surely the waves on the pool's surface indicated that there was a dragon in its depths? Gao Shiqi, recognizing that the emperor's question put him in a somewhat tricky situation, astutely answered by paraphrasing the classics on the ambiguity of the dragon: "The transformations of the dragon are impossible to predict—sometimes they appear, sometimes they are hidden. Sometimes a deep pool, on principle, would seem to contain a dragon, but in fact the pool might not necessarily have one." This apparently lighthearted exchange reveals a certain skepticism about the existence of real dragons, but a deeply ingrained awareness of their presence in the historical landscape.[56]

While real dragons remained elusive, the dragon's presence was clearly manifested through symbolic representations on the journey to Mukden. Tapestries embellished with dragons lined the road prepared for the emperor.[57] Images of dragons fluttered on the flags that marked the divisions of the emperor's armies—the Eight Banners devised by Nurhaci, each designated with a different-colored dragon. But it was upon entering Mukden that representations of dragons were most visible—in the form of soaring golden dragons made of gilded wood that were used as motifs on Mukden's palace architecture.

Dragons are the first thing that one sees at the Hall of Great Rule (Dazhengdian), the octagonal headquarters of Nurhaci's Eight Banner army and the most imposing structure of the Mukden palace complex. The entrance to the Dazhengdian is flanked by a pair of twenty-foot-long golden dragons that curl around the columns on either side of the entrance. The dragons' bodies are rendered as vivid three-dimensional sculptures and appear as entirely separate beings that have just alighted on the columns. The tails of the dragons wrap three times around the bottom of the columns, lower claws grasping the columns for support. The upper parts of the dragons' bodies then rear out into the air, their heads and shoulders looming rampant on either side of the palace entrance, their bared white teeth and sharp claws lunging in midair over the heads of those who enter. At the center of the octagonal inner space inside the Dazhengdian stood a "dragon throne," the back carved with dozens of interlocking bodies of writhing dragons, the arms draped with three-dimensional beasts guarding the emperor's sides. Behind the dragon throne was placed a massive screen carved with dragons and sprouting nine dragons at the top. A horned dragon head, staring straight ahead, emerges from the screen, directly behind the head of the emperor.[58]

More three-dimensional dragons stare down from their positions atop the main columns flanking the entrance of the Hall of Supreme Rule (Chongzheng dian), the main civil and ceremonial center of the palace residential complex located to the immediate west of the old Eight Banner headquarters. Inside the Hall of Supreme Rule, another striking pair of golden dragons have alighted on the columns of the throne dais (plate 1). These dragons are portrayed as if descending down from the heavens face first. Their tails wrap around the top of the dais, while one front claw grasps the bottom of the column as if halting the dragons' descent. Fighting against gravity, the dragons rear the front of their bodies upward, their suspended heads and front claws forming a 'dragons rampant' position on either side of the throne. While the Mukden palace complex was much smaller than Beijing's and sported far fewer dragons, these three-dimensional "flying" dragons form the most stunning symbols of the imperial palace complex: a literal manifestation of the power of the ancestors who established the Great Qing dynasty.[59]

In addition to the presence of visible, albeit symbolic, dragons in the architecture at Mukden, invisible dragons also coursed through the ground beneath the city. As the Kangxi emperor and his courtiers looked out over the landscape, they were aware of a numinous energy in the land, an energy that sometimes took the form of dragons. As felicitously phrased by Peter Carroll in his study of Chinese urban forms, "land was not mere geological earth but an animate environment."[60] The basic principles of traditional Chinese geomancy (*fengshui*) hold that *qi*, or the elemental energy-stuff of the universe, flows through discernible channels in the body of the earth in a form known as the "dragon vein" (*long mai*).[61] The main ridges of mountain ranges form the visible indication of the dragon vein, a central flow of *qi* that can stretch for hundreds of miles. Every region could have its own dragon vein, but a mountainous landscape that was the birthplace of emperors was particularly numinous, a place where the dragon vein below the earth had given rise to dragons that walked the earth. Capital cities that were built on or near such a dragon vein also represented the cosmic powers of heaven and earth, drawn together and mediated through the person of the Son of Heaven, the emperor, and manifest in an architecture that reflected and augmented the powers of nature harnessed by a righteous dynasty.[62]

The city of Mukden and its surrounding "greater Mukden" region resonated with powers of nature, but they were not limited only to the earth. The *Shengjing tongzhi* (Comprehensive gazetteer of the flourishing capital), an official history of Shengjing/Mukden published just two years after the second eastern tour, gives a sense of the powers imbued in the sky, stars, and atmosphere of the region.[63] Within the gazetteer's opening lines is the phrase: "For a ruler to achieve sagely virtue and sacred power and receive the Heavenly maps and registers, he must possess a place that has a unique concentration of *wang qi*" (the *qi* of kings, what Craig Clunas has fashioned as "kingly aura").[64] The Mandate of Heaven—the divinely bestowed privilege of ruling the empire—would come to those who arise from a land that has a naturally occurring concentration of cosmic power. Shengjing/Mukden, the land where Nurhaci established his rule, was just such a place, as indicated by its geo-astronomical location. The gazetteer continues its geomantic praise of the region by linking the land to the stars: "A region of ancient fortifications, its earthly location corresponds to the Celestial Mansions of 'Ji' (Scorpius) and 'Wei' (Sagittarius). In the embrace of the Long White Mountains, fast by Mount Wulu, flanked by the boundless sea, extending to the great desert, it is the concentrated dwelling-place of the force of Thunder, the most advantageous terrain under Heaven."[65]

The *Shengjing tongzhi* offers only the scantest description of the capital city itself but dwells at length on the relationship between the region's land and the heavens. It notes the sector of the heavens that governs the land, as well as the sacred mountains and great terrestrial features that surround the region. Even a symbol from the *Yijing* (Book of Changes), the thunder (*zhen*) trigram associated with the person of the emperor and the imperial dragon, is associated with the geography of Greater Mukden. These "cosmic demarcations" were ancient correspondences between segments of the starry sky ("celestial mansions," or *xiu*) and zones of the Chinese empire.[66] Not only would the correspondences locate terrestrial sites, marked as degrees from particular stars; any unusual heavenly occurrences in the assigned sector of the sky—the appearance of comets, suddenly bright stars, or shooting stars—portended important events that would take place in the sky sector's corresponding earthly location. Local observers reported and imperial bureaus recorded such celestial portents, along with other unusual natural occurrences in the region, including monstrous births, the appearance of strangely marked animals, surprising weather occurrences, and earthquakes.[67]

According to the Chinese texts consulted by the compilers of the gazetteer, the region around Shengjing was long known to have corresponded to the celestial mansions Ji and Wei, in the eastern sky sector under the Azure Dragon (in Western terms, these stars are in the general vicinity of Scorpio and Sagittarius).[68] Remarkable natural events were rich in this region, particularly when the region gave rise to a new dynasty such as the Khitan Liao dynasty in the tenth century and the Jurchen Jin dynasty in the twelfth century. In the first year of the Liao emperor Tianxian (926) white smoke blotted out the sun (an event that could be an indication of the "millennium eruption" of Mount Paektu—see chapter 3). That same year, a ring of fire was seen above the sky in Kaiyuan, a city fifty miles north of Mukden. Most marvelous were the multiple sightings of the "Yellow Dragon," an apparition associated with earthly power and the emperor himself. In 926, the appearance of sun-obscuring smoke was combined with the falling of a meteor and the appearance of the Yellow Dragon—signaling the Liao victory over the kingdom of Balhae. In 1167, the Yellow Dragon (along with falling stars and a beam of red light) appeared in the capital of Emperor Shizong of the Jurchen Jin dynasty (considered the forebearers of the Qing).[69] Kangxi's immediate ancestors also beheld numerous auspicious phenomena in the region, with comets, stars, and rays of heavenly light signaling Heaven's support for the exploits of the dynasty's founders.[70]

Verbiest may have been proud of his reckoning of the latitude of Mukden at N 41°56′, but his astronomy might seem a poor, one dimensional technique compared to the cosmological vision that linked the city's energy to the pulsating powers of nature. The Catholic father certainly knew that the thousands of Qing courtiers around him held other ways of linking the heavens to the earth: after all, the imperial Bureau of Astronomy he directed was also the bureau tasked with determining geomancy and divining portents for the empire.[71] Rather than being side notes to the important "scientific" task of calendar making, the cosmo-geomantic arts were in fact central to the work of the bureau and could, at times, become life-or-death matters.

Indeed, Verbiest himself had risen to his position as bureau director in the fallout of the *fengshui*-related "Calendar Case" that rocked the Qing court in the 1660s. In this deeply complex case involving mathematics, visions of the cosmos, and treason, Chinese literati in the court had accused the Bureau of Astronomy and its then director, the German Jesuit Adam von Schall (1591–1666), of miscalculating the cosmologically appropriate time of the burial of the Shunzhi emperor's infant son, an error that ultimately (according to the accusation) resulted in the untimely deaths not only of the boy's mother (the emperor's favorite consort, Donggo) but also of the Shunzhi emperor himself, who died the following year at the age of twenty-two. Kangxi, only seven years old when his youthful father died in 1661, was by 1669 old enough to intervene in the case and reinstated the Jesuits to the bureau leadership, although not before several officials in the bureau had been cashiered, imprisoned (the aging Schall died behind bars), and even executed. Through Kangxi's intervention, Verbiest's position within the Bureau of Astronomy was secured. The bitter fallout from the Calendar Case had inspired Verbiest to pen scathing essays criticizing Chinese prognostication practices, works that glowered with the indignation of a mathematician and a devout Catholic.[72]

It is important to remember that the European approach to natural knowledge was also deeply intertwined with ideas about the sacred. Even in his essays that decried Chinese use of astronomy and mathematics for the purposes of grave siting and fortune telling, Verbiest wrote of astronomy and mathematics as techniques that revealed the greater glory of God. Verbiest faulted Chinese for not reading ancient texts for their metaphoric illumination of underlying moral principles and criticized geomancers for interpreting texts such as the *Yijing* (Book of Changes) literally, as actual tools for predicting the future. In this they mistook man-made systems for a larger sacred truth. He criticized the practitioners of mantic arts not so much for the content of their work but for its theological failings, since the future can be known only to God. By claiming this ability for themselves, they were "usurping the power of the Lord who alone determines the fate of human beings."[73]

In spite of his deep distaste for Chinese geomancy, Verbiest ironically held partial responsibility for the very thing that made Manchuria a sacred place. The *fengshui* of the tombs of Kangxi's ancestors had been determined by Chinese scholars in the Bureau of Astronomy who were undoubtedly known to Ferdinand Verbiest: Yang Hongliang and Du Ruyu. The two men were originally sentenced to death for their role in miscalculating the auspicious burial time for Shunzhi's infant son, but their sentences had been commuted due to their role in successfully establishing the auspicious geomancy for the resting places of Kangxi's imperial forebearers—the very things that maintained Manchuria's sacred energies.[74] Now that Verbiest found himself in the land of the emperors' origins, he seemed intentionally to ignore the fact that those around him felt they walked on sacred ground imbued with the cosmological power of the dragon. Verbiest's silence about Mukden may, in fact, prove that he understood all too well the significance of the city. Given the sensitive nature of any potential association with idolatry, Verbiest's Vatican report only briefly mentions the activity that was the declared purpose of the entire eastern tour: ritual veneration at the tombs of Kangxi's ancestors.

Bodies of the Ancestors, Veins of the Dragon

Kangxi's most important task once he arrived in Mukden was to perform rituals at the tombs of his ancestors. Unlike all the other Qing rulers who are buried in massive collective mausoleum complexes south of the Great Wall, Nurhaci and Hong Taiji each enjoy separate individual mausoleums fast by their Manchurian capital. The mausoleum of Nurhaci (d. 1627), styled "Fuling," or "Mausoleum of Prosperity" (Manchu: Hūturingga munggan), was located on a mountainside seven miles east of the Mukden palace complex. Nurhaci's son, Hong Taiji (d. 1644), was interred at another grand tomb known as "Zhaoling" ("Shining Mausoleum," or "Mausoleum of Brilliance," Manchu: Eldengge munggan) four miles to the north. The third mausoleum, Yongling ("Eternal Mausoleum," Manchu: Enteheme munggan), contained the remains and relics of Kangxi's more distant ancestors and was located eighty miles east of Mukden at Nurhaci's original power base of Hetu Ala.[75] Together, these three tomb complexes would form the focus of the imperial tour's ritual activities.

The tomb of Nurhaci, constructed over the course of many years after the khan's death in 1626, is perched on the steep slope of "Column of Heaven" Mountain, facing south above the flow of the Hun River just outside of what was Mukden's city wall. This imperial tomb, like all sites designed for ritual workings in imperial China, is not simply a single mausoleum; it consists of multiple buildings composed around an expansive walled rectangular court-yard covering several acres. Its design mirrors the layout of a Chinese city, but it is a city for the dead. A "spirit path" leading to the complex, lined by massive statues of guardian animals, signals the beginning of sacred space. Today one first encounters a pavilion housing a granite stele inscribed with Nurhaci's exploits, but this was erected at Kangxi's orders after the second eastern tour. A rectangular walled enclosure anchored by turrets at the four corners marks off the castlelike mausoleum complex proper, which is entered through an arched gateway topped by an elaborate three-story-tall balustrade. The main buildings of the complex are relatively small but finely rendered; the central Hall of Abundant Blessings, the setting for ritual ceremonies to honor the spirit of the departed ruler, is set on an impressive three-layer marble plat-form, its sides and balustrades intricately carved with dragons rising from billowing waters. One hundred eight marble stairs (a number with cosmo-logical significance in Buddhism) approach the hall, flanking a deeply incised marble dragon path. Beyond the Hall of Abundant Blessings, behind another castlelike towering gate, lies the "Precious Abode," the unadorned crescent-shaped plot of land beneath which the ashes of Nurhaci and his empress are entombed in an underground chamber.[76]

Kangxi had come to the northeast to communicate with his ancestors, and protocol required visiting the eldest first. Arriving at Fuling on April 14, 1682, Kangxi entered the confines of the complex, and at the Hall of Abun-dant Blessings performed the major ritual addressing his great-grandfather Nurhaci. The speech, delivered by the emperor himself, informed the spirit of Nurhaci that the enemies of the dynasty—the treacherous rebels of the Three Feudatories—had finally been destroyed. Indicating his subordinate status in the presence of his great-grandfather, Kangxi began his speech by introduc-ing himself using his own personal name, Hoiwan ye. So sensitive was the taboo against the personal name of the living emperor being seen in print

or heard in public that the imperial chronicle that records the speech obscures the characters for this name with a black square. While he referred to
himself by his "first name," Kangxi addressed his great-grandfather using the
deceased ruler's full imperial posthumous title: "The Supreme Ancestor, Re-
ceiver of Heaven's Mandate, He of Expansive Cosmic Fate, Sagacious Virtue,
and Divine Power; the Creator of the Epoch and Initiator of the Ultimate; the
Benevolent and Filial, Sagacious in War and Magnificent in Rule, the Founder
of the Imperial Enterprise, the August Emperor on High." Kangxi informed
Nurhaci that since receiving succession, he had "succored the people in all directions." Only the "bandit" Wu Sangui—the Chinese general who abandoned
the Ming and swore allegiance to the Qing—resisted the dynasty by rebelling
from his base in Yunnan and spreading treachery and discord throughout
the provinces of the South. Kangxi's address specifically lists all the provinces
involved in the rebellion—Fujian, Guangdong, Guizhou, Sichuan, Hunan—as
if to give the dead Nurhaci a sense of the great geographical expanse of the
empire that his house had achieved and to vividly portray the overwhelming
threat to that empire posed by the traitorous Chinese general. Through numerous campaigns, Kangxi continued, the dynasty's armies were finally able
to "exterminate the villainous traitor and his chieftains." Even though Nurhaci
had been dead and buried for more than fifty years, Kangxi credited the victory to him: "Were it not for the numinous effective power [*ling*] of our ancestor, the chaos could not have been checked!"[77]

After two days of rituals at Fuling, the emperor and his court proceeded
north of the city to Zhaoling, where they conducted identical rituals at the
mausoleum complex of Kangxi's grandfather Hong Taiji. Again, Hoiwan ye
announced his presence, again recited the narrative of the treachery of Wu
Sangui, and again detailed the victory over the rebels. Hong Taiji's spirit must
have been particularly gratified to know of the destruction of the traitorous
"twice turncoat" Wu Sangui, since his own forces had fought against Wu in
the 1640s when the Chinese general was still a "loyal" servant of the Ming.
Kangxi again concluded his address by attributing his own victory to the
ling—the numinous effective power—of his revered grandfather, intoning the
same phrase of gratitude: "Were it not for the numinous effective power [*ling*]
of our ancestor, the chaos could not have been checked!"[78]

Kangxi's gratitude to his ancestors for wielding power in the present day
was not simply a rhetorical turn of phrase; it reflected a concrete belief that
the dead, like the gods, could exert direct influence over the affairs of the living. The historian Valerie Hansen has highlighted this power, *ling*, as a central

feature of Chinese religions. The term *ling* is frequently translated as "soul" or "spirit," but Hansen perceptively points to the use of the term to describe the degree of success or actual power a deity can effect in the human world.[79] This basic premise can be traced back to the most ancient written texts on the Chinese mainland, the "oracle bones" of the Shang dynasty (c. 1500 BCE) that were designed to augur the possible anger of the ancestors as causes for disease or crop failure. This belief permeated ritual life in traditional northeast Asia in everything from Confucianism to shamanism. The ancestors do not simply effect this power to benefit descendants out of the kindness of their hearts, however: their power must be cultivated, augmented, channeled, and accessed through appropriate ritual performances and utterances for the dead, and, most important, through careful positioning of the bodies of the dead within the energies of the earth itself.

This is the underlying purpose of *fengshui*: it is the observation and calculation of land formations to reveal the hidden powers within terrain that make the ancestors even more *ling*. The techniques of *fengshui* are numerous and varied, but the basic premise is to map symbolic meanings derived from ancient texts such as the Book of Changes onto the directionality and layout of a specific terrain. While *fengshui* calculations are employed to site the dwellings and business places of the living, the most important determination is the location of an auspicious site for the burial of the deceased.[80] Accurate readings and well-reasoned grave locations can benefit descendants for generations, while mistakes in readings or missteps in burial can produce a loss of fortune, personal calamity, and even early death. When the deceased in question was also a personage of political importance, the power of the grave site would not only affect the genetic offspring but also could involve the entire fate of the country.

It is no wonder, then, that in the texts generated by the imperial tour of 1682 we see a constant cross-referencing of the energy of the environment in and around Mukden, the tombs of the dynasty's founders, and the fate of the dynasty itself. Gao Shiqi describes Hong Taiji's mausoleum Zhaoling in geomantic-poetic terms: "The dragon coils fast by the vault of heaven, the phoenix soars thundering in the myriad directions."[81] The landscape of Nurhaci's resting place is a place where the "hundred waters meet and the myriad mountains converge to embrace the dynasty's august lineage; a place truly established by heaven and earth, an eternally sacred mountain, awe-inspiring in its divine power" (*shenling*).[82] The Kangxi emperor similarly describes the intersection of Fuling's architecture and the "airs" of its natural setting: "A numinous beauty of auspicious mist gathers around the watch towers; the morning vapors envelope the palaces; ten thousand mountains encircle to the north; while the hundred waters pour from the east."[83] Aesthetic appreciation of Manchuria's natural beauty is inseparable from an appreciation of the invisible numinous power of the landscape, a power that both emanates from and accretes around the bodies of the imperial lineage.

After Kangxi paid his respects to his great-grandfather and grandfather, there was one more set of ancestors who needed to know about the victory over the Three Feudatories. Two days' journey eastward from Mukden was Yongling, the tomb complex of Kangxi's most distant direct ancestors: Nurhaci's father Taksi (1543–1583), Nurhaci's grandfather Giocangga (1526–1583), Nurhaci's great-grandfather Fuman (d. 1542), and the most distant direct ancestor of the imperial line, Möngke Temür (1370–1433), who had lived and died more than four hundred years before the triumph of the Qing. The tombs of these near-mythical figures were established a few miles to the west of Hetu Ala, the original stronghold of Nurhaci's family and the birthplace of the armies and strategies that would eventually result in a new empire.[84]

As they advanced eastward from Mukden toward Yongling, the imperial entourage was once again traveling backward along the timeline of the Manchu conquest. On April 16, the imperial entourage passed Fushun, the Ming walled fortress captured by Nurhaci in 1618 in one of the first Manchu victories over Ming frontier defenses. In 1682, more than sixty years after the siege, the city was still in ruins, its walls crumbling amid the weeds and only a few families inside "living with ghosts as their neighbors." Passing Fushun was a rare moment when Gao Shiqi did not see thrilling victory in the landscape; instead, he found that viewing the city's present state was a "gloomy and depressing" experience. Gao added a somewhat uncanny or even spooky observation: even though the day was calm and the sky was clear, as the entourage passed the ruins, a mysterious strong wind blew and unfurled the military insignias of the mounted horsemen.[85] Natural forces seemed to acknowledge that the Qing was returning to its original *bantu*, or territory, but the landscape provided a simultaneous silent testimony to the human suffering wrought by the Manchu conquest.

The next day the expedition arrived at the plain of Sarhu, the spot where the east-to-west-flowing Hun River converged with the Suzi River that flowed in from the southeast. This intersection was the site of a pivotal battle between Nurhaci's forces and the Ming armies (along with their Mongol and Korean allies) that had been sent to destroy the Manchu power base at Hetu Ala in

1619. The terrain at the site played an important part in the battle. The narrow valleys of the two rivers are lined with rocky hills that rise abruptly to heights of about a thousand feet. At a strategic point where the rivers converge, the Jurchen had established a defensive fort on Mount Wahonmu, known in Chinese as "Ironback Mountain."[86] As the Ming troops under general Du Song approached from the west, Nurhaci used the terrain to his advantage. He had temporary dams built upriver to partially hold back the flow of the Hun and amassed his forces to lie in wait above the river on Ironback Mountain. As Du Song led his forces across the Hun at night, Nurhaci ordered the dams broken. The river swept away half the Ming forces, and Nurhaci's troops ambushed the remaining troops from their mountain fortress.[87] Seeing this famous battlefield for the first time, Gao Shiqi mused about the reasons for the rise of the Qing. Experiencing the mountains and passes in the northeast, Gao could see the extreme difficulty that Nurhaci and Hong Taiji faced in establishing the dynasty and grew in his appreciation of their strategies. By reading the landscape, Gao concluded that their remarkable military might was nothing short of miraculous: accomplished not through human effort alone but with the assistance of Heaven.[88]

Kangxi, too, was clearly aware of the historical significance of the Sarhu landscape. As he had several times before when encountering significant battlefields, the emperor ordered his bannermen into formation and conducted a massive hunt on the land where his ancestors had slaughtered the troops of a dying dynasty. In a poem commemorating the landscape, Kangxi imagined his glorious ancestors engaging the battle of Sarhu in the form of a soaring dragon:

> From over the castle walls, the dragon leapt and soared into the clouds;
> Swiftly the imperial halberds and banners conquered Liao.
> Here before Ironback Mountain, the thrilling victory was won,
> Flying on the wild whirlwind, the vast journey accomplished in an instant.[89]

At Sarhu, Kangxi celebrated the "dragon's" victory with the deaths of hundreds of deer, rabbits, badgers, and foxes, along with the carcasses of two tigers.

After seeing the spot where Nurhaci had decisively defended his power base from a massive attack, the tour pushed farther east to arrive at the homeland itself: a series of Aisin Gioro–dominated settlements in the valley of the Suzi River. This area had not always been the origin of the family: For the centuries prior to the rise of Nurhaci, the clan that was to become the Aisin

Gioro had moved numerous times, pursuing opportunities, making alliances, and attacking enemies. In the early 1400s, historical records note a Möngke Temür who led his "Odoli Jurchens" first into the northeastern frontiers of the Chosŏn kingdom (in today's North Korea) and then into the area around the Pozhu River (known today as the Hun River), a tributary of the Yalu that flows forty miles to the north of the Korean border. Their peripatetic exploits in northeast Asia finally found them muscling into the very perimeters of Ming power in Liaodong. Granting them imperial investiture as the lords of the "Jianzhou Jurchens," the Ming court recognized and hoped to control the growing power of this group led by the Aisin Gioro clan.[90]

By the sixteenth century, the Aisin Gioro had moved away from the Yalu and created a homeland along the Suzi River. Called the Suksuhu River, or Fish-Hawk River in Manchu, the Suzi lay in a narrow plain that was lined to the north and south by two gently sloping parallel mountain ridges. This placid valley was where generations of the Aisin Gioro clan were born and raised. In 1559, Nurhaci was born in Hulan Hada, his grandfather Giocang-ga's redoubt on "Chimney Peak," a 2,200-foot mountain with a pronounced pointed summit just two miles to the south of the Suzi River. The khan later established his own settlements in spots just a few miles away, first in Fe Ala (Old Flat Hilltop), and then at Hetu Ala (Broad Flat Hilltop) near the southern bank of the Suzi.[91] This mountain-lined river valley was also where Nurhaci's people were buried. In 1556, just a few years before Nurhaci was born, his great-grandfather Fuman died and his ashes were interred on the slopes of Niyaman (Central) Mountain, a thousand-foot-high hill located a half mile to the north of the Suzi. Over the ensuing years of the later sixteenth century, wives, siblings, uncles, and aunts of the clan were buried in the same location, anchoring the family to the land.[92]

While generations created a thriving society there, the Suzi valley became a center of a struggle for power in northeast Asia. Möngke Temür's descendants were wedged both politically and geographically between the Ming in Liaodong, Chosŏn Korea, and a host of other Jurchen tribes and Mongol rivals. In the 1580s, these rivalries ultimately led to battle against another Jurchen leader fighting with the backing of Ming commanders at Mount Gure, located twenty-five miles to the west of Hetu Ala. In the battle of Mount Gure, Nurhaci lost both his father, Taksi, and his grandfather Giocangga. After prolonged negotiations with his rivals, Nurhaci finally recovered the body of his father, but nothing remained of his grandfather, who had been burned alive in the battle. Nurhaci interred his father and symbolic relics of his grandfather at the foot of Niyaman Mountain, in the same soil where Fuman and others were buried. Twenty-five years later, the deaths of his father and grandfather would top the list of the Seven Grievances (1618), Nurhaci's war cry against the enemies of the family that would ultimately be hailed as the event that triggered the birth of the Qing dynasty.[93]

These central figures in the history of the Qing were entombed at Yong-ling, but the site did not rise organically from the homeland's original burial ground. That the bodies of these men rested together in this one place was part of the intentional creation of an auspicious "imperial lineage" out of a family of tribal headmen. Some of the ancestor's remains had been moved several times, and some didn't even exist. As Nurhaci's forces penetrated westward into Ming territory, he moved both his capital and his ancestors' ashes with him. In 1624, the remains of Nurhaci's father and grandfather were removed from Niyaman Mountain to Nurhaci's temporary capital of Liao-yang, leaving just Fuman's remains at the old family burial ground. Ten years later, Nurhaci's son Hong Taiji commanded that a crown and royal robe rep-resenting the distant ancestor Möngke Temür be buried next to Fuman at Niyaman in order to cement a connection between the old head of the Odoli Jurchen and the newly ascendant leader of the Manchu people. In 1648, just four years after establishing the Qing capital in Beijing, the new court be-stowed posthumous imperial titles upon the distant ancestors, thus turning the four deceased tribal headmen into emperors. Finally, in 1657, once the Beijing Bureau of Astronomy confirmed that the site of the old graves at Hetu Ala had good *fengshui* (in a set of decisions that would eventually lead to the Calendar Case), construction began on an official imperial tomb complex at the base of Niyaman Mountain, which was given the auspicious Chinese name Qiyun shan, or "mountain that launches destiny."[94] The remains of Taksi and Giocangga were excavated from their mausoleum in Liaoyang and moved one hundred miles back to the Suzi River valley to join the remains and rel-ics of Fuman and Möngke Temür. The new tomb complex of the four gen-erations, containing a mélange of improvised objects and ashes, was dubbed Yongling, the Everlasting Tomb. A four-hundred-year-old lineage cobbled together from far-flung ancestors was now permanently sunk into the land at a site that would represent the sacred root of the dynasty.[95] For nature to take on the power of the imperial dragon required a tremendous expenditure of human effort.

As Kangxi's entourage approached the Yongling tombs, the landscape grew more and more heroic in the eyes of the observers. From Sarhu, Yongling was a day's ride to the southeast through rugged terrain. The altitude increased by over a thousand feet. Even though it was late April, the changeable Manchurian weather had suddenly grown cold and cloudy, and when the expedition arrived at Yongling on April 18, a steady snow was falling. From a *fengshui* perspective, the placement of the tombs within the landscape is perfect, with folded mountains rising behind the mausoleum and the Suzi River coursing in front. While it enshrines four "emperors," the complex is on a far more intimate scale than the monumental Qing tombs south of the Great Wall. Near the entrance to the main courtyard stands a line of four small pavilions, each containing stele lauding the accomplishments of Taksi, Giocangga, Fuman, and Möngke Temür. One modest ceremonial hall at the center of the complex sufficed for all four spirits, and behind the hall, one crescent of earth covers the underground burial chamber of all four emperors and their consorts.[96]

As the Kangxi emperor moved through the landscape approaching the tomb, he sensed a singular energy:

Peak upon peak, river upon river,
A royal *qi* pervades and protects Yongling;
The myriad mountains coil into a crouching tiger,
The soaring peaks launch the dragon into flight.
Clouds envelop the ancient garden's trees and bridges
Snow embraces the graves of my ancestors rising before the imperial road.
As the Zhou rulers moved to Qishan, so did my ancestors move here to begin the Great Enterprise—
To follow in their footsteps fills me with dread and awe.[97]

For Kangxi, the land where his ancestors were buried contained a dragon's numinous power, a power that had given his ancestors the strength to conquer nations. In this poem, Kangxi references Qishan, the site in present-day Shaanxi province that was the capital of the ancient Zhou dynasty (1046–256 BCE). This area, to the west of the present-day city of Xian, was seen as the center of the earliest classical Chinese civilization, the origin of all tradition and excellence in imperial rule.[98] The mountains in this vicinity were not only considered the origin of the dragon vein for the Zhou dynasty—they were the origin of the dragon vein for all of China. Using the language of *fengshui* and dragons, the Kangxi emperor likened his dynasty to the greatest dynasty of Chinese civilization, suggesting that the land of his ancestors possessed an energy as sacred as anywhere in China proper. Having just waged an exhausting massive war to unite the empire, Kangxi felt the anxiety of maintaining the ancestors' accomplishments. A supernatural energy was needed to continue in their footsteps, and this energy resided within the land of Manchuria, intensified and channeled by the sacred bodies that lay buried beneath its soil. In this poem from 1682, Kangxi uses the technology of *fengshui* to wrench the fate of the entire empire from its previous Chinese geographical moorings and anchor it solidly within the environment of Manchuria.[99]

Embodied Knowledge of the Perilous Frontier

On April 19, 1682, the imperial tour left the ancestors' tombs and turned to the north. Gao Shiqi records that the emperor "desired to conduct a review of the border region," in order to "witness firsthand the hardships experienced by the ancestors as they established the dynasty."[100] Kangxi undoubtedly wished to inspect the progress of his border defenses against persistent Russian military incursions in the northeast, but his desire to "witness the ancestors' hardships firsthand" was more than just a rhetorical turn of phrase. Their destination was the settlements of Girin Ula, located almost two hundred miles to the northwest of Hetu Ala on the banks of the Sungari River. The trip to Ula would satisfy both the military strategies of the emperor and his desire to know, through direct bodily experience, the true nature of Manchuria.

At this point the imperial tour again left the main established routes for traveling in the region and followed a shortcut. Even before he left Beijing, Kangxi anticipated the need for a route between the Suzi River valley and Ula that could avoid the need to backtrack to Mukden and ordered local Manchu officials to reconnoiter a new path. Kangxi warned that the going in that part of the country would be rough, and emphasized that the bannermen who planned the route would have to possess superior mapping skills and an ability to "record the lay of the land."[101] Gao Shiqi describes the travel in this section along the resulting path as a sort of bushwhacking process through the "myriad mountains," following a trail created only by the passing of horses that had gone before. He noted that the spring warmth was melting the snow and that the horses' footing became less and less secure on the muddy mountain paths.[102]

In spite of the dangerous terrain, Kangxi's zeal for hunting seemed only to increase as he moved further north. The remoteness of the territory certainly created an environment full of game: but once again the hunts seemed to correspond to sites of major battles fought by the ancestors. In this more northerly landscape, the former enemies were not just the Chinese armies of the Ming but also the armies of rival Jurchen and Mongol groups who had stood in the way of the Aisin Gioro rise to power. As they passed by Hada town, a site where Nurhaci's armies had defeated armies of the Hada tribe in 1599, Kangxi launched a hunt that bagged two tigers. More tigers were killed at the nearby "Tiger Cave," north of a site of one of Nurhaci's major victories against the Ming. The tour continued working its way northeast, passing in and out of the Willow Palisade near the point where the branch delineating Manchu territory and the branch delineating Mongol territory converged. This was the original territory of the Yehe tribe, a onetime ally of Hada in their struggle against Nurhaci.[103] In spite of an injunction against hunting pregnant deer proclaimed by a benevolent Kangxi on the occasion of his twenty-ninth birthday, the body count for tigers reached its height in this region, ranging from one to five each day.[104]

For the last two weeks of April, Kangxi drove his men hard through the old Hada and Yehe homeland. The warming weather made the terrain increasingly unmanageable. For almost two hundred miles, the entourage rode across a monotonous expanse of rolling 1,500-foot hills. Gao Shiqi described how the trail had turned into a combination of mud, fallen leaves, and melting snow ranging from one to three feet thick. The horses lost their footing or became stuck in the mire. Riding became treacherous: most riders opted to dismount and led their steeds, growing more exhausted and mud encased with each passing step. In the valleys between hills, the travelers had to slog through half-frozen swamps choked with rotting weeds; the stagnant rust-colored water soaked their boots and stained their clothing red. To make matters worse, it began to rain, a rain that was to dog the party for the next several weeks. Of the road to Ula, a miserable Gao Shiqi recorded: "We rode in the rain for twenty, thirty miles a day . . . as soon as one mountain ended, another began; exhausted from climbing the peaks, I was unaware of the coming of the night or the arrival of dawn."[105] The court of the Qing empire had descended into a true frontier, more remote and wild than any terrain they had ever experienced, and the emperor gave his court no respite. Kangxi, it seems, was consciously achieving his goal of "witnessing firsthand the hardships experienced by the ancestors."

Eighteen days after leaving the comforts of the Mukden palaces and forty-seven days after leaving Beijing, the court finally arrived at their furthest destination, the outposts of Girin Ula on the banks of the Songhua (Manchu, Sungari) River. The area had been the stronghold of the Ula (Big River)

Jurchen, a people conquered by Nurhaci in the early 1600s after the exchange of women between the two groups produced interfamily drama but failed to produce peace.[106] At Girin, the Songhua River has wound over 150 miles from its eastern origins on the slopes of Changbaishan and executes a graceful semicircle curve before flowing northward into the broad central Manchurian Plain. From this point, the Songhua was navigable all the way to the Heilongjiang River over five hundred miles to the north. The Qing administration established a fort at Girin Ula in the 1660s that included a shipyard to produce the warships needed to defend the Songhua region from Russian incursions. About fifteen miles to the north of Girin Ula, where the river splits into several small rivulets, lay the outpost of Butha Ula (Chinese, Dasheng wula), where the Qing established its administration overseeing the complex network of hunting and foraging enterprises that collected the bounty of the northeast—pearls, fur, ginseng, fish, honey—for consumption at the Beijing court.[107] In 1677, Kangxi had the northeast's military headquarters transferred from Ningguta to Girin Ula, increasing the city's banner population and ensuring the centrality of the place to the military strategies of the empire. In spite of Girin's importance, it was still a sparsely populated frontier outpost, and the population of the imperial tour may have outnumbered the population of the town itself. With its rude wooden balustrade and log barracks, Girin Ula could hardly have presented a comforting sight to the exhausted nobles and courtiers of the court who arrived there in the spring of 1682. Nevertheless, recognizing its importance as the military nerve center of the frontier region, the emperor staged a triumphant arrival ceremony and entered the town to the accompaniment of trumpets and drums.

Both Gao Shiqi and Ferdinand Verbiest record the remarkable actions of Kangxi as he arrived near the banks of the Songhua River.[108] According to their accounts, Kangxi stepped down from his sedan chair (Verbiest has him dismounting his horse), kneeled on the riverbank, and prostrated himself on the ground to venerate the sacred Long White Mountain (Chinese: Changbaishan; Manchu: Golmin Šanggiyan Alin) in the distance. Verbiest records the emperor bowing three times; Gao notes instead that the emperor performed the "three prostrations and nine knockings," the full *ketou* (kowtow) that the emperor (as the Son of Heaven) performed only for the most sacred of objects in the universe, including Heaven and Earth. The emperor's kowtow was, once again, an intentional marking of the significance of the northeast. It was also a remarkable bodily performance to an invisible ideal: Changbaishan, at 150 miles away from Girin Ula, is not visible from the city to an observer at the ground level (see fig. 3.2). From his position, the emperor could certainly see a ring of impressive mountains around the city, including the 2,500-foot-high ridges of Dragon-Pool Mountain (Longtan shan) to the east and Vermilion Bird Mountain (Zhuque shan) to the south. These mountains, while beautiful, were not the sacred Golmin Šanggiyan Alin, the mountain where, according to the Aisin Gioro origin myth, the ancestor of Möngke Temür, Burkuri Yongson, was born to a heavenly maiden who had descended to earth. The mountain's spectacular peaks, crater lake, and waterfalls had been the stuff of legend, and the Kangxi emperor had, just five years before, sent an expedition to find the mountain and confirm its location (see chapter 3). On this imperial tour, however, the Kangxi emperor would be satisfied with bowing in the general direction of an unseen object. There was to be no pilgrimage, no imperial ascent of the sacred mountain. Even for this emperor who would later personally trek thousands of miles through deserts in pursuit of his enemies and climb multiple sacred mountains in China proper, the Long White Mountain was seemingly too remote and would remain a mountain of the mind.[109]

The rain returned the next day. Rain kept the imperial entourage in the Ula area for almost a week, giving them time to contemplate their surroundings at length.

Both Gao and Verbiest clearly saw Girin Ula as a geographical and cultural frontier. The European saw the region primarily as a liminal space within a global geography. For Verbiest, the area around "Kirin" was unimaginably remote, "the uttermost spot of the inhabited world."[110] The eastern tour had already passed through uninhabited mountains teeming with tigers, but Jilin was wilder than any landscape he had ever experienced. Verbiest had heard that there were Qing outposts even further away, out there in the vaguely defined stretches of wilderness: "Niucuta [Ningguta] is distant seven hundred Chinese *stadia* in a north-easterly, or rather more northerly direction." Informants told Verbiest that a forty-day journey northeast beyond Ningguta would lead to the sea, a location that Verbiest surmised to be the "Strait of Anian," the elusive Northwest Passage between North America and Asia.[111] Verbiest felt himself to be at the edge of the world, on a remarkable journey that led to places as remote and exotic as those traversed by other European explorers—indeed, in his estimation, he was just a little more than a month's journey away from a passage to the New World—and yet he had done all of this not on any explorer's expedition but in the company of the emperor of China. The aged father had traveled together with the leader of a world civilization to one of the farthest frontiers on earth.

Gao Shiqi also sensed Girin Ula as a bewilderingly remote frontier. Like the Han Chinese exiles who journeyed to the deep forests of Ningguta twenty-five years earlier, at Girin Ula, Gao turned away from the military and political history that had been the mainstay of his understanding of the landscape and turned instead toward natural history. He spends several pages describing the environment of Ula. The mountains were mostly black pine forest that produced an incredible number of pine nuts. The region produced ginseng and pearls, the rivers were full of unusual fish like skate and sturgeon, eagles and falcons soared in the skies, and the forests abounded in deer, bears, boars, and sables.[112] Gao offered long passages on each of these precious natural resources, with particularly detailed observations on the medicinal root ginseng:

> Those who pick ginseng say: Ginseng sprouts in the middle of spring, mostly in shady moist places. At first the young plants are small, about four inches long, with one prong and five leaves. After four or five years, the plant has two prongs and five leaves, but still has no flower buds. After ten years, the plant grows a third fork. Older plants develop four forks, each with five leaves and a bud in the center: these are commonly called "Hundred Foot Clubs." In the fourth month a small flower blossoms, as small as a kernel of corn, with silk-like stamens and a purplish-white color. After the autumn arrives, the plant develops seeds, in clusters of seven or eight, similar in shape to soybeans. The seeds start out green, turn red as they mature, and then fall away from the plant.[113]

This detailed natural history of ginseng is presented as knowledge garnered from "those who pick ginseng," the rugged mountain bannermen whose industriousness Gao so admired. Gao notes that Girin Ula was home to hundreds of newly recruited banner troops of the "New Manchus," fierce hunters from east of the Songhua River who were accustomed to harsh frontier life and "rode their horses up and down the steep mountains as if they were flying."[114] These bannermen lived a rugged life, busy all year round. Besides their farming duties to ensure they had enough to eat, in the summer the men collected pearls, in the fall they picked ginseng, and in the winter they trapped sable. But a further examination of the text reveals that Gao learned about ginseng not from these "native informants" but from the silk-bound books in the imperial library. Gao's description of the medicinal root is lifted almost verbatim from an eleventh-century herbal, the *Bencao tujing*, by Su Song. It is possible that Gao discovered this passage in the more recent sixteenth-century work *Bencao gangmu*, by Li Shizhen, since Li cited Su Song at length in his own definitive natural history of ginseng. The eastern tour may have arrived at "the uttermost spot of the inhabited world," but there were still familiar texts to be deployed in the description of the new. At the extremes of the empire, history had given way to natural history, but texts had not entirely ceded to observation.

Gao's books hinted at even more remote places beyond the great forests, beyond Ningguta, lands that were vaguely known by the Jurchen and the Mongol dynasties of several centuries ago but difficult for him to access since the written records in Chinese were sparse and vague. After scouring the official *Jin shi* (History of the Jin Dynasty) for information, Gao even turns to Wu Zhaoqian as a source for the geography beyond Jilin, using the poet's writings to envision the place where the Songhua and Heilongjiang Rivers meet and then roll out to the sea. Gao speculates about "what's out there" in the same way as Verbiest, and while he does not mention the New World, it is obvious that his geographical imagination is equally as fuzzy. Ultimately, Gao, like the Han Chinese exiles who came before him, admitted that knowledge of this frontier is incomplete: "Using various texts to investigate these places, we find that the information on geography and distances is not particularly consistent. At present, this is all that we can understand through texts: the problem awaits further investigation."[115]

*

All observers seemed to agree that Ula was the most extreme environment they had ever encountered, but the Kangxi emperor stamped his mark on the frontier through the same technique he had used at many other sites along the tour: the capture and killing of wildlife. At Girin Ula, Kangxi's hunting strategy turned from tigers to "dragons"—from the king of the forests to the king of the rivers. On May 6, the emperor mounted a fishing expedition to a spot on the banks of the Songhua River twenty-five miles north of Butha Ula (at Lengpeng) in search of the massive Amur sturgeon, or kaluga (*Huso daricus*) (Chinese, *huang*; Manchu, *ajin*).[116] Valued as a delicacy for its roe, the kaluga was also valued as one of the hunters' most daunting adversaries. These fish were truly the dragons of the river: adult fish could weigh over a thousand pounds and attain a length of fifteen feet. Verbiest sensed that the emperor pursued the kaluga with more fervor than any other game, and he even suggested that "it was principally for the amusement of catching these fish that the emperor went to Ula."[117] But on this occasion, the emperor fought a war with nature, and nature won. The incessant rainfall and melting snows had turned the Songhua into a raging torrent that swept away the imperial fishing nets and nearly capsized the emperor's boats. The terrified Gao Shiqi wrote that the crashing of the river's waves was like the "sound of an army of a hundred thousand men" that made his heart stop in fear.[118] The emperor returned to Dasheng Ula empty handed, while Gao acquired a new appreciation of the power of the "Water Dragon" (shui long).[119]

Not satisfied with the "dragon-hunting" expedition at Lengpeng, Kangxi threatened to take his entourage even further into the wilds: Gao Shiqi reports that on May 6, the emperor desired to see the primeval forest, the Lamu Wuji (Manchu, Namu weji, or "oceanic forest") that lay between Jilin and Ningguta. Gao described this forest as a dark, dreaded place where few ventured, with "trees as dense as scales on a dragon, teeming with streams, impenetrable and bewildering."[120] Perhaps Kangxi wished to continue to "experience the hardships of the ancestors": after all, Nurhaci had won victories over the Ula and Weji tribes in these forests seventy-five years before in the early 1600s.[121] But because of the incessant rain and the risk of flood, the idea was scuttled. One wonders whether the entourage hadn't breathed a sigh of relief at being spared the trek even further into the wilderness.

The emperor finally abandoned his enthusiasm for pushing into the wilds and on May 7, 1682, the expedition began its journey southward back to Beijing. Further hardships awaited them on the return trip. In spite of his tremendous power, the emperor of China could not control nature, nor could his power over a massive organization shield him from nature's dangers. The skies had poured rain for more than three weeks, and the tributaries of the Songhua that had been small streams when the tour encountered them on the way north had turned into raging rivers. River crossings became moments of life or death, since bridges were washed out and boats shattered against rocks. Travel became impossible: at times the emperor and his massive entourage had to wait hours on end for storms to stop in order for them to move. The emperor himself frequently had to dismount and wade through knee-deep mud. It was at this point that the eastern tour turned deadly. Members of the entourage were swept away in the floodwaters, pack animals and men disappeared in muddy bogs that waited at the slightest misstep off the designated path. Boats carrying the young prince and relatives of the blood came close to capsizing. Both Verbiest and Gao Shiqi recorded the extreme hardship they endured; at times, both men feared for their lives. Verbiest noted that "old officers who had followed the court for more than thirty years" told him they had never suffered as much as on this journey through what was ostensibly their native land.[122]

Nevertheless, even in the midst of this extreme hardship, the emperor still found other, more subtle ways to demonstrate his mastery over nature. After weeks of rain, one night the storms cleared, revealing a star-filled heaven. Although they were exhausted, hungry, and muddied, the emperor saw an opportunity in the Manchurian sky. He asked Verbiest to sit with him and "name for him all the constellations which appeared above the horizon." In a vignette that seemed to truly touch Verbiest, Kangxi produced "a little map of the heavens" that the old priest had given him many years before. Putting to use the knowledge that Verbiest had passed along to Kangxi in his youth, the emperor used the map to "find the hour of the night by the southern star." Verbiest reported that the emperor seemed to be "pleasing himself by showing to everyone the knowledge he had in these sciences."[123] As Catherine Jami has noted, Kangxi often displayed his power to his courtiers through "performance" of mathematics and science.[124] It is remarkable that when faced with death and extreme hardship at the hands of nature, the Kangxi emperor attempted to reassert his mastery over heaven by naming the stars.

No doubt because of the extreme conditions, on the route back the tour did not retrace their rugged shortcuts through the mountains between Mukden and Ula, but instead tacked close to the established road that passed from Ula in a straight southwest line through the flat grasslands of Yehe, past the towns of Kaiyuan and Tieling, and then back to Mukden. The entourage stayed in the old capital to rest for a few days: here Gao Shiqi noted in passing that the mid-May weather in Mukden had finally warmed up enough that he could "gratefully retire his fur coat."[125] One would think that the tour had come full circle and the route home would be clear and fast, but Kangxi still had more terrain to inspect and more history to relive. From Mukden, instead of following their original (and more direct) route back into Liaoxi, the tour stayed to the east of the Liao River and traveled south to visit Liaoyang, the temporary capital of Nurhaci's enterprise from 1621 to 1625. The tour then continued south, where the emperor enjoyed a day amid the scenic Buddhist temples and steep peaks of Qian Shan—the Thousand-Fold Mountains. Keeping a stiff upper lip while his emperor climbed mountains, Gao Shiqi briefly noted that, once again, it was raining.[126]

As if to put a historical bookend on this journey to tell the ancestors of his victory over Wu Sangui—the bloody Revolt of the Three Feudatories that finally ended the conquest begun with Nurhaci's battles of consolidation one hundred years before—Kangxi performed two final tasks involving the dead. First, he traveled to Haicheng, a town just twenty-five miles from the top of the Bohai Gulf, to visit the tomb of Shang Kexi (1604–1676). Shang was one of the Ming generals tasked with defending the northeast who switched his allegiance to the Qing in the early years of the conquest and was subsequently rewarded with his own "feudatory" in Guangdong after the Qing victory in 1644. When the other two Ming generals turned Qing feudal lords of the south Wu Sangui and Geng Jingzhong rebelled against the Qing, Shang Kexi refused to take part. Shang had died just a few years before the tour commenced and was buried in his family's lands in Liaodong. Kangxi visited the tomb and offered a libation in memory of his dynasty's loyal Chinese general.

While traveling in the same region, Kangxi also wrote to his grand secretary Mingju, noting that he had seen an "extreme amount" of unburied bones and desecrated graves scattered across the Liao borderlands between the old Jurchen lands and the outposts of the Ming. He ordered Mingju to have the bones collected and buried. With this gesture, Kangxi thus erased the visible traces of one hundred years of violence from the landscape.[127]

Eighty days after their initial departure from Beijing, the imperial tour finally arrived back at the capital. While the immediate imperial entourage all returned safely, we do not know how many cooks, grooms, porters, or soldiers may have lost their lives on the journey. In poetry, the "land where the dragon arose" may have been the sacred domain of ancestral tombs and the Long White Mountain, but it was a land of impenetrable forests, raging rivers, and mud, a place where a dangerous, wild nature threatened human existence.

Kangxi noted the impact of this experience in a letter to his beloved grandmother, the Lady Borjigit, who had remained back in the palaces of the Forbidden City. Written while still hundreds of miles from Beijing, Kangxi confessed to his grandmother that the imperial tour had made him finally realize the great difficulty experienced by his ancestors in "opening the frontier" (kai jiang).[128] Kangxi's grandmother (personal name "Bumbutai" [1613–1688]) was a wife of Hong Taiji, the son of Nurhaci who had fought fiercely against the Ming and led the Qing armies all the way to the Great Wall. From an elite family of Khorchin Mongols, she had been married to Hong Taiji at the age of twelve and had given birth to Kangxi's father, the Shunzhi emperor, in Mukden when she was twenty-five years old.[129] A dowager empress at the age of seventy, Kangxi's grandmother was by this time his closest direct link to the past of the conquest leaders. That Kangxi used the wording "opening the frontier" in his letter to Bumbutai is telling. Kangxi is remarking how parts of Manchuria seemed a world away from Beijing, a sophisticated urban center that was on its way to becoming the largest city in the world. We can also interpret this comment as an awareness of the multiple borders contained within the very nature of Manchuria. The land of his ancestors' exploits, in Liaoxi, around Shenyang, and into the region around Hetu Ala, was fairly familiar, home to cities and sacred tombs. Pushing further north, however, into the forests around Girin Ula and along the Sungari River, had exposed Kangxi to lands that were "other." Not only were they homelands for tribes and confederations other than that of his own lineage; they also physically appeared and felt more wild, less controlled, more frontierlike. There, every river crossed and every mountain scaled represented remote hardships. Kangxi does not specify the location of this "frontier" his ancestors "opened." But like the Han Chinese exiles who had ventured to Ningguta twenty-three years before, Kangxi's journey across the terrain had left him with a sense of shifting, ill-defined boundaries in the Manchu "homeland." With the conclusion of his second eastern tour, this knowledge of Manchuria as both a homeland and a frontier was imprinted within the very bodies of the emperor and his court.

Conclusion

Part military reconnaissance, part religious pilgrimage, part literary-historical reenactment, the 1682 eastern tour reflects the unique multivalent way of understanding the nature of Manchuria in the early Qing. Participants contributed knowledge according to their own areas of expertise: history and natural history, poetry and astronomy, cartography and *fengshui*. Through a grueling two-and-a-half-month journey across multiple terrains, the emperor and his servants, including Manchu generals, Chinese literati, and European astronomers, celebrated the invisible sacred aspects of Manchuria while simultaneously finding visible empirical evidence for that sacrality. While they imprinted the landscape with meaning, at the same time the nature of Manchuria established itself within their bodies, a knowledge written in blood, mud, and exhaustion. The making of a Manchu homeland was a deeply physical process, one rooted in death and violence.

By paying attention to bodies and movements across a specific terrain, we are better able to grasp the Qing's multisensorial and multiethnic modes of creating place. A close reading demonstrates that as the tour traveled east, it was going backward in time. Under Kangxi's direction, the eastern tour was an embodied history lesson, and the nature of Manchuria was a history book. The bannermen and members of the court were unwitting historical reenactors, camping on what had been battlefields, creating military formations and unleashing violence on animals where once the enemy had been human. The goal of this embodied history lesson was clear. From the beginning of internecine warfare in 1583 to their entry into Beijing in 1644, the armies of Nurhaci and Hong Taiji had faced sixty Manchurian springs with their drenching rains, mud, and raging rivers, and yet they had overcome their foes to establish the dynasty. By placing the dynasty's bodies in the same environment and subjecting them to the same environmental extremes, the vigor of the ancestors could be both experienced and their power achieved. The tour imprinted meaning on the land, but at the same time, the land imprinted meaning on bodies, creating a cultural identity rooted in physical potential.

This combination of history, hardship, and violence was but one aspect of Qing knowledge creation: quantification was also necessary to achieve a finer-grained and accurate understanding of the nature of Manchuria. For this goal, the scientific skills of the Catholic father Ferdinand Verbiest were of great importance to the empire. His ability to generate numbers from observations matching the land to the heavens could provide the Qing court with standardized, transferable knowledge of the earth that could retain its accuracy regardless of the user's language or position, as long as the techniques employed were the same. In his role as technical adviser to the eastern tour, Verbiest had not only provided the emperor with the names of stars and an etching of a sea lion: for locations where the time in town allowed for observations and calculations, he generated a list of relatively accurate latitudes, placing Shenyang at N 41°56′ and Jilin at N 41°20′.[130] Where weather and exhaustion thwarted systematic observations, Verbiest also generated a list of the exact distances between rest points along the journey. Verbiest appraised the usefulness of his work, suggesting that "it would be worth the trouble of someone with greater leisure than I can command to make out from this list . . . a map of the province of Leauton [Liaodong]."[131]

For the Manchu court, such knowledge was crucial for its survival. Since the 1670s, the mapping of the northeast had become an intense concern of the Kangxi emperor in the face of the growing Russian military threat to the Manchu homeland.[132] Geographical information-gathering projects in the northeast had begun even before the 1682 eastern tour and would continue with increasing urgency after it. Within a year of the tour's return, Kangxi ordered the official survey of a route from Girin Ula to the Heilongjiang

(Black Dragon River, or Sahaliyan Ula—Black River—in Manchu, Amur in Russian) more than five hundred miles to the north. With these logistics in place, Kangxi's forces launched two sieges against the Russian stronghold at Albazin on the banks of the Heilongjiang, a spot almost seven hundred miles away (overland) from Girin Ula. These attacks were led by Sabsu, the Manchu officer whose martial prowess had been lauded by the poet Wu Zhaoqian. The Russians finally surrendered, and in 1689 the Qing and the Romanov courts, assisted by a new generation of European Jesuits in the employ of the Kangxi emperor, negotiated the Treaty of Nerchinsk, establishing peace on the northeast Asian mainland.[133] Even after the settlement of the Treaty of Nerchinsk, Qing mapping expeditions continued apace, and Kangxi sent multiple teams of Manchu and indigenous trackers into the far northeast to probe the rivers and mountains well to the north of the Amur in the last years of the seventeenth century.[134] With the arrival of Jesuits from the court of Louis XIV to the Qing court at the beginning of the eighteenth century, Kangxi would launch an empirewide mapping project using European and Chinese mapping techniques that began in the Manchu homeland. Verbiest's measurements taken during the hardships of the eastern tour can be seen as a precursor to the great enterprise that resulted in the famed "Jesuit Atlas" some twenty-five years later.[135]

While these mapping projects demonstrated the Qing commitment to the recording of accurate empirical data about the earth, this impulse toward empiricism was not accompanied by any disenchantment of the earth. Both the precise calculation of geographical coordinates and the precise calculation of geomantic fate comfortably coexisted in Qing knowledge of Manchuria's nature. The eastern tour was a pilgrimage-expedition to discover data and the sacred powers of the land. Ferdinand Verbiest was the only member of the entourage who felt unease with the enterprise. Rejecting the idea of a numinous Manchuria, Verbiest maintained his own sense of enchantment, but his science instead pointed toward God.

While Kangxi's military objectives dictated that the rivers and mountains of Manchuria required accurate mapping, there was one other important part of the terrain that needed to be mapped: the path of the dragon's energy. All along the path of the eastern tour, the dragon was a constant companion: hidden in pools and rivers, emblazoned on banners, columns, and thrones, shining in the stars of the sky, lurking in the thunderclouds that brought the rains. The emperor and his courtiers such as Gao Shiqi frequently imprinted this imperial dragon on the environment with poems that pictured dragons soaring over castle walls or dragons that lay coiled within the mountains. The dragon was sensed most strongly in the lands where the remains of the ancestors were buried: resonating in the landscape around Fuling and Zhaoling outside of Mukden and visible in the contours of the mountains around Yongling, especially in the gentle curve of Niyaman Mountain or Qiyunshan—the Mountain That Launches Destiny—at Hetu Ala.

Established as fact in the late seventeenth century, the dragon vein of Manchuria grew in importance over the succeeding years. As the knowledge of this dragon vein increased, so did its size and impact: the dragon vein's path expanded and engulfed the entire terrain of Greater Mukden like a surging unstoppable wave, as portrayed in this description from the 1730s:

> Respectfully observing the terrain around Fuling, [we note that] its "dragon" and that of Yongling are but different branches of the same root. The dragon arises from where the tall mountains touch the heavens and auspicious *qi* is remarkably concentrated. Traveling from the western sections of Changbaishan range, it moves along a path to the north of the Hun River in the same direction as the waters. Starting with the towering 10,000-fathom peaks, it leaps over the myriad mountains; it accumulates to hundreds of cubits, imbuing all the high and low points in the terrain. It stretches from horizon to horizon, hurtling from the depths to the heights, causing the dragon to take flight and the phoenix to rise. To the farthest directions there is not one peak the dragon vein does not encompass, from the interior to the ocean, not one flow it does not receive. Through it, the utmost profound blessings of heaven and earth are enjoyed by all: thus the concentrated numinous power of the mountains and valleys nurtures both the civilian and martial aspects of imperial rule with divinely sagacious, everlasting blessings. The land marked by such a dragon is indeed the most excellent of all lands![136]

In this depiction (used to provide the epigraph to this chapter), the dragon vein is deeply embedded in the ancestors' tombs, but the surrounding environment was so imbued with the energy of the dragon that "there is not one peak it does not encompass, not one flow it does not receive." This was the ultimate auspicious environment, "the most excellent of all lands," so full of the numinous power of the dragon that the hills and valleys could not contain it all.

This geomantic-geographic understanding of Manchuria as the land where the dragon arose was made real through the Kangxi emperor's second eastern tour. The dragon was both a visible beast and an invisible energy, manifest in the body of the emperor and in the depths of Manchuria's rivers, present in architecture and in the clouds. The dragon was simultaneously biological, astronomical, spiritual, and geographical, like Qing knowledge of nature itself.

The only thing missing in this complex vision was concrete knowledge of the exact location of the dragon vein's origin, the place where even the ancestors' tombs derived their strength: the sacred Long White Mountain where the progenitors of the Aisin Gioro lineage had descended from heaven. At Girin Ula in the spring of 1682, Kangxi had made a tremendous demonstration of venerating this mountain, but he had prostrated himself in the general direction of an invisible landform. Five years earlier, Kangxi had begun a series of expeditions to find and map the White Mountain, a project that would continue on for decades. Manchuria may have been the land where the dragon arose, but tracing the course of the dragon from the sacred mountain to the ruling house in Beijing would require further acts that merged the mind, the body, and the spirit.

SI(GH)TING THE WHITE MOUNTAIN

Locating Mount Paektu/Changbai
in a Sacred Landscape

The peak on the horizon is highly visible.
—YI-FU TUAN, *Space and Place*

In the spring of 1677, the twenty-three-year-old Kangxi emperor uncovered a troubling geographical problem. Only a generation after the Manchus had left their northeastern homeland and conquered China, no one seemed to know the location of the Manchu's sacred origins, the Long White Mountain (in Manchu, Golmin Šanggiyan Alin). This was the mountain where a seed dropped by a divine magpie impregnated the heavenly maiden Fekulen, who then gave birth to the hero Bukuri Yongson, the progenitor of the imperial Aisin Gioro clan.[1] For Kangxi, the Manchu's power to vanquish enemies and sustain a dynasty was directly born of the special relationship between his family and Golmin Šanggiyan Alin. How could it be that no one at court had ever seen it? To remedy this problem, the emperor dispatched an expedition from the capital to journey northeast in search of the mountain.[2]

Kangxi's expedition sought to gather empirical evidence of the mountain through seeing. However, the closer the expedition got to the mountain, the more the mountain stubbornly evaded sight. Ultimately, the only way the Qing expedition could see the mountain was by praying to the mountain's spirit. Other northeast Asian elites who traveled to the mountain throughout the seventeenth and into the eighteenth centuries, including those who approached from the Chosŏn kingdom to the south, also had to resort to prayer and ritual to see it. Visual knowledge of the Long White Mountain relied on mastering knowledge of the unseen world.

Today we take for granted a certain kind of seeing, facilitated by technology. Leaving aside the immense efforts required to create, compile, and deliver these images, the mountain is easy to see for anyone with a computer: there is no need physically to climb the mountain, and no prayers are needed to make it visible. A glance at a Google satellite image of northeast Asia readily shows the whole of Golmin Šanggiyan Alin (called Changbaishan in Chinese, or Paektusan in Korean) positioned within its larger environment, its slopes rising like a white-capped dome from a dense green forest at the top of the Korean Peninsula. If we zoom in, the image signals to us that the mountain's white cap is actually the cone of a massive volcano. The satellite can even show us that there is a stunning crystal-blue caldera lake nestled in the center of the volcano. Contour lines encoded on the map inform us that the peaks range between eight thousand and nine thousand feet above sea level, making it the tallest mountain in this part of the world. If we switch to a three-dimensional view and zoom in from the north, we can see a large waterfall—the origins of the Songhua River—pouring over the edge of the crater and falling two hundred feet into the valley below. Tourists to the mountain today can climb to the top and take photographs of the stunning caldera lake (plate 2). As the cover image of this book demonstrates, this mountain is truly a breathtaking visual spectacle. Why, then, were historical observers unsure of the mountain's location, and why did they have such a difficult time seeing it?

This question becomes even more pressing when we consider that this "White Mountain" was and still is an extremely important symbol in northeast Asia. In its guise as Changbaishan (Long White Mountain), the mountain has become the ultimate symbol of northeast China: the *baishan* (White Mountain) of *Baishan heishui*, the nature-based metonym for Manchuria: White Mountain/Black Water. The Manchu rulers of the Qing empire revered the mountain as the source of their cultural identity. When Japanese forces occupied Manchuria in the 1930s, Changbaishan became a nostalgic rallying cry for Chinese nationals seeking to recover the region as an integral territory of the Chinese republic. After 1949, the new People's Republic lauded the mountain as a site of anti-Japanese resistance led by the Chinese Communist Party.[3]

This very same mountain, known in Korean as Paektusan, or Whitehead Mountain, is also considered the ultimate symbol of Korea. The evolution of this symbolic importance is described most thoroughly in English within the work of Andre Schmidt, who has detailed how the idea emerged in the late nineteenth century that Paektusan was the home of Tan'gun, the mythical progenitor of the Korean people.[4] Today, the mountain is of particular political importance to North Korea, featured on the national emblem of the Democratic People's Republic of Korea (DPRK) and lauded as the birthplace of Kim Jung-il. It is also revered in South Korea as a symbol of the potential for a unified nation. The mountain is said to be the origin of the vast north-south mountain range, the Paektudaegan, that runs almost the entire length of the east coast, forming the spine of what would be a unified peninsula were it not for the tragedy of Korea's division.

Politically, this significant mountain is split in two. Today, the border between the PRC and North Korea runs down the middle of the crater lake, effectively dividing the territory of the mountain in half between the two countries (plate 3). Many mountains around the globe are split between political entities—the list includes Mont Blanc in the Alps (between France and Italy), Ojos del Salado in the Andes (split between Argentina and Chile), and, of course, Mount Everest in the Himalayas, which is split between the PRC and Nepal.[5] Though less famous than these mountains, the division that sunders this mountain in northeast Asia is perhaps the most profound. Because of the border, it is impossible for anyone to take a hike around the beautiful crater lake. Scientific study is relegated to one side of the mountain or the other. Heated arguments have occurred between China and Korea over the disposition of the mountain's environment and cultural legacy. Even selecting a name for the mountain—"Long White" or "White Head," *Changbai* or *Paektu*?—is a politically fraught decision.[6] A rich scholarship details how the mountain became split between the two polities that became China and Korea: a long and contentious history of boundary making that is still a deeply sensitive topic in the region today.[7]

This chapter takes a step back from the boundary question and tries to understand how knowledge of the mountain was created through the sense of sight. Asking how the mountain was perceived by those who sought it focuses our attention on the intersection of human senses and the physical presence of the mountain itself. Before the mountain could be sited within borders and dealt with as a political possession, it needed to be sighted—seen and comprehended within a physical landscape. This was no easy task. This chapter starts off by asking how such a famous mountain could be lost—or at the very least, why expeditions were required to find it—and why it was so difficult to see. Today, aerial and panoramic photographs of the mountain are easily reproduced and circulated, serving as political symbols and objects of tourist desire. We assume that these images are the reality of the place. But if we take away the aerial perspective and approach the mountain from the reality of the surrounding terrain, we find that its physical configuration, combined with the way the mountain is situated in its environment, makes it difficult to grasp. Indeed, until one actually ascends to the peaks and gazes down at the spectacle of the caldera lake, the mountain can evade visual comprehension, giving, in the words of one early twentieth-century climber, "the impression of something broad and confusingly arranged."[8] It is with this paradox of an object at once visually spectacular and visually elusive—in essence, an invisible volcano—that my inquiry begins.

Even in scholarship that tries to complicate vision and landscape, seeing mountains is often assumed to be a relatively straightforward task. Mountains are, after all, "High Places," the ultimate symbol of visibility.[9] One of the foundational texts of spatial studies, Yi-fu Tuan's *Space and Place*, even begins its chapter "Visibility" with a universal experience of "admiring a famous mountain peak on the horizon," the mountain being an object that naturally "looms large" and is "highly visible."[10] This perspective on mountains as automatically visible presupposes two things. First, it assumes a mountain shaped in the classic inverted V, its peak towering above the low ground that surrounds it. Second, it assumes a singular perspective of an imaginary universal spectator, unencumbered by position, location, or even a physical body. Indeed, the very act of siting/sighting a landscape has been portrayed (following a drawing by Le Corbusier) as "a giant disembodied eye . . . that directs its gaze toward the distant ridge of recumbent mountains seen on the horizon" (fig 3.1).[11] This chapter seeks to place this hovering eye back in the body, place the body back on the ground, and understand how embodied vision, traveling through space, created politically and spiritually charged knowledge of Manchuria's nature.

Clues about the embodied quest to see the White Mountain can be found in the narratives written by those who journeyed to the mountain. This chapter is anchored in these narratives, beginning with the earliest extant description of a journey to the mountain undertaken by the Qing court in 1677 and continuing with detailed narratives of ascents by elites from Chosŏn Korea in the early 1700s. These narratives are important moments that made the mountain known beyond its confines and helped turn the mountain into an important cultural and political symbol. I use a close reading of these narratives to generate a spatial history of the mountain, one that takes into consideration what can and cannot be seen, and acknowledges that different people may see things in different ways.

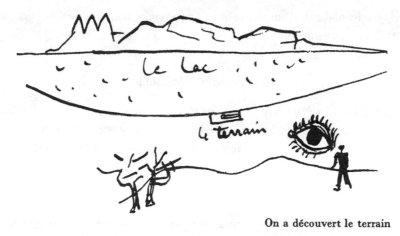

On a découvert le terrain

Fig. 3.1 A disembodied eye "discovers the terrain." Le Corbusier, *Une petite maison, 1923* (Zurich: Girsberger, 1954), 7. © F.L.C. / ADAGP, Paris / Artists Rights Society (ARS), New York 2021.

Siting the mountain required more than just sight. Several of these expeditions were mapping quests, led by metropolitan elites who wielded mathematical formulas and scientific instruments. As such, they are part of the well-studied phenomenon of early modern East Asian polities bringing their territories "onto the map" as part of political projects of centralized governance.[12] Their quest for a technologically mediated vision of the mountain is a central part of this story. A close reading of the expeditions to the White Mountain, however, indicates the importance of local guides, men who led metropolitan explorers across both the physical and spiritual terrains of the journey.[13] Drawn from populations who lived in close proximity to the volcano, they navigated the terrain with no need for augmented vision, their knowledge of the mountain born of exquisite skills gained through movement. They also viscerally sensed the mountain as a sentient being, a spirit that rewarded good or punished evil through the agency of wind, storms, and fire. Within the narratives of White Mountain exploration, this embodied sensing of the invisible was a crucial requirement for si(gh)ting the mountain.

Through examination of seventeenth- and eighteenth-century narratives of those who sought it, this chapter brings a portrait of the White Mountain into view. Making this portrait required different senses, including the visual and the kinesthetic, and multiple technologies, from telescopes and quadrants to incantations and animal sacrifice. Retracing this history must be done while standing on the ground, for it requires us not only to understand the topography of the land but also to imagine the embodied experience of people as they moved through time and space, desiring to see.

The Invisible Volcano

Today, Mount Paektu/Changbai seems to sit placidly while tourists clamber up its slopes and descend into the caldera for a glimpse at its beautiful lake. Few seem concerned with the identity of the mountain as presented through science: in this view, Mount Paektu/Changbai is an active volcano with a long history of violent eruptions. In this version, the mountain is understood as having been created by several nearby geological faults and by the movement of the West Pacific Subduction Zone along the coast of Japan.[14] Over the past two million years, successive waves of magma flows from the volcano formed a basalt shield that covers over seven hundred square miles, creating a massive plateau that is both part of the mountain and also the stage upon which the mountain sits. From the center of the shield's swell, the volcano's cone rises abruptly, creating an object recognizable as a mountain that reaches a height of over nine thousand feet above sea level.[15]

At one point, however, the mountain was much taller. Scientists estimate that around 946 CE, Mount Paektu/Changbai experienced a catastrophic eruption ten times larger than the famed 1883 eruption of Krakatoa. The cone collapsed, forming a massive basin at the center. The resulting caldera lake, fittingly known as the Pool of Heaven or Heavenly Lake (Tianchi in Chinese, Chŏnji in Korean), is over three miles wide and more than half a mile deep—a silent remnant of the extreme violence that created one of the most spectacular mountain landscapes in the world.[16]

This idea of the mountain as a volcano, while seemingly a singular reality expressed in number, is a product of evidence gathered from around the world, interpreted in order to see into deep time, and held together by the remarkable coordinating practices of science.[17] To create their understanding of the Millennium Eruption, vulcanologists not only examine the mountain/volcano itself—they gather evidence from far-flung locations, some thousands of miles away. Dating of the eruption is discerned by analyzing the composition of Greenland ice cores, which contain tephra (material blown out of volcanoes) that was borne to the Arctic on wind currents and preserved within glaciers. Analysis of patterns of tephra in rock layers found as far away as Japan indicates that the Millennium Eruption sent what scientists call a "Plinian" plume over fifteen miles into the air, higher than the Hiroshima mushroom cloud. It is estimated that this eruption blew off more than twenty-three cubic miles of material from the cone, reducing its height by as much as a third, and thus explaining the mountain's characteristic long, low profile.[18]

Given the magnitude of this event, it is remarkable that no specific written record of the eruption has come down to the present. No eyewitness recorded the eruption the way that Pliny the Younger detailed the famous eruption of Mount Vesuvius in 79 CE.[19] Scholars have identified one line from the fifteenth-century Korean text *Koryŏ sa* (The Koryŏ history) as possibly capturing the eruption, but the chronicle simply states that "the drums of heaven sounded." Remarkably, the text bears no mention of a volcano, an eruption, or even the mountain itself.[20] These vague references remind us that the White Mountain is not at all like Mount Fuji or Mount Vesuvius: familiar volcanic mountains near major metropolitan regions, imbued with distinctive conical peaks that can be seen from miles away.[21] By comparison, the White Mountain is remarkably removed from metropolitan human observers, located far away from urban centers on both the Korean and the Chinese side.

The mountain's location helped render its physical qualities invisible in the textual record even as its legend grew. Texts compiled in China proper from around the time of the Millennium Eruption describe the mountain as a shining white dwelling place of a benevolent bodhisattva where animals coexisted in peaceful harmony. The Jurchen Jin dynasty (1115–1234), whose territory extended over much of Manchuria and North China, rose to power not long after the eruption and worshipped the mountain as the birthplace of its founders.[22] Koryŏ, the unified kingdom that began to control the Korean peninsula from 918 CE (just thirty years before the eruption), also claimed Mount Paektu as the origin of its ruling house and lauded the mountain as the source of all the mountains in the kingdom. Some have even interpreted Koryŏ texts to suggest that Tan'gun, mythical founder of the Korean people, began his great enterprise at Paektusan. It is also tempting to speculate that local oral folktales of a destructive fire-breathing dragon dwelling on the mountain emerged from the time of the Millennium Eruption, although it is impossible to trace these origins in historical texts.[23]

By the early modern period, a vague but standardized physical description of the mountain emerged in Chinese geographies. Locating the mountain "in the land of the Jurchens," the fifteenth-century Ming Unified Gazetteer (*Da Ming yi tongzhi*) depicts Changbaishan as one thousand *li* long and two hundred *li* tall (a *li* is about a third of a mile), formed of five peaks with a lake in the center that is eighty *li* in circumference. Three rivers, the Yalu, the Tumen, and the Songhua, were said to originate from its slopes.[24] This description is repeated in multiple sources, the "Five Peaks and Three Rivers" trope standing in for the mountain itself. It was not until the seventeenth century that dreams of a distant sacred mountain were supplemented by the desire of political elites to see, measure, and walk the mountain itself. Scholars have noted that early modern Europe experienced a moment when geography, once a textual genre, became "intensely visual," driven by a desire to "make visible and viewable that which had hitherto been unseen."[25] The same impulse in East Asia was about to render the sacred White Mountain visible.

Sighting the Manchu Mountain

The Qing court's 1677 quest to locate Golmin Šanggiyan Alin is the first record we have of an actual expedition to the Long White Mountain. Several scholars have fruitfully examined the narrative of Umuna's expedition.[26] The Qing imperial chronicle from that year preserves the edict in which the Kangxi emperor ordered his fellow Aisin Gioro clansman, a member of the imperial guard named Umuna, to travel to the outpost of Girin Ula and from there to find the route to the sacred mountain. A narrative of Umuna's journey was published in a Chinese anthology in the late seventeenth century, and Manchu versions appeared in eighteenth-century court-sponsored compilations.[27] But the whole affair can seem rather odd. How could the Manchus have lost their most important mountain in the first place? And why would Kangxi insist on *seeing* the mountain, when textual descriptions of the miraculous mountains had sufficed for centuries in the past? Examining Umuna's narrative of the expedition from a spatial perspective reveals how vision-based empirical evidence emerged as an imperative for Qing imperial rule.

Kangxi's order to Umuna holds an indication of the expedition's motivation. Kangxi ordered Umuna to "see and understand" the mountain in order to "facilitate the judicious execution of the sacrificial rites." Kangxi certainly wished to tie the mountain more closely to the imperial family for domestic political reasons: although the mountain was sacred to all Manchus, it was, after all, *his* family that claimed direct descent from its slopes.[28] Kangxi was also concerned about the growing military threat of the Russians to the north as well as the economic threat of Korean ginseng poachers who infringed on the imperial monopoly over this extremely expensive medicinal root that grew in the region of the mountain.[29] But the Kangxi emperor was also not satisfied with leaving Golmin Šanggiyan Alin as a "mountain of the mind," hidden from his imperial view. Influenced by Chinese practical-learning studies (*shixue*) and tutored in mathematics and astronomy by European Jesuit advisers, vision and empirical evidence had become central elements in Kangxi's approach to ruling the Qing empire. Whether manipulating cones and spheres for a geometric proof, wielding astrolabes to plot the heavens, or ordering mapping surveys of his entire realm, Kangxi used direct observation of nature as a way to perform his power.[30] The expedition to find the Long White Mountain can be seen as part of this impulse.

But in seeking the mountain, Kangxi did not just seek the location of a strategic geological formation. He was also in search of a god. He desired empirical evidence of the mountain's *ling*—its numinous, sacred power, its ability to exert invisible agency in the affairs of men—and sent Umuna on a journey to bring back this proof.

Umuna's team departed Beijing in early June 1677 and traveled northeast for seven hundred miles before arriving at Girin Ula, the Manchu military outpost located on the banks of the Sungari River. There Umuna met Manchu military leaders familiar to us from the first chapter: Bahai, the commander of Manchu troops at Ningguta and Wu Zhaoqian's Manchu patron, and the young officer Sabsu, who would later go on to defeat the Russians at Albazin. With the help of Bahai and Sabsu, Umuna interviewed the bannermen of Ula and Ningguta to find someone with knowledge of Golmin Šanggiyan Alin's whereabouts. To their surprise, the same lack of knowledge that frustrated Kangxi at the court in Beijing was displayed by the warriors of the Manchu homeland: "We proclaimed the imperial edict before the generals, then inquired among all the troops stationed at Ula and Ningguta, and among the hunters living in the settlement at Ula, but no one knew where Mount Changbai was. Everyone simply said, 'I've seen it from a distance.'"

That a high-ranking member of the court could read the emperor's edict (a rather involved process that involves kowtowing before the document and proclaiming it in an august cadence) and receive nothing but shoulder shrugs in response is quite astonishing. Golmin Šanggiyan Alin was the sacred birthplace of the emperor's lineage and the tallest mountain in northeast China. It would be a bit like the Greeks not knowing the whereabouts of Mount Olympus, or the Japanese losing track of Mount Fuji. If the Manchus, or more specifically, the lineage that gave rise to Nurhaci, the founder of what became the Qing dynasty, were from the region of the Long White Mountain, as we learn today, then one might imagine the young Nurhaci in the mid-1500s growing up with a towering, white-capped volcanic mountain in the background, this image of his ancestral birthplace shaping his consciousness and his ethnic pride. How could the mountain be lost?

A combination of semantics and geography helps explain this conundrum. First, there is some linguistic confusion: in Chinese and in English, both the mountain itself and the range it is situated within go by the same name: Changbaishan. The situation is a bit like having a Mount Rocky located within the Rocky Mountains. Nurhaci's power base, Hetu Ala, was in the valley of the Suzi River, over 170 miles (as the crow flies) to the west of the volcano. The terrain of Nurhaci's home town is surrounded by the low foothills of the Long White Mountain range, but *the* Long White Mountain is nowhere in sight. Other Jurchen tribes, the Yalu, Neyen, and Jušeri, had lived closer to the mountain before the conquest, but Nurhaci conquered those tribes very early on in his campaigns of unification, and they became among the first Jurchen to have their regional identity subsumed under a Manchu banner identity.[31] The formation of the Manchu military machine meant the displacement of tens of thousands of families, who were moved to centralized locations far from their original homes.[32] By the mid-seventeenth century, the conquest of the Ming capital, Beijing, added hundreds of miles of distance from the region and totally transformed the lived environment for most Manchus. By the late seventeenth century, then, when the Umuna expedition was launched from Beijing—now on its way to becoming the world's largest city—Golmin Šanggiyan Alin was a remote, alien environment, out of sight, invisible.

The rustic bannermen who stayed behind in the northeast after the conquest were certainly more familiar with the local terrain than the urban Manchus from south of the Great Wall. Tasked with supplying the Beijing court with its beloved furs and wild game, these bannermen spent much of their lives hunting and foraging in the forest. But as David Bello has shown, in the late seventeenth century, even these frontier activities had become spatially circumscribed, heavily regulated, and centrally managed.[33] It is not surprising, then, that Umuna's questioning led to blank stares. Finally, an old hunter-forager named Daimubulu who was originally from the Ehe Neyen River region told Umuna how his father described the journey to the mountain: "If you go hunting at the foot of the Long White Mountain and carry the deer home on your shoulders, it takes three nights on the road, and by the fourth day you can make it home."[34] The Beijing team had to translate this fragment of embodied local knowledge—a sense of distance and direction measured by the heavy steps of a tired hunter, recalled decades later—into forms that could guide the journey of the imperial expedition. While the exact location of the mountain was still uncertain, Umuna at least now had a direction—toward Ehe Neyen, southeast of Girin Ula.[35]

The hunters responsible for foraging and trapping in that general area could guide the way, but questions of logistics remained. Umuna reasoned that rivers could provide a possible route to the mountain. Girin Ula was on the banks of the Sungari River, and the Golmin Šanggiyan Alin was known as the origin of the Sungari. Why not simply follow the river to its source? But local hunters warned that finding the mountain was not that simple. The Sungari at Jilin is the product of the confluence of multiple smaller rivers that emerged from various places on the vast slopes of the mountain. These rivers flow in confusing twists and turns, crisscrossing an east-west expanse of over a hundred miles. They agreed that the Ehe Neyen River was certainly in the vicinity of the mountain, but they warned that navigating the river upstream could be circuitous and dangerous. In the Manchu language, the Ehe Neyen was, after all, quite literally the "Bad" Neyen River: full of rapids and confusing turns. The Sain Neyen, or the "Good" Neyen, was easier to navigate but would not lead to the mountain.[36] Knowing that the river would eventually lead to the mountain but hoping for a more direct route, the team split up, with some traveling via river, and others traveling on horseback overland.

The logistics of the journey demonstrated just how remote the mountain was perceived to be. Even though the seven-hundred-mile journey from Beijing to Girin Ula had taken just a little over two weeks, Umuna's team called for horses and provisions enough for a three-month round-trip journey. Fearing that the horses would expire carrying such immense loads, the riverine force was tasked with transporting food and grain by boat to a location near the mountain where they could be claimed later to sustain the return trip. Bahai stayed behind in Girin Ula and assigned the younger colonel Sabsu to assist the imperial expedition. In 1677, Sabsu was still several years away from becoming the celebrated general who would defeat the Russian armies at Albazin along the Amur River. This perilous expedition to find the Long White Mountain was Sabsu's first major assignment, and in it we find hints of the rugged glories that were to mark his later career.[37]

Sabsu's detachment of two hundred soldiers and hunters took the more direct overland route, but the expedition's progress was painstaking. They avoided the perils of traveling by water, but their movement was still slowed by the numerous rain-swelled rushing rivers they had to cross: a familiar impediment to travel in mountainous eastern Manchuria. In Umuna's narrative, each of these small rivers bear individual names in Manchu, appearing today like long-disappeared Native American place-names on an Anglicized American landscape: the Undehen, the Kirsa, the Hontoho, Burkan, Hoifa, Fa, Bakta, Narhūn, Joronggo, the Wanuhu, the Fodoho.[38]

Most striking, however, was the fact that as they moved forward, the mountain proved impossible to perceive. According to Umuna's narrative, dense forests went on "as far as the eye could see" and obscured lines of sight. Sabsu ordered his soldiers to cut down trees as they advanced, but this still afforded no view of the mountain. Finally, Sabsu had to order his trackers to climb hills that rose above the treeline in order to catch occasional glimpses of their goal. Even as they got closer, the mountain proved elusive and continually cloaked itself in clouds and mist.[39]

Before we dismiss these details as mythologizing narrative, we must consider that the very physical embodiment of Mount Paektu/Changbai makes it truly difficult to see. The mountain was known as "Long White" in Manchu and Chinese for a reason: the Millennium Eruption reduced the mountain to a long, low profile in relation to its broad horizontal expanse. On the approach from the north, the massive basalt shield of the stratovolcano rises very gradually from a height of around 2,300 feet above sea level to a height of 3,300 feet at the base of the cone, resulting in an incline of only 2 percent over the course of sixty miles. Much of this long, low slope is covered in dense forest consisting of mixed deciduous and broadleaf evergreen trees that give way to tall Korean pines as the altitude increases.[40] From this extended plateau, the cone of the volcano rises to a relatively modest height in relation to its horizon-filling breadth.

A modern viewshed analysis helps us to imagine Sabsu's dilemma (fig. 3.2). Viewshed maps are generated by GIS software to indicate the lines of sight experienced by a viewer standing upon a specific terrain. The algorithms take into consideration Earth's curvature, differences in elevation, and even vegetation to generate a representation of the lines of sight between an observer's position and a visual target, or, alternately, as with the map in figure 3.2, generated by GIS specialist Jeff Blossom, to estimate the visibility of an object from various places in a given terrain. In the past ten years, archaeologists and historians have used viewshed maps to help us understand everything from the distribution of taverns in Pompeii to what Robert E. Lee could see at the Battle of Gettysburg.[41] While this particular map is based on contemporary geographical data and not historical maps, the results are highly suggestive. Due to the variable terrain, the curve of Earth's surface, and the height (or lack thereof) of Mount Paektu/Changbai, from low positions on the ground, the

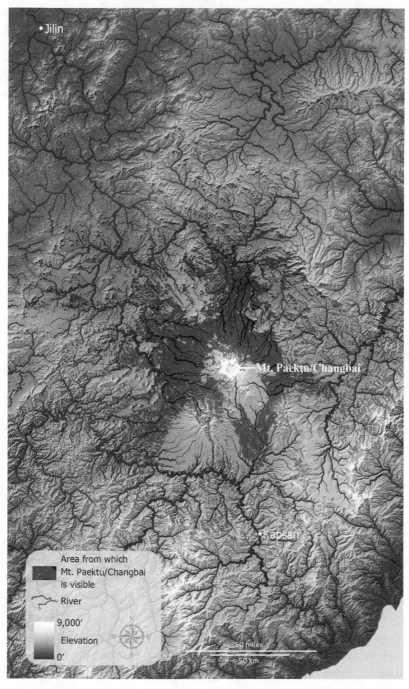

Fig. 3.2 Viewshed map of the approach to Mount Paektu/Changbai.
Map by Jeff Blossom. See plate 16 for color version.

mountain is visible only from a distance of approximately ten to twenty-five miles, and then it is visible only from a limited number of vantage points. This viewshed map helps us to understand the dilemma faced by Umuna and other explorers who sought the mountain: as the traveler approaches the mountain, the mountain refuses to appear in view.

Not only does the geometry of the Long White Mountain and its surrounding environment block lines of sight; the climate also prevents clear vision. As we saw in our discussion of the climate of Ningguta, the Changbai mountain range forms a barrier that stalls the advance of the seasonal monsoon as it moves from southwest to northeast across East Asia. The heavy snowfall in the winter months is accompanied by perpetual drizzle the rest of the year: data gathered in the late twentieth century show precipitation at the mountain's peak as high as seventy inches per year, roughly equivalent to the average rainfall in Hawaii or Louisiana.[42] The gradual rise of the mountain, the tall trees, low clouds, and even the curve of the earth all conspire to conceal the mountain from the gaze of those who seek it.

According to Umuna's account, it was only through the power of the mountain itself that the expedition succeeded in seeing it. Sensing they were near but surrounded by a dense fog, members of the expedition knelt and read the imperial edict directing them to find the mountain on behalf of the Son of Heaven. As if in response, the clouds dispersed, and the mountain allowed itself to come into full view, its white peaks capped in glory. As we shall see, this is a common element in almost all early modern narratives of ascents of Mount Paektu/Changbai: a prayer or ritual supplication to the invisible spirit of the mountain was required before the mountain would reveal its physical presence to the sight of man.

As they ascended to the top of the mountain, the expedition was met with a spectacular sight. Heaven Lake was a pure and brilliant blue, with rippling waves that were "astonishingly beautiful." The peaks surrounding the lake were so steep that they "give the impression that they are on the verge of collapsing, and to look at them incites fear." The awe the men felt at seeing the mountain did not deter them from their task of "visual inspection" for the emperor. Umuna's report includes measurements of the mountain's topography that contradict the received knowledge of the Ming gazetteer and actually make the mountain smaller. The mountain was one hundred *li* tall (not two hundred *li*) and the lake was thirty to forty *li* in circumference, not eighty *li*. Three major rivers did not flow away from the lake, but multiple streams sprang from many places on the mountain's slopes. The report also attempted to site the mountain in broader descriptive terms relative to its surrounding landscape. Seen from afar, the whole of the mountain appeared long and vast, while the summit itself appeared round and white topped. The mountain dwarfed everything around it: when standing at the summit, a bear spied at the edge of the lake appeared as a speck. Gazing from the summit outward to the surrounding terrain, the numerous mountains visible in the distance all seemed tiny in comparison.

Their visual inspection completed, the expedition descended the mountain, and as they descended, the miraculous mountain once again withdrew itself from sight. The expedition "turned to gaze once again at where Mount Changbai had been, but suddenly it disappeared, enveloped by clouds and mist."[43]

Siting the Mountain in Chinese Space-Time

Umuna returned to Beijing with a more accurate understanding of the mountain's location along with eyewitness testimony of the mountain's miraculous spiritual power. He also bore with him a poem about the mountain written by the exiled literary genius, Wu Zhaoqian—a lengthy elaborate *fu* (rhapsody or prose poem) on the Long White Mountain penned as a gift to the emperor himself. While the expedition had sought to position the mountain within the terrain of the Manchu homeland, the exiled Wu Zhaoqian sought to situate the mountain within the cultural terrain of Han Chinese civilization—and by doing so, he sought to gain the favor of the Manchu emperor.

By the time Umuna rode in to the rustic fort at Ningguta in the summer of 1677, Wu had been living in exile there for almost twenty years. The brash young poet whom we met in chapter 1 was by then a middle-aged man approaching fifty, with a wife and family of grown children of his own. The years had not made the harsh Manchurian winters any easier, and Wu longed to return to his homeland in the south. To survive, Wu had ingratiated himself with the upper echelons of Manchu society at Ningguta and had frequently written paeans to the military prowess of Bahai and Sabsu. It is most certainly through his contact with the Manchu commanders that Wu became aware of the emperor's determination to find and worship the Long White Mountain. The occasion of Umuna's expedition inspired Wu to create a *fu* dedicated both to the mountain and to the Kangxi emperor.

In his Long White Mountain Rhapsody (*Changbaishan fu*), Wu Zhaoqian unleashed his poetic talents, writing an epic that manifested all the attributes of the genre: overwhelming visual description and detail, extravagant language, erudite allusions, and an intended audience consisting of the Son of Heaven. The poem begins with an effusive prose introduction equating Mount Changbai with the most essential icons of Chinese culture:

> Mount Changbai is the Sacred Mountain of the Easterly Quarter. As the
> Jin minister Yuan Hong once said: "The East is the origin of all things, and
> Mountains are the abode of the Gods." Mount Changbai is the awe-inspiring
> origin of our country's dynasty. Nurturing the realm, it is esteemed for ten
> thousand years and extends righteous rule to the four corners of the earth.
> Verily, this mountain protects the imperial realm and brings good fortune to
> the imperial house. Like Youchao's Mount Shilou or the palace of the Yellow
> Emperor, it shines brilliant glory upon all future generations.[44]

With this introduction, Wu claims Mount Changbai as the "Dong Yue," or the Sacred Mountain of the East. This move makes Changbai into *the* most sacred mountain in all of China—in effect dethroning Taishan in Shandong province, known for over a thousand years as the "Dong Yue," the paramount sacred mountain. Yuan Hong's original quote, from his fourth century *Annals of the Later Han*, is actually from a passage discussing Taishan, not Changbai, but Wu makes a stunning reappropriation and moves the center of Han culture to this Manchu mountain.[45] Wu goes on to suggest similarities between Changbai and the abodes of the mythical founders of Han Chinese culture: Youchao, who taught primitive man how to build shelters, and even Shaodian, the father of the Yellow Emperor, the revered founder of Chinese civilization.[46] With these lines, Wu takes this mountain, once on the periphery of Han civilization, and sites it both at the pinnacle of Chinese sacred geography and at the root of Chinese mytho-historical time. The contrast with the gloom-ridden representation of Manchuria's environment in the poems of an earlier generation—Wu Weiye's crawling vermin and river monsters, Qian Qianyi's goblins and demons—could not be more startling. Wu Zhaoqian, the Han poet from the center of Chinese culture, from a line of Ming loyalists and exiled by a Manchu emperor, validates the Manchu possession of the Mandate of Heaven, proclaims the sovereignty of the Manchus over all of China, and pleads for an end to his banishment—all with a poem describing a mountain:

> It stands alone, venerated among all the sacred mountains. Viewing its great
> presence: it is // sheer and sharp, steep and soaring // vaulted and towering,

massive and remote // secluded and distant, stretching continuously // elevated and vast, wild and immeasurable // its never-ending expanse cloaked in mists, its soaring peaks hurtling to the sky.

Like all landscape *fu*, Wu Zhaoqian's *Rhapsody* presents a distinctive way of seeing. Working in an almost cinematic way, it structures its description of the mountain in progressive scenes like a camera gradually panning over a terrain. The poem begins with an overall description of the mountain, a wide-angle view that attempts to encompass the entire mass with its sweeping language. From the initial wide-angle panoramic view, the poem's focus then zooms in and begins to describe the mountain in stages or scenes from its base to the summit. Each of the ten "scenes" comprises a densely layered description of either a specific locale or a category of natural phenomena—with separate sections dedicated to rivers, trees, birds, and beasts.

The first scene is the "great forest of black pine" that lies draped over the foot of the mountain, stretching as far as the sky in all directions. This is a primeval forest, shrouded in mists, deep, dark, and impenetrable. The poet then describes the birds found in the forest—eagles, hawks, ravens, falcons, and other birds of prey that soar and swoop on the winds, "fierce-minded, golden eyed, frost draped on their wings." Spotted deer, wolves, bears, tigers, and wild boars abound in the forest, as do black sable, brown wolverines, speckled martens, and even sea lions—claws flashing, eyes burning, padding and pausing on the ice, "so many types of animals that it is difficult to enumerate them all." The narrative eye then moves up the mountain in stages to reach the summit, where "myriad peaks unite into jade and soar" and "a thousand cliffs of peeled jade congeal and rise." At the top of the mountain lies the Lake of Heaven, Tianchi. A separate section then describes the waterfall that pours out from Tianchi, spilling over the edge of the peaks, "leaping and bubbling, galloping and foaming . . . jade-green, kingfisher-green, clashing and swirling, flying and churning."

The last sections of the poem praise the powerful spiritual qualities of the mountain: "It is said that heavenly beings walk about at the summit, and those who dare to shoot deer on the mountain get lost in enveloping clouds and fog." The mountain is shrouded in a "primordial mist" that protects its spiritual potency; the profuse gathering of yin and yang in and around the mountain imbue it with a dense "spiritual power" (*ling*). In the twelfth and final section, the poem reaches its climax with a celebration of the link between the mountain and the ruling house. Because of its remarkable physical attributes and its rich spiritual properties, Mount Changbai "spreads its magnificence to the entire imperial realm." Although it lies in a distant corner of the empire, the mountain stands "streaming an unassailable golden essence, bestowing good fortune upon our Qing for a million years—forever firm, unchanging."

With this poem, Wu Zhaoqian attempts to capture the mountain in its entirety and deliver it to the emperor, presenting the essence of the mountain itself in a way that a dry report from a Manchu courtier never could. At the same time, he culturally transforms the mountain, wrapping it in all the trappings of a glorious Chinese history. Descriptions of the northeast's wild sable and soaring eagles coexist with tales from classical Chinese mythology and quotations from ancient China's most revered statesmen and scholars. Wu, the Han Chinese expert on the nature of Manchuria, created a place at once quintessentially Manchu and Chinese. Wu sites the mountain within multiple dimensions: uniting plants, animals, nature, spirit, terrain, and man, forming a joint Manchu-Han universe.

Indeed, the southern Han literatus even identified himself with this Manchu sacred mountain. Like Changbaishan, Wu saw his talent as pure and beautiful, and yet they both languished in undeserved obscurity. Both Wu and the mountain awaited the emperor's validation to achieve the recognition they deserved:

> Changbai, hero of the Northeast, towers over the frontier lands,
> It borders on the distant summery seas
> and ranges as far as the autumnal wastes;
> White clouds stretch across it for thousands of feet
> and from it, two rivers leap into the sky.
> How can the sacrifice to the mountain be performed?
> It must wait the journey of the imperial cortege.

Wu hoped that Kangxi would be moved by the poem and grant the loyal exile the right to return. But the poem did not have the desired effect. The poem was not enough to inspire the filial Kangxi emperor to overturn a decision made by his father two decades before.[47] Wu languished in exile for three more long Manchurian winters, until supporters in Beijing finally managed to raise enough money to purchase his freedom. Wu returned to the south in 1683, just one year after the Kangxi emperor's eastern tour, but he died the next year—nostalgic, it is said, for the land of the White Mountain.[48]

Siting the Mountain on the Border and on the Earth

Umuna's expedition and Wu Zhaoqian's poem were part of a larger imperial project of bringing the Long White Mountain firmly "onto the map" of the Great Qing empire.[49] When Kangxi kowtowed to the mountain at Girin Ula while on the eastern tour of 1682, he was bowing to an invisible place, but his vision of the mountain was already informed by Umuna's 1677 expedition and Wu Zhaoqian's rhapsody. To poetry and eyewitness description, Kangxi would add another layer of knowledge: Kangxi was determined to make the mountain visible through cartography. To do this, the emperor had to site the Long White Mountain vis-à-vis the neighboring state of Chosŏn: an act that required using European technology and Manchu strategy to extend the imperial gaze to the other side of the mountain.

Many scholars have examined Kangxi's penchant for mapping his dominions.[50] As the Manchu homeland, the northeastern regions were among the most intensively mapped regions of the empire. As we have seen, the most urgent reason for mapping the northeast was military. The encroachment of Russian forces in the Heilongjiang and Songhua River regions in the mid-seventeenth century was an existential threat that arose almost simultaneously with the rise of the Qing empire itself. Securing the Manchu homeland from the Russian threat required a clear grasp of these far-flung territories, along with knowledge of routes for moving and supplying troops. Indeed, the elaborately planned simultaneous riverine and overland expeditions from Girin Ula to Mount Changbai in 1677 could be seen as a sort of training run for an emerging transport system that would ultimately enforce Qing military dominance in the northeast.

Golmin Šanggiyan Alin stood at the center of a different but related borderland mapping project in the northeast: a decades-long search for cartographic accuracy about the terrain dividing the Qing empire and its vassal kingdom, Chosŏn Korea. The research of Andre Schmid, Seonmin Kim, and Nianshen Song has greatly deepened our knowledge about the contentious border between China and Korea.[51] For centuries, Jurchen and Korean peoples had traded, coexisted, and sometimes clashed in the mountainous region along the Tumen (Korean, Tumen) and Yalu (Korean, Amnok) Rivers, long considered the waters that separated Korea from China (indeed, the name "Yalu" means "division" or "boundary" in Manchu).[52] Lore had it that the Yalu and Tumen Rivers both emerged from the sacred White Mountain, but exactly where the two rivers emerged in relation to the mountain was a mystery to both courts. By the late seventeenth century, the Qing court could no longer tolerate this lack of precise knowledge. From the late seventeenth through the early eighteenth centuries, the Qing dispatched multiple surveying teams to probe the area around the mountain by following the Yalu and the Tumen upstream from either side. These surveys were not well received by the Chosŏn court, who feared ulterior motives. Chosŏn assumed that the Manchus were not particularly interested in the mountain itself but were instead reconnoitering a possible pathway for retreat to the Manchu homeland through Korean territory in the event of the collapse of their imperial ambitions in China proper. The Koreans proved uncooperative vassals and repeatedly obstructed Qing exploration of Chosŏn's northern borders.[53]

Chosŏn even managed to thwart the Kangxi emperor's most comprehensive imperial mapping project, the *Huangyu quanlan tu* ("Overview Maps of Imperial Territories," sometimes called "The Jesuit Atlas"). These mapping expeditions employed French Jesuits of the Académie royale des sciences—the "King's Mathematicians" who had been sent with the support of Louis XIV to the Kangxi court to spread science and further the missionary work of the Catholic Church. Ferdinand Verbiest had long since passed away, but the spark he provided in the European cartographic arts flowered with the new arrivals of the Society of Jesus. The French Jesuits introduced the latest Parisian cartographic equipment to Beijing and tutored the emperor and his courtiers in mathematics and surveying techniques. Between 1708 and 1712, Kangxi dispatched Jesuit advisers, together with large teams of Manchu surveyors, to map his empire, from Outer Mongolia in the north to Hainan Island in the south, and from Tibet in the west to Manchuria in the east.[54]

In 1709, the team of Pierre Jartoux (1669–1720), Jean-Baptiste Regis (1663–1738), and Xavier Fridelli (1673–1743), along with over a dozen Manchu personnel, set out to map the northeastern domains. These regions had been mapped before through the efforts of Manchu explorers and their indigenous guides, but the Europeans and Jesuit-trained Manchus would partake in an empirewide project that would combine standards of latitude and longitude together with Chinese measuring techniques and Manchu tracking skills.[55] Following the routes traveled in the previous century by Chinese exiles and Manchu emperors, the team first journeyed to the old Manchu power centers of Hetu Ala, Mukden, Girin Ula, and Ningguta, and then pushed north to explore the middle and upper reaches of the Amur River, going as far as what today is Russian territory north of the city of Khabarovsk—a distance of over 1,200 miles. But when on their return the team approached the Yalu and Tumen Rivers, representatives of the Chosŏn court halted their progress. The team probed the opposite ends of the east-west river border on the Qing side: they managed to visit Fenghuangcheng, where the Yalu empties into the Yellow Sea, and they also mapped the area near Hunchun where the Tumen River empties into the East Sea. But as the Manchu-European team sought to travel along the south banks of the rivers inland toward the Long White Mountain, the Chosŏn court detained them, arguing that Westerners were forbidden to enter the Chosŏn kingdom. Even though Chosŏn was a vassal state of the Qing, the Qing recognized the Chosŏn's right to ban foreigners, and there was little that the multinational surveying team could do to counter this decree. The maps these expeditions produced left the Kangxi emperor frustrated by impediments to the imperial gaze.[56]

The next year, however, the murder of a Qing merchant by Koreans on the north bank of the Yalu finally gave the Kangxi emperor a diplomatic pretext to see the other side of the mountain. Kangxi dispatched the Girin Ula hunting battalion official named Mukedeng to Korea to investigate the site of the murder, but his real mission was to site the sacred mountain.[57] Mukedeng (1664–1735) was an excellent choice for the assignment. As the "Master of the Hunt," Mukedeng oversaw the bureaucracy that supplied the Qing court with its game meats, wild honey, ginseng, and furs, and was thus familiar with the rough terrain of the northeast. But Mukedeng was more than just a brawny forester—he was also an accomplished court official who was conversant with European technologies. Father Matteo Ripa (1682–1746), an Italian Jesuit who served as a painter and engraver in Kangxi's court, described Mukedeng as someone who was well known to him and his friend ("da me ben conosciuto e mio amico") and mentioned that Kangxi had ordered Mukedeng to receive instruction from his confreres in surveying techniques.[58] Under Louis the XIV's mathematicians, Mukedeng learned how to handle quadrants, telescopes, and compasses, determine terrestrial distances, and take readings of the height of the sun and stars to calculate latitude and longitude in the field. He would be the emperor's eyes to see on the other side of the mountain and calculate, finally, its position on the globe. It took two attempts and over two and a half years of cajoling, bushwhacking, and negotiations, but in 1712, Mukedeng finally achieved the summit of the mountain from the Korean side. From the Korean officials who obsessively recorded Mukedeng's every move, we know that he carried with him telescopes, compasses, and "an instrument for measuring the heavens and to measure the land." According to the historical geographer Ma Menglong, this was most likely a combined geometric square and quadrant, based on Parisian models and produced in the imperial workshops in Beijing.[59]

Mukedeng's 1712 expedition is best known for marking, with a tombstone-sized stele, the location that he determined to be origin of the Tumen River, and thus the site of the border between the Qing and Chosŏn. But the mountain's waterways once again turned out to be too tangled and confusing, and Mukedeng identified the wrong riverbed. This erroneous site turned out to be several miles south of the summit, thus technically removing the most important part of the mountain from the territory of Chosŏn, much to the disappointment of the Koreans.[60] If we turn our attention away from the location of the border and instead focus on the location of the mountain itself, Mukedeng's most significant accomplishment was determining the latitude and longitude of the mountain: 42°00′ N, 127°11′5″ E.[61] Gazing at the sun and the stars while standing on the mountain, Mukedeng generated a sequence of numbers. European, Manchu, and Han Chinese court technicians then plugged these numbers in to a field of other numbers arrayed across a plane of paper in order to fix the mountain's position on the surface of the earth. The blank space on Kangxi's map that had frustrated him for years could now be filled with the precise site of the sacred mountain.

In spite of the central importance of siting the mountain for the Overview Maps of Imperial Territories, Golmin Šanggiyan Alin makes a rather modest appearance in the Mukden/Shengjing sheet of the atlas, nestled almost imperceptibly in the far southeast corner of the region. The map indicates, in a subtle way, the uniqueness of the landmark. While other mountains are represented by standard-sized inverted Vs (the universal generic marker for mountains), the Long White Mountain is indicated as a larger cluster of mountain symbols, the inverted Vs slightly rounded and a bit taller than those around it. But when we view the northeast area of the map of the "complete realm," the mountain almost disappears, hidden in a tangle of rivers and mountain ranges. For all of the effort the Qing expended to see the spectacular mountain, when sited, it simply became another data point on a global universal grid.

Siting the Mountain in Qing Spiritual Space

Even before the mountain was sited within the political geography of the empire, Kangxi made sure the mountain was given a significant place within the sacred geography of the empire. This was accomplished in two ways. The first strategy was to establish a formal ritual of worshipping the mountain.

This entailed the creation of a temple space where representatives of the court could meet with the spirit of the mountain and conduct significant bodily movements in its honor. Kangxi also sited Golmin Šanggiyan Alin within the empire through a combination of cartography and *fengshui* logic that traced the sacred energies of the empire to the northeast in what Stephen Whitman has called an "auspicious chorography." Together, these techniques attempted to capture the Long White Mountain within the visible confines of the invisible energies that constituted China.

As we have seen, since ancient times the sacred mountains of the five cardinal directions had formed an important foundation for thinking about the space of China. Each of the five sacred mountains (*yue*) had their own large temples where ritual sacrifices sponsored by the court would take place.[62] At Kangxi's order, the Long White Mountain would also become the object of imperial veneration, but sacrifices were not held at the mountain itself. Deeming the mountain too removed from any settlement, the Qing conducted rituals over one hundred miles away at Girin Ula, the banner garrison along the Sungari River where Umuna had begun his search in 1677. Instead of worshipping the mountain while standing on the mountain itself, the Qing chose to worship the mountain while looking at it from a great distance. The ritual itself was known as the *wang ji*: literally, "Gaze into the Distance Sacrifice."[63]

At first, the *wang ji* was performed under a large tent temporarily erected for this purpose, but in 1733, Kangxi's grandson, the Yongzheng emperor (r. 1722–1735), ordered a permanent temple built for the god of Mount Changbai on top of Girin's Xiao Baishan ("Little White Mountain"). Xiao Baishan is an undistinguished knoll that looks nothing like its namesake, but the site was quite picturesque. The temple was fairly modest in size, the main hall only five columns across, but the architecture was exquisitely detailed. Perched at the top of a hill above the Songhua River, the temple had a commanding view of the Long White Mountain and was surrounded by lily of the valley, a delicate, fragrant flower used to make the incense for Manchu rituals.[64] Here, on this quiet, flower-covered hillside, the sacrifices to Mount Changbai—standing mutely in the distance—continued until the end of the dynasty.

The sacrifices were held twice a year, once at the spring equinox, once at the fall equinox. Before the ritual, a deer, a boar, and an ox were sacrificed as offerings for the god, and over two dozen ritual vessels were prepared to hold the meat, wine, and incense. The most important vessel was a box covered by a yellow silk cloth. Inside lay the spirit plaque of the mountain god, a wooden tablet with the name of the god written in Manchu and Chinese. This piece of wood, small enough to be carried respectfully in human hands, was to become the site (*wei*) that could accommodate the spirit of Manchuria's tallest and most powerful mountain.

We can think of the ritual as a space where the mountain and humans communicated through movement. On the day of the ritual, officials from Mukden and Jilin purified themselves and donned their court robes.[65] The doors to the temple were opened, and the entourage slowly proceeded into the main hall, bearing the ritual items. As in all court rituals, movements were slow and deliberate, and participants moved their bodies through the ritual space in specific directions according to a strict litany. Within the center of the hall stood a simple altar, unadorned but large enough to accommodate the multitude of offerings to the mountain. The ceremony began with the welcoming of the god. As the master of the ceremony intoned instructions, the official in charge of incense slowly moved forward and set the incense burners on either side of the altar, lit the incense, opened the small wooden box, and set the spirit plaque of the mountain in the center of the altar. Traveling across space in the blink of an instant, the spirit of Golmin Šanggiyan Alin now occupied the tablet, the small piece of wood imbued with all the sacred energies of the massive mountain.

As the presiding official called out, the officials slowly lowered their bodies to the stone floor and performed three prostrations (or "kowtows") before the tablet, each one a kneeling accompanied by three knockings of the head upon the ground: the ultimate embodied form of reverence performed only for the emperor and other heavenly beings. One official approached the altar on his knees, presented the spirit with a bolt of yellow imperial silk, then performed the three prostrations and returned to his position. Another official approached the altar with a ritual wine vessel and set the vessel at the center of the altar directly in front of the spirit tablet. Another official approached the altar, prostrated three times, and presented the text of the ritual prayer to the mountain. Kneeling before the tablet, the official intoned the ritual prayer for the mountain god. Once the reading was finished, the official replaced the text in its silk-lined box upon the altar, kowtowed three times, and retreated to his original position.

After presenting the plaque with more wine (and kowtowing each time), the master of ceremonies announced the sending off of the god. All the officials performed three prostrations and nine knockings in unison to see the god home. With bows and prostrations, the officials then approached the altar and solemnly removed the offerings of silk, wine, meat, and the spirit plaque, and escorted them all to the storehouse. The master of ceremonies pronounced the ceremony completed, and all retreated and closed the doors behind them. The spirit of the mountain returned south to Golmin Šanggiyan Alin, leaving the fragrance of half-burned incense lingering in the dark hall.

*

The sacred energies of the mountain were also celebrated through maps that revealed that which could not be seen. As early as the 1650s, through calculations performed by court geomancers, the pulse of the imperial dragon had been established coursing in the earth around the tombs of the early Qing founders at Hetu Ala and at Mukden. The origins of this dragon vein lay further to the east, at the place where Bukuri Yongson, the Aisin Gioro progenitor, was born of a heavenly maiden atop the sacred White Mountain. Through surveying expeditions, the geographical coordinates of the mountain were entered into the vast data pool of thousands of other coordinates sited at mountaintops, cities, plains, and rivers throughout the empire. The result was that the sacred mountain seemed to disappear into a maze of mathematically determined points and lines that represented the world on paper.

Although Kangxi's maps did not seem to emphasize the Long White Mountain, Kangxi nevertheless used his maps' new modes of aerial-view visuality to cement the mountain's spiritual significance—a significance that had profound political import. After examining evidence from his numerous expeditions and taking in the broad sweep of his empire through his state-of-the-art overview map of the empire, the emperor penned a statement that entirely reoriented the sacred geography of the Qing, creating what Stephen Whiteman has insightfully termed "the auspicious empire."[66] In a long and technically brilliant essay, the emperor established that the "dragon vein" of Taishan, the premiere sacred mountains of China proper, actually originated with the Manchu's sacred Long White Mountain.[67] As we have seen, in East Asian cosmography, the "dragon vein" was thought of simultaneously as the visible elevations of land formations, especially mountain ranges, as well as the invisible *qi* energy embodied within those formations. Kangxi criticized Chinese scholars who relied on textual knowledge alone and assumed that the dragon vein of Mount Tai began in central China, near the cradle of Han Chinese civilization. Kangxi's cartographers, by producing precise maps of the empire, had allowed the emperor to detect empirically the line of *qi* energy and find its origins in the Manchu homeland: "We have studied with precision the earth's form, deeply researched the [*qi*] channels of the earth, dispatched men who have navigated the seas and conducted geographical surveys. We therefore know that the Dragon Vein of Mount Tai actually originates at Mount Changbai."[68] In a visually vivid essay written from a bird's-eye perspective of mountain ranges afforded him by his new European-inspired maps, Kangxi traced the progress of the dragon vein from Mount Changbai, following its main branch through the land of Mukden and then southwest along the Liaodong Peninsula. Continuing southwest, the dragon vein plunged under the Bohai Gulf and reemerged on the Shandong Peninsula, where it finally accumulated its energies and became Taishan, "its towering peaks twisting majestically into the sky, first among the five sacred mountains."

For millennia, Mount Tai had been esteemed as the most sacred mountain in China, but Kangxi had replaced it with the mountain that had given birth to his own ancestors. Nowhere in Kangxi's essay did he acknowledge the work of Zhang Jinyan, the Ningguta exile who suggested as far back as the 1660s that the Long White Mountain belonged in the ranks of China's most sacred mountains. Nor did he acknowledge the poetry of the long-suffering Chinese exile Wu Zhaoqian, whose *Changbaishan Rhapsody* elevated the Long White Mountain over Mount Tai. Kangxi had clearly been dreaming of reconfiguring the empire's sacred geography since he sensed the dragon's energy at Yongling during the second eastern tour and wrote a poem equating Manchuria's earthly pulse with that of the august Zhou dynasty. But thanks to advanced cartography, now Kangxi had an airtight case. Through mathematics and ritual, Kangxi could demonstrate both the natural coherence of his empire and the supernatural source of power of the Aisin Gioro line. He did this by constructing a hybrid form of imperial knowledge: one that combined the data from geographical surveys with the sacred power of natural forms. By siting the mountain upon the surface of the earth, Kangxi subjected the environment of his empire to number, not to remove its valences of enchantment but to enhance them.

Paektusan as a Korean Mountain

At regular intervals, the god of Golmin Šanggiyan Alin not only traveled north, summoned by the wine, meat, and prayers of the Manchu court, but also traveled south—as the spirit of Paektusan—to enjoy the sweet offerings presented to it by Korean officials of the Chosŏn kingdom. Known in Korean as "Whitehead Mountain," the mountain had long been revered by local people, and flashes of this reverence emerge in Korean texts dating as far back as the thirteenth century. It was the repeated probing, mapping, and worshipping of Golmin Šanggiyan Alin by the Manchu court in the seventeenth and eighteenth centuries, however, that prompted the political and intellectual elites of Chosŏn to situate and elevate the mountain they called Paektusan above all others within the physical, political, and spiritual landscape of their own kingdom.

Paektusan is the tallest mountain on the Korean Peninsula, but the landscape of Korea is over 70 percent mountainous, and Paektu is one mountain among many. By world standards, Korea's mountains are not particularly tall, but the profusion of ranges over the peninsula as well as their topographical prominence (the difference in elevation between the mountain's peaks and the surrounding land) make them the most characteristic element of Korea's landscape. The distinctive jagged-edged Taebaek (Great White) range extends down the east coast of the peninsula, famous for its scenic mountains such as T'aebaeksan and the Kumgang (Diamond) Mountains (all at around five thousand feet high).[69] The lower half of the peninsula is bisected by the Sobaek (Small White) range, with its renowned mountains, including Chirisan (six thousand feet) that anchors the southern end of the range. The highest mountains on the peninsula are in the north, where the landscape is dominated by the Kaema Plateau—the "roof of Korea." On a contemporary topographical map, this region appears as a brown-and-white-tinged raised patch that covers the center of North Korea.[70]

Paektusan rises from the northernmost rim of this remote highland. Because the Kaema Plateau is already at five thousand feet and many of the surrounding mountains are between seven thousand and eight thousand feet, Paektusan's altitude of nine thousand feet is not topographically prominent. Moreover, the shield of the volcano—or the platform from which the cone arises—is even more elevated on the Korean side of the current border, as the tephra and lava from the Millennium Eruption was carried by monsoon wind currents in an easterly direction.[71] This natural history of faults, eruptions, and global wind currents makes the mountain singularly elusive when approached from the Korean side. As beautifully captured in this photograph taken in North Korea by the New Zealand photographer Roger Shepherd (plate 4), Paektusan is not as visually pronounced when viewed from the south but appears instead as an indistinct crest, a wave atop an ocean of other waves that merge with the sky.

Moreover, from the perspective of Korea's population centers on the coasts, the mountain is positioned in a remote, sparsely populated northern frontier. Hundreds of miles of forests, rivers, and mountain after mountain separate it from the more populous political centers of the south, including the Chosŏn capital and center of Yi dynasty intellectual life, Seoul. Because of these factors, the mountain now lauded as the most sacred of the Korean people required considerable effort for the Korean people to see.

The peninsula's mountains, collectively speaking, have been a cornerstone of Korea's culture. Many were considered "famous" because they were the site of Buddhist temples, the birthplace of notable men, the hideout of eccentric hermits, and above all, the home of mountain spirits, worshipped in local shamanistic rites or even perhaps by the royal court itself.[72] To find Paektusan within these traditions is not difficult, although there is some controversy about what constitutes its earliest textual manifestation. Paektusan was clearly revered by the Koryŏ kingdom (918–1392), the first unified kingdom to control the entire peninsula. Five centuries before the Kangxi emperor made his geomantic claims about the physio-energetic connection between Changbaishan and the sacred mountains of China proper, the Koryŏ dynasty claimed Mount Paektu as the originating point of the Paektudaegan (the Great Paektu Range): the visible line of mountain ranges that descended down the eastern coast of the peninsula bearing the qi (Korean ki) energies that sustained the unified kingdom.[73]

It was during the Koryŏ period that the legend of Tan'gun, the mythical founder of the Korean people, was first recorded—a legend that would become closely identified with Paektusan. The legend holds that Hwanung, the Heavenly Prince, left heaven to live on earth on Korea's beautiful Mount Taebaek (Great White Mountain). A tiger and a bear approached Hwanung, requesting that he transform them into human beings. Hwanung gave the animals nothing but garlic and herbs to eat and ordered them to seclude themselves in a cave for one hundred days. The bear persevered through this hardship and became a woman, whereas the tiger failed and remained a tiger. Hwanung took the bear-woman as his wife, and their child, Tan'gun, was born on the mountain. Tan'gun went on to establish a city and found the nation that would become Korea.[74] While there is disagreement among scholars about the location of this "Great White" mountain, Paektusan has for generations been assumed to be the birthplace of Tan'gun.[75]

Mount Paektu was undoubtedly significant to Koreans for centuries, but the last dynasty, the Chosŏn (1392–1910), saw an intensification of this significance, spurred by the political and military rise to power of the Manchu Qing dynasty. The early years of Chosŏn were marked by perpetual dealings with the Jurchen peoples on its northern border, whom the Koreans disdain-fully called the "Wild Men."[76] Two hundred years later, the descendants of the former Jurchen "Wild Men" had become the rulers of China and suzer-ain overlords of the Korean kingdom. The ascendancy of the Qing—whom they still referred to in private as barbarians—was a shock to Chosŏn, and it prompted a new celebration of the kingdom's own cultural achievements and natural landscapes. Qing exploration and mapping of the border area helped to make Paektusan a focus of a new geographical nativism. At the same time, cosmopolitan connections between Beijing and Seoul fostered by frequent tribute missions meant that Chosŏn's elites shared in many of the same intel-lectual trends as their Qing counterparts. Chosŏn's *sirhak* (practical learning) scholars, such as Yi Ik (1681–1763), Park Jiwon (1737–1805), and Sŏ Myŏng-ŭng (1716–1786), participated in the same global intellectual exchanges that influenced the *shixue* (practical learning) trends in China, including a drive toward empiricism and an interest in Western technologies of astronomy, mathematics, and surveying.[77]

They also shared an obsession with Paektusan. Beginning in the eigh-teenth century, Chosŏn's scholars produced a flood of treatises that probed the position of Paektu in Chosŏn territory and recast its significance in his-torical, spiritual, and spatial terms.[78] Most *sirhak* scholars explored the moun-tain from the comfort of their studies by perusing texts, but several climbed the mountain themselves and wrote about their journeys, producing the first extensive eyewitness accounts of Mount Paektu/Changbai not only in Korean history but also in the history of the world.[79] Ultimately, these narratives also demonstrated the crucial role that local knowledge played in creating a new vision of the supreme national mountain.

Local Knowledge in Si(gh)ting Mount Paektu

Ironically, one of the first Korean narratives of an ascent of Mount Paektu, Hong Se-t'ae's "Paektusan gi" (A record of Whitehead Mountain), centers on the 1712 Qing expedition led by Mukedeng as related by one of the Korean translators who accompanied the Manchu official on his expedition.[80] Two individuals stand out in "Paektusan gi": one is Mukedeng, described as a vigorous but haughty military man who brought telescopes, quadrants, and compasses on the climb. The other main character in the narrative is the ex-pedition's local guide, a humble but wily ginseng picker from the borderland known simply as "Aesun."[81] The narrative captures two different approaches to creating knowledge about the mountain. The Manchu Mukedeng, dispatched from the metropole Beijing, would know and see the terrain primarily through the mediation of surveying equipment. The local man Aesun, in contrast, knew Paektusan through embodied experience of the terrain, flora, and fauna. More importantly for the survival and success of the expedition, Aesun also possessed a direct knowledge of the mountain's powerful god.

Hong's narrative begins in the spring of 1712, when the Chosŏn court personnel tasked with accompanying the Qing court's "murder investigation" journeyed "several thousand *ri*" from Seoul to the northern territories. They rendezvoused with Mukedeng and his entourage along the Yalu River, then traveled to Samsu prefecture, north of the garrison town of Kapsan on the Kaema Plateau. The Manchus and Koreans set off in what they thought was the direction of Paektusan, but they could see it only by making exhausting and time-consuming climbs up the tops of other mountain ridges. Only from these vantage points did their goal appear indistinctly perched on the hazy horizon, "like a white cow resting at the far edge of a field." Mukedeng used his telescope to try to get a better view of the mountain, and dejectedly proclaimed it more than three hundred *ri* (one hundred miles) away. In spite of the presence of advanced vision-enhancing technology, the expedition could proceed only with the knowledge of a local guide. At Hyesan, an outpost at the point where the Yalu River's path turns almost due north to Paektusan, the expedition hired a man named Aesun, a Korean-speaking local ginseng forager who was said to know the way to the mountain. When asked to lead the way, Aesun feigned ignorance of the mountain—not surprisingly, since Qing regulations forbade picking ginseng in the region and Chosŏn regulations made the very act of crossing into Qing territory a crime punishable by decapitation. Mukedeng laughingly assured Aesun that he would be pardoned for all of his past "crimes," and the ginseng picker was finally coaxed into being the expedition's guide.

As they moved north, Aesun's intimate and accurate knowledge of the mountain became readily apparent. Aesun had no need for a telescope: without seeing the mountain in the distance, his steps were guided in the right direction by his knowledge of the rivulets, knolls, and caverns along the way. Aesun also knew the mountain's flora and fauna, leading the party to the hiding places of game and gauging the progress of the ascent by examining the size and shape of the trees. Aesun could even sense the weather: in the last leg of the ascent, the climbers ignored the ominous clouds and humidity gathering around them, but Aesun insisted they pause, because any increase in altitude would mean that the party would encounter snow, high winds, and freezing temperatures. Aesun's abilities combined skillful vision with other embodied sensibilities: a gauging of time, distance, and altitude—a hunter or forager's way of feeling the landscape acquired through mindful and meaningful dwelling.

Aesun also had expert knowledge of the mountain's supernatural landscape. The gathering storm clouds were a clear indication that the god of Mount Paektu was displeased by their approach, and Aesun insisted that the party stop and conduct a ritual prayer to placate the spirit before their final ascent. Mukedeng scoffed at the idea because of his superior rank: he insisted he was not a common deer hunter or ginseng gatherer, but a representative of the Son of Heaven with no need to supplicate a mere mountain god. The Korean interpreters warned Mukedeng that those who skipped the prayer might not survive the climb. The haughty Manchu finally acquiesced to the rites, and after the ritual was completed (so the narrative goes) the clouds miraculously dispersed, revealing a starry firmament overhead.

Once the party reached the summit (absent the out-of-shape Korean officials who could not make the difficult ascent), the Koreans and Manchus alike marveled at the spectacular beauty of the mountain and the lake. At the summit, Mukedeng used his equipment to estimate the heights of the peaks and measure the circumference of the lake—but he also demonstrated his new command of the supernatural terrain, leading a prayer to pacify the spirit of the mountain who had roiled the waters of Heaven Lake with a terrifying sound like thunder as the expedition approached. After this remarkable encounter at the summit, Mukedeng went on to complete his more mundane cartographic tasks: finding the source of the Tumen River, establishing the border between Qing and Chosŏn territories, and determining the latitude and longitude of the mountain.

In later years, Korean patriots used Hong's narrative to decry the injustice of what they saw as Mukedeng's inaccurate determination of the Chosŏn boundary. But the narrative can also be read to illuminate the differences between local knowledge of the mountain and that of the elites who traveled from distant political centers. Both Manchus sent from Beijing and Koreans sent from Seoul were neophytes to the mountain, unable to gauge distances or find paths in spite of their command of maps, treatises, and telescopes. The unlettered local guide, the forager Aesun, was the one who possessed accurate knowledge of the mountain's geography, flora, and fauna.

Aesun can be seen as a practitioner of what Tim Ingold has called a "sentient ecology," a "perceptually skilled agent, who can detect those subtle clues in the environment," a possessor of "skills, sensitivities and orientations that have developed through long experience of conducting one's life in a particular environment."[82] Importantly, Aesun's skills included an exquisite awareness of the mountain as a sentient being and knowledge of how to appease the spirit with prayer and offerings. Indeed, without the embodied knowledge and spiritual skills of the local guide, the cartographic goal of the expedition would not have been achieved. Behind the much-lauded numbers of latitude and longitude determined by the European-trained Mukedeng stood the power of a northeast Asian mountain god and the expertise of the god's local acolyte.

A later Korean ascent narrative, "Yu Paektusan gi" (A record of travels to Mount Paektu), written by the scholar Sŏ Myŏng-ŭng (1716–1787), tells a similar story of an intersection between metropolitan science and local knowledge, but this time the cartographic technician was Korean, not Manchu. Sŏ Myŏng-ŭng was an astronomer and mathematician who was one of Chosŏn's leading *sirhak* intellectuals. Proud of his command of the Chinese classics but also open to "foreign learning," he is best known for combining an expertise in Western science with a framework derived from traditional Confucian studies.[83] In 1765, Sŏ balked at accepting a royal assignment to the court of King Yŏngjo (r. 1724–1776), and as a punishment, he was exiled from Seoul to the northern prefecture of Kapsan. Just as the Qing exiled intellectuals to Ningguta in the heart of their former homeland, so too did the Chosŏn court send intellectuals into exile in the northernmost reaches of the kingdom. From there, these intellectuals had a hand in turning Paektu into Chosŏn's most sacred mountain.

According to his "Record,"[84] Sŏ left from Seoul's eastern gate in June 1767 and set upon the northern road to exile, a three-hundred-mile trek that brought him from the cosmopolitan capital near the central west coast (at 125 feet above sea level) to Kapsan on the rugged Kaema Plateau (at 5,000 feet above sea level). There he found himself in a remarkably different environment, both physically and intellectually. During the later Chosŏn, southern literati tended to view the denizens of the north with contempt, deeming even the educated of the region to be rugged frontier folk who were less refined than their southern counterparts.[85] Rather than despair, however, Sŏ looked at the bright side: because of the misfortune of exile, he had a golden opportunity to see the famed Whitehead Mountain. Like Mukedeng before him, Sŏ staged his ascent from the outpost of Samsu, where the party hired two local men to serve as guides. From Samsu, the guides assured Sŏ, it would be a four-day journey north to the mountain, keeping the Amnok (Yalu) River on their western flank.[86]

As they traveled through the rugged countryside, Sŏ perceived his surroundings almost entirely as a military landscape, and such a landscape demanded quantification. Sŏ chided himself at times for simply enjoying the scenery when he could be taking cartographic measurements in order to improve his country's defenses. He noted the size and strength of remote garrisons and fire-signal beacons, ever aware that these outposts were created to guard against the "barbarian"—never mentioning the Manchus by name but alluding to the potential threat they posed. The Korean mathematician proved himself master of the same technologies that Qing surveyors had used several decades before, and at night, he determined his coordinates by taking readings of the stars with a quadrant he had directed a local carpenter to make. Sŏ noted that his readings produced a latitude for his position of slightly less than 42 degrees, a figure, he noted, that was similar to that of Shenyang, the old Manchu capital located on the Liaodong Peninsula 250 miles to the west.[87] Sŏ used his mathematical expertise to site his position on the globe and to situate Paektusan within the geography of a greater northeast Asian landscape.

Sŏ soon discovered that his technological knowledge was no substitute for the embodied knowledge of the physical and spiritual terrain possessed by his local guides. As in Hong Se-t'ae's account, the metropolitan elites wished to quickly achieve the summit in spite of the gathering darkness, but local guides insisted on waiting until morning for safety's sake. The local guides also insisted on conducting rituals to the mountain spirit. Like the Manchu official Mukedeng fifty years before, some of Sŏ's Korean companions looked askance at the local ritual, mocking it as an "absurd attempt at creating a temple in the wilderness." But the mathematician embraced the ritual and even agreed to write the prayers himself.[88] Addressing the mountain in reverent tones, Sŏ entreated Mount Paektu to "look down from on high" and "accept the sincere offering." He also added a personal touch, including a *sirhak*-style lecture for the mountain god that sited the mountain within the sacred geography of East Asia: "There are thirty-six sacred mountains, and among them, Kunlun is the progenitor of them all. Among the Chinese, there is none who would not climb Kunlun to see its magnificence, and Kunlun would never think to hide its magnificence from them . . . Our nation's Paektusan is like China's Kunlun. . . . As the Spirit of Kunlun would not hide itself from the people of the Central Country [China], so Paektusan's Spirit should not be stingy with the people from the Eastern Country [Korea]."[89]

Sŏ's geographical prayer (combined with the ritual offerings of his local guides) was apparently effective, and the next morning, a cloudless day greeted the travelers. Once the party reached the summit and looked down at the beauty of Heaven Lake hundreds of feet below, the rational geographer found he was too awestruck (or overcome by vertigo) to move. In his stead, the local guides confidently descended into the crater and explored the shoreline of the lake, making sketches and using a borrowed compass to site the positions of the surrounding peaks. That evening, the elated members of the party played instruments and sang songs into the night in celebration of their successful ascent. Sŏ did not join the revelries and instead spent his evening trying to get an accurate reading of his position by siting the stars with his quadrant.[90]

Sŏ's "Record" demonstrates the way that Chosŏn elites combined cartographic measurement and a sense of the sacred in order to site Paektusan in the physical and spiritual terrain of Chosŏn. Sŏ observed the land around Paektu with "national security" concerns in mind, and did calculations in the field in order to generate accurate understanding of strategic positions. In spite of Sŏ's mastery of cartographic science, it was the embodied knowledge of the local guides that ensured the success of Sŏ's ascent. Just as important was the local guides' mastery of techniques to appease the mountain deity. Sŏ recognized the efficacy of the mountain's power as a sentient being and joined his guides in offering prayers to its spirit. We are not sure if Sŏ's inability to function at the summit was a manifestation of vertigo or a spiritually inflected sense of awe: perhaps it was a combination of both. It is clear that at Paektusan, the mathematician sensed a presence that both incorporated and went beyond the mountain's visible form.

In the late eighteenth and nineteenth centuries, Chosŏn's elites expressed their awareness of the mountain's numinous presence not only through formalized ritual but also by combining sacred sensibilities with the number-based language of cartography. Korean maps of this period sited the sacred mountain in relation to the kingdom's terrain and also positioned the entire kingdom within the mountain's sacred energies. To do this, they made invisible energies visible and imbued those energies with the power to transcend political boundaries.

Siting Chosŏn in Paektusan's Spiritual Space

In 1767, not long after Sŏ's ascent, the Yi court incorporated Mount Paektu into the roster of Chosŏn's sacred mountains and established official ritual sacrifices to the spirit of the mountain. This daring decision to make Paektusan the *pukak*, or "Northern Premier Mountain," was the result of a long court debate that centered on siting the mountain within multiple dimensions of Chosŏn: the political, the historical, the geographical, and the sacred. The debate began with King Yŏngjo and several of his courtiers linking the mountain to the founder of the Yi dynasty, Yi Sŏnggye (1335–1408), whose family had emerged from the northern province of Hamgyŏng. They argued that every "stream and mountain" of that province emerged from Paektusan, and thus the mountain was the "auspicious birthplace of the glorious spirit of our dynasty." Some conservative courtiers protested this idea as geographically imprecise, stating that Yi Sŏnggye's hometown was well over a hundred miles from Paektusan. Others pointed out that the border fixed by the Qing set much of the mountain outside of the kingdom, and state rituals could not be conducted on a mountain outside a state's territory.[91]

Ultimately a compromise was found, one that acknowledged political borders but suggested that the spirit of the mountain could transcend them. Mukedeng's stele marked the border on the southern slopes of the mountain, meaning that at least part of the mountain was in Chosŏn territory and therefore could be worshipped by the court. Moreover, it was not necessary to conduct the rituals on the mountain itself. The Qing, after all, sacrificed to the mountain from a place more than a hundred miles away. While the Qing conducted "gazing from afar" rituals from the north, Chosŏn could similarly conduct "gazing from afar" rituals from the southern side.[92] The invisible line constituting the political border would neither obstruct the worshippers' line of sight, nor would it impede the travel of the sincere spiritual energy of the court's prayers.

When time came to find a site for the temple, a familiar character dominated the discussion: Sŏ Myŏng-ŭng had returned to Seoul from his exile in the north and had taken up the position of vice minister of the Bureau of Rites. With Sŏ's input, the temple was placed on a remote mountaintop in Kapsan near the outpost of Unhŭng that became known as "Gazing at the Mountain Peak," a location specially sited for its direct-line view of Paektusan approximately fifty miles to the north.[93] The temple was a rustic T-shaped structure with a small main hall and an external veranda, set on a raised stone platform, and, as later documents mention, it was often surrounded by unkempt grounds and fallen trees because of the extreme weather and remoteness of the northern territory. The temple contained a fifteen-inch-tall spirit tablet inscribed with the name of the mountain—an abode for the mountain god when summoned to enjoy the sacrificial offerings.[94] On the spring and autumn equinoxes, Kapsan officials would conduct the sacrifices to Paektusan, following a litany mandated by the court in Seoul. Although the setting was rustic, the prayer to the mountain recited by officials embodied formal cadences, as reflected in this prayer to Paektusan from Chosŏn court ritual:

> Peaks extending as high as Heaven,
> Honored by all the sacred mountains;
> For ten thousand generations, our Eastern Land
> Has basked in your vast merits.
> Dwelling in your prosperity, we present myriad offerings,
> Seeking your boundless blessings
> With our utmost sincerity, this we pray.[95]

While Chosŏn rituals marking the sacrality of Mount Paektu were carried out on a remote mountaintop in the rustic north, Chosŏn's cosmopolitan elites in Seoul proclaimed their adulation of the mountain through visual representations in maps. The later Chosŏn geographical renaissance resulted in an explosion of cartographic expression. The era saw a profusion of hand-drawn and individually inked maps capturing the entirety of the peninsula or focused on individual provinces. Many of these maps are exquisite works of art, vividly colored and imaginatively rendered.[96] Increased emphasis on empirical evidence and Western cartographic methods through the eighteenth and early nineteenth centuries resulted in the increased accuracy of later Chosŏn maps, but along with accuracy came another consistent element: an obsession with the position and significance of Paektusan in the geo-body of Korea. Indeed, the maps of this period can be read not only as ways of representing the mountain within Korea but also as ways of siting the entire kingdom of Korea within the sacred energies of the supreme mountain. Most remarkably, even as techniques of cartographic science were taken up by Chosŏn elites, the sacred nature of Paektusan was not diminished but seemed in fact to be intensified even as maps pursued a goal of accuracy.

Several scholars have traced the cartographic transformation of Paektusan's position on the Korean Peninsula.[97] The earliest maps from Chosŏn's Yi dynasty (c. 1450) displayed a squat, rectangular peninsula with an ill-defined northern border region. Conflicts with Jurchen tribes over the course of the century inspired more detailed attention to the north, and with it, the distinctive appearance of Paektusan perched at the top of the peninsula. While Korea's highly mountainous terrain is represented with generic inverted-V symbols in these early maps, Paektusan alone gets its own personal portrait—painted with jagged towering peaks and a round lake in the center. The mountain is disproportionately large, its sacred uniqueness indicated by the use of vivid color: its peaks tinted with white, the lake a brilliant blue—a standard form of representation that would continue from the fifteenth century well into the nineteenth century.

While these maps site Paektusan at the northern fringes of the kingdom, they also begin to place the kingdom within a network of the mountain's energy: the Paektudaegan, or the great spine of Paektu.[98] In these maps, the Paektudaegan is represented as a jagged line of vibrant green that starts at the base of Paektusan and winds its way down the eastern edge to the center of the south. Conceived both as a system of mountain ranges and as a line of *ki* (*qi*) energy, the Paektudaegan on maps presents a geomantic vision of a nation nurtured by a life force that surges from Mount Paektu through a network of mountain arteries. The historian Gari Ledyard, while acknowledging the obvious geomantic (*fengshui*) intent of such maps, suggests that the lines actually captured the true contours of mountain ranges, rivers, and drainage basins, and could "dramatically reveal the overall character of the landscape in images that common sense cannot easily generate."[99]

These lines of *ki* energy were even compatible with developments in cartographic technology. Indeed, the interest in empiricism and measurement that emerged with the *sirhak* movement in the eighteenth century did not put an end to the numinous magic of Paektusan; instead, this period saw cartographers' deepening commitment to representing the mountain's sacred energies even as they sought to capture accurately the visible contours of the earth. The most prominent map maker of the early eighteenth century, Chŏng Sanggi (1678–1752), was one of the first Korean cartographers to employ a standard scale and empirical measurements, using an odometer and plane scale to accumulate data on distances and altitudes of places he visited personally throughout the kingdom.[100] His maps were the first to represent clearly the more angular shape of the peninsula and detail the north-south peregrinations of the Tumen and Yalu Rivers. Yet for all their accuracy, these "Chŏng Sanggi-style" maps are shot through with dramatic *ki* channels or ridgelines that emanate from an easily identifiable Paektusan, and the mountain is still rendered in a painterly style, with Heaven Lake nestled amid towering purple peaks.

Following traditional East Asian cartographic conventions, these maps do not represent land beyond the borders of the area in question: terrain ends with borders and gives way to blank space. Focused as they are on the Korean kingdom's territory, the maps do not represent Qing territory, but they do include Mount Paektu: the mountain floats above the peninsula, suspended in an ethereal pure blank space to the north of Chosŏn. The fact that the Qing has claimed the mountain is represented by the positioning of Paektusan within the blank area that would be Qing space. Nevertheless, in these maps, a fine but vigorous thread of *ki* energy descends south from the mountain, linking Paektu to the peninsula like a tether to a balloon or like a fetus to its mother via an umbilical cord. As seen in one nineteenth-century map of South Hamgyŏng province (plate 5), this mode of representing Paektusan continues long after the impact of European mapmaking: maps still feature a floating Paektusan and its umbilical cord of *ki*, even amid a field of precisely rendered latitude and longitude.

This biological or obstetric analogy, while strange sounding, is not terribly far fetched. Perhaps the most astonishing rendition of the relationship between Paektusan and its surrounding land can be found in an early nineteenth-century map of Kapsan, the original "home county" of the mountain (plate 6).[101] At first glance this map does not look like a map at all; it appears to be a drawing depicting some sort of organic tissue—a web of blood vessels, a network of neurons, or perhaps even a placenta. Floating in space nearby is a remarkable green kidney-shaped object, surrounded by white appendages rendered like the tendrils of a strange sea plant or the fimbriae of a Fallopian tube surrounding an ovary. The labels on the map reveal this numinous "embryo" to be Mount Paektu itself, envisioned from directly overhead: the dark green kidney shape is the caldera lake, while the wavy white fronds are the peaks surrounding the crater. The blue "vessels" are rivers, flowing and branching in a field of rounded green lobes representing mountains. The main "artery" is the Yalu River, and square cartouches along its tributaries indicate settlements. Two lines tether the mountain-embryo to the network-landscape: a thicker green artery emerges from the center like an umbilical cord but goes off into empty space: this is the Songhua River, which leaves the plane of the map and disappears into Qing territory. The other tether is a thin red line that meanders through Korean territory from Kapsan, Samsu, Unhŭng, and Hyesan, past the three ponds at Samjiyon, and then approaches the mountain-embryo from its right flank. This line is the actual path to the mountain itself, created by the knowledge of local guides and mapped by *sirhak* scholars such as Sŏ Myŏng-ŭng. Like those scholars and their guides, this map grasps Paektusan not as an inanimate mountain of rock but as a living landscape, one shot through with pulsing energies that both emerge from and nurture a sentient being. The same can be said of the entire peninsula. The sentient being in question is both the mountain and the Kingdom of Chosŏn, linked through a numinous energy that political boundaries could not sever.

Conclusion: The Visible, the Invisible, and Transcendent Visions

While little known in the West, within East Asia, Mount Paektu/Changbai is a well-known "environmental spectacle," a famous symbol of an ostensibly pristine wilderness celebrated and marketed in both the PRC and Korea.[102] The "reality" of the mountain is typically presented from a singular viewpoint—from atop its towering peaks or from an aerial shot, looking down at the brilliant blue of Heavenly Lake. This image is reproduced and consumed in tourist brochures, documentaries, books, and murals found everywhere from government meeting rooms to restaurants. The low-profile view of the mountain's broad white expanse even appears on the national emblem of North Korea; and its slopes are immediately recognizable as the backdrop for heroic images of that nation's leaders. Perhaps nowhere else is the image of a singular mountain so closely associated with a country (with the exception, perhaps, of Japan's Mount Fuji), and perhaps no other image is so closely associated with Manchuria as a place.

This chapter has demonstrated that this spectacular image of Mount Paektu/Changbai is not a given. What is taken for granted as the "reality" of Mount Paektu/Changbai was produced over hundreds of years through a process that involved the very intentional pursuit of a certain type of seeing: empirical eyewitness examination. Beginning in the seventeenth century and continuing through the eighteenth, intellectual trends and the desire for military precision meant that elites such as the Kangxi emperor desired vision as proof over textual evidence: they were no longer satisfied with leaving this storied place as a mountain of the mind. Ordered to visually inspect the mountain, expeditions set out to find it, but this task was far from simple. Metropolitan elites journeyed hundreds of miles from urban political centers and overcame tremendous physical hardships, all the time struggling to see a mountain that evaded their sight. Placing expeditions within their physical environment and understanding their senses as they moved through space show that the mountain remained frustratingly invisible—a hidden spectacle.

Sometimes the act of seeing required the deployment of different senses in order to augment the sense of sight, or even the ability to sense that which could not be seen. Local ginseng foragers and hunters who guided expeditions saw the mountain in a different way, with skills that developed through constant interaction with the environment. Their "dwelling" within the mountain, learning to react to the land, plants, rivers, and animals in order to survive, provided them with a multisensory knowledge that allowed them to simultaneously see, read, and feel the terrain. While metropolitan elites relied on augmenting their neophyte vision through technology, the guides' own augmented sight proved a superior way of seeing, even if what they saw was ultimately invisible. This augmented sense included an awareness of the mountain's spirit, a kind of awe in response to environmental cues that was at once haptic, visual, and internal.[103] This sensing of the sacred was essential to the eventual siting of the mountain, as essential as the telescope or the quadrant.

It is important that we not reduce East Asian cartographic enterprises to rationalizing strategies of territorial management. Empirical observation and cartographic accuracy are always intertwined with belief, and measurement coexists with myth. Siting the White Mountain was never just a question of creating maps with accurate borders. Those who climbed it strove to situate the mountain not only within physical space but also within the invisible and unmeasurable dimensions of the spiritual world.

Even when the invisible mountain finally came into view, it was politics, and not the environment, that ultimately limited vision: the majority of government-sponsored climbers were confined by national borders to one slope or the other. We have ample evidence that local climbers routinely ignored these borders (at the risk of execution if caught), but most seemed to be all too aware of the power of states to circumscribe the experience of the mountain and its environment. As a result, most of the perspectives we have considered in this chapter—whether Qing or Chosŏn—site the mountain solely in relation to the territory of their own polity. Some may have mentioned distances to foreign cities or made vague references to rivers or mountains beyond the border, as we saw with Sŏ Myŏng-ŭng's musings on Paektusan's location. Yet most site the mountain in a corner of provincial terrain, as in Qing maps of Shengjing, or, as in Chosŏn maps, which represent the mountain as a free-floating growth on the top of the geo-body, the southern slopes linked to the nation but with all other sides surrounded by empty space.

✳

There were, however, some observers of Mount Paektu/Changbai who were not concerned about siting the mountain within a political formation and who needed neither maps nor even the sense of sight to journey to the mountain. These were the shamans who lived in local communities near the mountain whose rites, songs, stories, and dances sited the mountain in a way that transcended time, compacted space, and ignored borders entirely.

Members of the Manchu Šikteri (Sinicized as Shi) clan of Xiao Han village in Jilin province possess rituals that focus strongly on the presence of the White Mountain. Today the Shi shamans live about forty miles north of Jilin City (Girin Ula), but clan genealogies suggest that five hundred years earlier, the clan had lived much closer to the mountain in the area of the Hoifa (Huifa) River. After the conquest of the Hoifa region in the early seventeenth century and the consolidation of the banner system, members of the Shi clan were placed into the Plain Yellow Banner and relocated north to the Girin Ula area.[104] The clan rituals and tales collected by anthropologists and folklorists in the 1980s are most likely fragmented remnants of older traditions. The songs of the Shi clan shamans are sung in Sinicized Manchu, the original language having died out in these locales long ago, but hints of past environments remain embedded in their words.[105]

Although today the mountain is too far away to be seen from their village, songs of the Shi clan shamans clearly envision myriad spirits—of animals, heroes, and ancestors—who dwell in silver and gold palaces atop Golmin Šanggiyan Alin's snow-covered peaks. The shaman's songs are intensely visual, providing vivid descriptions of the spirits, the physical attributes of the ancestors, and, most of all, vivid descriptions of the mountain and the landscape around it. The rituals begin with the shaman summoning the spirits to join him, calling to them with songs that include oral maps to guide their journey.

The mountain itself is manifest in the persona of Cooha jenye (the Military Immortal), the spirit of the mountain who travels as a warrior-chieftain leading a great army:

> Cooha jenye, he who dwells on the Long White Mountain, on the myriad ridges, past the nine peaks, the eight ridges, the seven ranges; our founding ancestor, the god of the Long White Mountain.
> Red faced Cooha jenye, riding a fire-dragon steed, commanding the myriad heroic spirits, brilliantly shining, embarking on righteous battle; with forty strong men riding massive steeds and twenty braves marching on foot, descend along the Nisheha River!
> From your dwelling in the mountain palace, the vast palaces high upon the golden peaks—descend along the silvery Sungari River and come to us!

Deified ancestors of the shamans going back seven generations, together with their assistants (now heroic acolytes), are called to make the journey as well:

> The First Great Master, Sucungga, the great hero spirit who commands the revolving sun and moon; holding the great bronze mirror, he who dwells in the silver and gold palaces along the spring of the three peaks of the Long White Mountain, descend along the Hoifa River!
> The Second Great Master Tasha, along with the Hero Serengtai, holding the iron whip, they who roam the White Mountain, who dwell on Alaci Peak, descend along the Neyen River!
> The Third Great Master, Saicungga, with the Hero Jakta, holding two swords, and the Hero Huyan, holding a bronze trident; they who dwell on the Long White Mountain, descend from the sky along the Sere River!
> The Fourth Great Master, Ferguwecuke, accompanied by the Hero Cahabuku, holding an iron cudgel, and the Hero Toholo, holding a three-pronged trident, descend from the Long White Mountain along the Sungari!
> The Fifth Great Master, Kiyangkiyan, who dwells on the five-layer peak, the Hero Sirgici, who dwells on the Long White Mountain, descend along the Dergi River.
> The Sixth Great Master Fokuce, who lives beyond Seheri Peak, and the Hero Salabuku, holding a golden spear, they who dwell on the Long White Mountain . . .
> The Seventh Great Master Mukefayanga, the Spirit of the Shula River, who commands all the people,
> Those who dwell on the summit of the Long White Mountain, descend!

Other songs call upon the spirits of the forest animals to join the shaman—the bear, the tiger, the fox, the eagle—and even the spirits of snakes and dragons:

> Great Eagle,
> He who covers the light of the sun and moon
> when he spreads his enormous wings in the sky,
> Anchun Eagle Spirit,
> > Descend from the peaks of the Long White Mountain!

Golden-Tongued, Silver-Tongued Eagle who lives on the Long White
    Mountain,
        Descend along the Hoton River!

Water-Bird Spirit who lives on the eastern slopes of the Long White
    Mountain;
where also dwells the White Bird Spirit Shulumanni
and the Wasteland Bird Spirit, together with the Magpie,
        Descend from the Long White Mountain along the Sungari!

Gold-forged Fire Dragon
who dwells on the summit of the Long White Mountain . . .
        Descend and follow the Lalin River, the Fulgiyan River

Flying Tiger Gods who dwell upon the Golden Cliffs of the Nine Peaks,
the Female Tiger Spirit, the Male Tiger Spirit, the Striped Tiger Spirit,
the Great Black Tiger Spirit, the Golden Tiger Spirit,
        All those who dwell on the Long White Mountain,
        Come down from the craggy summits!

The eighty Fox Spirits, the ninety Wolf Spirits,
        All those on the Long White Mountain
        Descend from the silver cliffs!
The spirit of the Black Bear; the Spirit of the priceless Sable;
        Fly over the Long White Mountain along the banks of the Nishiha
            River,
The Eight-Yard-Long Python Spirit, the Nine-Yard-Long Snake Spirit
        Descend from the blue sky along the Nishiha River;

The spirit of the Golden Spark is on his way
Those on the Long White Mountain
The golden-feathered Eagle Spirit, the silver-feathered Eagle Spirit
All who dwell on the mountain's summit, descend![106]

Although the people of Xiao Han village cannot see the mountain, the
presence of the mountain is simultaneously envisioned as a fierce warrior-
god and a spectacular natural formation with luminous peaks of silver and
gold. The naming of myriad spirit-beasts of the mountain invokes the natural

abundance of the region's forests. The shaman calls down dozens of animals—birds, mammals, reptiles, even dragons—in rhythms and repetitions reminiscent of Wu Zhaoqian's 1677 *Changbaishan Rhapsody*, the poem capturing the totality of the mountain written for the Kangxi emperor. In his *fu*, the Chinese poet presented a wild, remote mountain that was home to a superabundance of wildlife, where the shriek of the sable and the cry of the eagle resounded off the icy cliffs. The shaman songs similarly reflect a vision of the area as teeming with wildlife, but it is far more than a distant nostalgic shorthand. The shaman's ritual literally manifests the intimate and immediate connections between humans and nonhumans: as the animals' spirits are summoned, they inhabit the body of the shaman, whose dance in turn manifests the winged flight of the eagle or the lumbering gait of the bear.

The shaman's songs also place the mountain within a significant past, but this past is concretely local. Unlike the vision of the White Mountain promoted by the courts in Beijing or Seoul, the mountain is not connected to the miraculous founder of dynasties, such as Bukuri Yongson, or the mythical founder of a national race, like Tan'gun. Nor is the mountain inserted into the hoary textual past of Chinese civilization, as Wu Zhaoqian attempted. Instead, the mountain is the dwelling place of the family's own personal ancestors—Sucungga, Saicungga, Mukefayanga—men entirely unknown to the outside world, but within the living memory of some, ancestors who are now powerful spirits who dwell in the mountain, living both in the distant past and in the present, but always in the local.

The shaman summons the animals, gods, and ancestors to journey from the Long White Mountain to the humble village, and the songs narrate specific directions. With these oral maps, the shaman calls upon the spirits to fly along pathways defined by rivulets that flow from the mountain's slopes—the Nishiha, the Hoton, the Fulgiyan. These names recount the waters that crisscross the expanse between Jilin and the mountain once traversed by Umuna and his local guides in the first Qing expedition to the mountain in 1677. The repetitive invocation of Manchu names for rivers creates a sonic cartography, charting a landscape that mirrors natural terrain and at the same time makes a temporal connection to a cultural past that has all but disappeared from the landscape.

As the spirits fly along the rivers to the summoning shaman, the ritual condenses space and transcends time. The Nishiha, the Hoton, and the Fulgiyan do not flow to Xiao Han village but instead flow to settlements that were home to the Shi family over five centuries ago. No matter. The spirits effortlessly traverse hundreds of miles, collapsing space and bringing humans immediately into the presence of a remote and unreachable place that many of them will never see. The songs imagine a place simultaneously invisible yet conjured into vision. The body of the shaman manifests this presence of the mountain until it is time to guide the spirits home:

> Tonight this humble shaman has
> Sought your blessings,
> The sound of my songs and drums reached far and wide.
> I burned incense for all of the heroes:
> Now, each of you, fly back to the forests and mountains,
> Depart and fly back to the faraway peaks;
> All together, return!
> Return to your palaces.
> This shaman may be old, but his body is pure and clean.
> This humble man,
> seeks your blessing,
> Prays for peace,
> Prays for your blessing;
> When it is time
> we will once again
> burn the incense and perform the ritual
> for you.[107]

FLOWERS ALONG THE AMUR

Making Sense of Plant Diversity
on the Amazon of Asia

How the smallest blade of grass takes interest when one
relates it to the universal order!
—Augustin de Candolle, 1829

Every blade of grass and every tree possesses its own principle,
and should be examined.
—The twelfth-century philosopher ZHU XI, quoted in an epigraph
to Joseph Needham's *Science and Civilisation in China*

In the summer of 1855, Karl Maximowicz found himself walking through a field of flowers along the banks of the Amur River. The young man had just graduated with a degree in botany from Dorpat University in Estonia and was working as a collector for the Russian Imperial Botanical Garden. Maximowicz arrived on these far eastern shores after a twenty-thousand-mile ocean journey from St. Petersburg that took him across the Atlantic, around the southern tip of South America, and across the entire breadth of the Pacific Ocean from Chile to the Straits of Tartary. While the botanist scoured the land for novel plants, warships of the Russian Imperial Navy were gathering along the Sakhalin Island coast and plying the Amur River itself—an ominous presence in what was Qing territory. But at that moment, Maximowicz did not seem concerned by any impending conflict between the two great empires. Instead, his attentions were focused on the plethora of brilliant flowers blossoming in profusion along the river.

Astonished by the botanical diversity that lay before him, the young man tried to capture this stunning sight in words:

> *Vicia pallida* and *Pseudorobus* penetrate through in all directions, and against the reddish shimmering silky sheen of the *Imperata* they give the whole prairie a blue-tinged, iridescent appearance . . . Also conspicuous are *Polygonum divaricum*, with whose blossoms the meadows are here and there dyed white; *Aster tataricus*, in rare luxuriance, and with their large clusters of bright-red flower-heads a great ornament of the prairie; also *Galatella dahurica*, *Biotia discolor*, *Veronica sibirica* and *grandis*, *Glossocotnia*, *Enpalorium kirilowii*, *Serratula coronata*, *Bupleurum scorzontraefolium*, *Gentiana triflora*, *Paeonia albiflora*, together with a few scattered *Asparagus sieboldi*, complete the scene.[1]

Maximowicz, a Russian subject of German heritage, rendered his descriptions of the meadow landscape in German phrases that bordered on poetry ("röthlich schimmernden Seidenglanz / in der Blüthe die Wiese weisslich färbend / ein grosser Schmuck der Prairie"). His description of the overall scene expresses awe, even bewilderment at this picture of diversity, but the names of the flowers provide the scene's anchor, the Latin genus and species names fixing each plant within a specific order.

This profusion of flowers along the Amur was absolutely central to Maximowicz's task of classifying all of the river's plants within a universal system—or rather, within a system that the botanists in Russia, Europe, and the United States deemed universal. In 1859, Maximowicz published his *Primitiae florae amurensis*, a remarkable tome expressing the "first harvest" (*primitiae*) of the Russian project of making sense of plant life along the two-thousand-mile Amur River. Through multiple processes, Maximowicz had managed to translate his rapturous vision in the field into stable categories on the page. The book contains names and standardized descriptions of over 900 species of plants, including 112 new species identified by the botanist and his Russian colleagues: plants whose names to this day bear some version of the name "Maximowicz."

Many of the plants that the Russians discovered and named, however, were neither unnamed nor unknown. The same plants named after Maximowicz were familiar to Qing elites, who named them through references to Manchu court language, pharmaceutical manuals, and the Chinese classics. Chinese settler-farmers knew some of these plants as things used for survival on the fertile but freezing northern prairies. And Maximowicz's plants were known to the many indigenous groups of the region, including Dagur, Orochen, and Nanai, the "small peoples" of the Amur. These groups had their own categories, their own understanding, their own names for the plants to which Maximowicz had laid claim.

The Amur River itself also went by other names and was known in different ways. Originating in the west from the Mongolian Plateau near Lake Baikal, it flows eastward through more than two thousand miles of grasslands, forests, and wetlands in northeast Asia until it empties into the Pacific Ocean across from Sakhalin Island. Many who lived along its banks called it some version of "Black River": in Mongolian, Hara Muren; Sahaliyan Ula, in Manchu. In Chinese, the river is the Black Dragon River, or Heilongjiang.[2] The Chinese term *heishui*, meaning "black water," is a poetic name for the river, and when coupled with "white mountain," the phrase *baishan heishui* forms the nature-based metonym for the entire region. Indeed, the Amur and the tributaries that flow into it from the west, north, and south form a watershed so vast that in some ways we can think of Manchuria as all the lands touched by the Amur and its many waters.

The new imperialism of the nineteenth century split this massive watershed in two. Since the 1689 Treaty of Nerchinsk, the border between the Russian and Qing empires had been set along the Stanovoy Mountain Range, hundreds of miles north of the Amur's main flow. The growth of mass armies, steam transport, and globe-spanning empires in the nineteenth century pulled the Amur into a new world context. In the 1840s and 1850s, the Qing empire was forced to cede territory to European powers after its defeat in the Opium Wars. Taking advantage of Qing weakness, Russia pushed its border with China from the northern edge of the river's drainage basin to the clearly etched line of the river itself. In this process, the Qing lost over two hundred thousand square miles of its territory, and Russia gained a vast fertile land that some imagined as the "Siberian Mississippi."[3] The river may have become a border that split the region between two empires, but the ecosystems that the massive river nurtured did not adhere to those divisions. Instead, plant life spread and swirled around and across the border according to its own logic, manifesting a spectacular diversity along the river's course.

This chapter examines the multiple ways that humans made sense of the Amur basin's abundant plant life at a significant time in the nineteenth century: a moment when the Qing lost much of its Amur territory to imperial competition and a moment when agents of Western scientific networks first explored the river's course. Against this backdrop of large-scale events, we follow individual observers—including a Russian botanist, a pair of Manchu officials, and an indigenous hunter—as they quietly walked among fields of flowers. Here along the riverbank, we may get a more vivid understanding of how human senses engaged with the environment. For some, the goal of this engagement was to discover the unique local qualities of useful plants; for others, it meant fitting each blade of grass into an imagined "universal order."

This chapter probes how these ways of ordering nature intersected with the projects of empire that began to engulf the region in the nineteenth century. Within this endeavor there is a certain element of East-West comparison: some of our observers on the Amur hailed from European centers of learning along the Baltic Sea, while others emerged from scholarly circles in Beijing. Direct comparison between Western and Chinese approaches to plants has certainly been of central concern to the history of Chinese natural knowledge. Some of this scholarship has attempted to evaluate Chinese knowledge about plants within classical texts with direct reference to the "universal truth" of Western science, while others have argued that direct comparison is fruitless.[4] Recent scholarship on botany during the Qing has instead begun exploring the complex networks and labor centered on plants in the vast Qing empire, tracing out the "multiplicities of cultures, languages, empires, forms of economic activity, and forms of political organization" that produced botanical knowledge.[5] I take up this strategy, focusing on specific actors and their contexts without attempting to evaluate the botanical knowledge of one through the standards of the other.

A focus on how these actors trained their senses upon one specific environment ensures that all methods of knowledge making are similarly situated within our inquiry. As Erik Mueggler observed in his work on the making of natural knowledge in another Asian borderland (Yunnan), British botanists and their indigenous guides were equally confronted by "the problem and promise of difference," and each reacted to the extraordinary proliferation of plant life in ways that demanded "intense empirical engagement."[6] For plant explorers along the Amur, this "intense empirical engagement" clearly took different sensorial forms: scrutinizing the sexual parts of tiny flowers, judging different sizes of millet grains, distinguishing vegetable from hallucinogenic forms of artemisia in the field. Rather than evaluate these different forms of knowledge according to their degree of proximity to a standard of Western science, this chapter suggests that we distinguish them by the degree of bodily sensorial engagement between plant and human—array them, as it were, on a scale organized according to the depth and robustness of interspecies entanglement. This entanglement had implications for the place of local knowledge within each schema. For the indigenous people of the Amur, human being was inseparable from plant being. Qing elites strongly sensed the landscape made up of things that nourished and healed human bodies, and acknowledged (with some frustration) that such knowledge often resided exclusively with indigenous informants. Maximowicz sought to distill the human from the plant, to translate fields of flowers into botanical samples: objects that could stand on their own, designed to appear free of their environment, free of human use, and thus devoid of local meaning. Western science hoped to create a universally valid system through this denigration, marginalization, and erasure of the local. Different empires worked with different logics.

As we join Maximowicz for a rapturous moment in a field of wildflowers or glimpse the deep connection between indigenous people and plants of the Amur region, it is also my hope that the reader might experience the past environments of northern Manchuria in a more visceral and vivid way. This frigid and remote world was home to a vibrant multiplicity of life. Highlighting the Amur environment in the nineteenth century illuminates a crucial moment before diversity—both of knowledge and of the flora itself—shifted radically as development, migration, and machines helped transform the land from a riotous sea of flowers to a sea of straight-line, monocropped farms.

Flowers along the Amur, then, have a great deal to tell us about the knowledge of nature both at the edges of empires and on the edge of a radical transition. To understand the significance of the Amur's flowers, we first need to understand the river, its diverse environments, and the diverse people who lived there.

The Black River and "Amuria"

The Amur River is the world's tenth-longest river, but it is relatively unknown in the West. If asked where the Amur is, a student of modern geography might simply point to the vast northeastern border that separates Russia and the People's Republic of China. The broad arc formed by the Argun and the Amur sweeps up from the Mongolian steppe, creating the PRC's distinctive northernmost hump. The river border then moves east-southeast for four hundred more miles until it comes to a sharp angled intersection with the north-to-south-flowing Ussuri River at the city of Khabarovsk. If the overall shape of modern China has (in a somewhat undignified manner) been likened to a

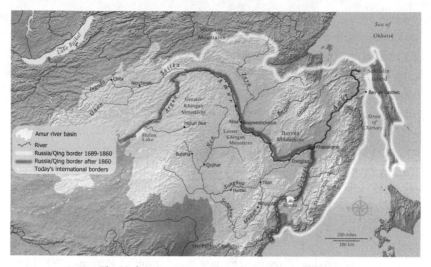

Fig. 4.1 The Heilongjiang/Amur River basin. Map by Jeff Blossom.
See plate 17 for color version.

chicken, then the Amur forms the cock's comb, its head, and pointed beak. A focus on the shape of China alone neglects the fact that the Amur does not end with this "beak" but rather continues to flow almost due north through what is now Russian territory for another four hundred miles until it empties into the Sea of Okhotsk across from the northernmost stretches of Sakhalin Island. Here the Amur takes waters that originate from deep within the Eurasian continent and merges them with the Pacific Ocean.

A map of the Amur watershed (fig. 4.1) shows a dozen major rivers with a bewildering variety of names coming together from the west, north, and south to create the "Black River." The Amur begins through the convergence of multiple streams rising east of Lake Baikal, including the Ingoda, the Onon, the Shilka, and the Argun. Other tributaries, including the Zeya and the Bureya, flow down from the northern Siberian Stanovoy Mountains. Multiple major tributaries flow in to the Amur from the south, including the Nen and the Mudan Rivers. Because of the shape of the terrain, however, most of the southern tributaries empty first into the massive Songhua River, the river that originates from Mount Paektu/Changbai on the Korean border and then flows from east to west through the middle of Manchuria. Where the north-to-south-flowing Nen River meets the east-to-west-flowing Songhua, the river makes a rather odd dogleg turn to the northeast and flows hundreds of miles until it merges with the Amur at Tongjiang. At this intersection of Manchuria's two largest rivers, the reason for the name Black River becomes apparent as the silt-filled "yellow" waters of the Songhua flow into the clear, dark waters of the Amur. If we calculate the length of the Amur to include its Central Asian headwaters, the river is over 2,700 miles long, ranking it between the Congo and the Mekong as one of the longest rivers in the world. Altogether, the Amur and its tributaries form a drainage basin of over 716,000 square miles, an area larger than the state of Alaska, or approximately the size of Western Europe.[7]

Given the vast size of this region, it is not surprising that the Amur basin encompasses many different environments. The upper Amur emerges from the dry, windswept hills of the Transbaikal steppe, then flows through the lush pine and larch forests of the Greater Khingan Mountain Range. Massive prairies stretch to the north and south of the middle Amur, with flatlands carved by the Nen, Zeya, Bureya, and Songhua Rivers. Forests rise again to the north and south of the river in the hills formed by the Lesser Khingan or Bureya Mountains. Low-lying lands predominate at the eastern meeting of the "Three Rivers" (Sanjiang)—the Songhua, Ussuri, and Amur—a confluence that forms a system of freshwater wetlands that were once among the largest in Asia (a place that is the subject of chapter 8). Beyond these wetlands, the Sikhote-Alin mountains on the Pacific coast rise to almost seven thousand feet, their dense forests watered by monsoon rains. The flora and fauna encompassed by the Amur basin make the region among the world's most biodiverse, and the river is dubbed by some "the Amazon of Asia."[8]

Before the twentieth century, the Amur River basin was also remarkably ethnically diverse, home to multiple indigenous groups who herded, hunted, farmed, and fished along the river and its tributaries. The major groups known today in Russia and the PRC as "small people" or "national minorities" include the Ewenki, Oroqen, Dagur, Solon, Nivkh, Udege, and Nanai.[9] The Ewenki and the Oroqen lived along the Greater Khingan Mountains and north, engaging in hunting and herding in a wide range from the fringes of the Mongolian steppe through the northern reaches of the upper Amur. The Dagur and Solon ranged over a two-thousand-square-mile territory in the plains along the Shilka, Zeya, and upper Amur, and south into the Nen River valley and the middle ranges of the Amur, where they engaged in livestock herding and agriculture. The Nivkh and Udege were among the groups who lived in the forests and estuaries of the Ussuri River to the mouth of the Amur and on the northern part of Sakhalin Island. Of greatest importance to this study, the Nanai people (also known Hezhe/Hezhen in Chinese, Hejen in Manchu, and Goldi in Russian) lived in the middle and lower ranges of the Amur, Songhua, and Ussuri river valleys. Their livelihood depended not only on hunting but also on fishing—to outside observers they were so closely associated with fishing that they were known in Chinese as "Fish-Skin Tartars" for their creative use of salmon skin to make clothes and daily-use implements.

These peoples and the river occupied the zone between the two largest land-based empires of the early modern period, the Qing and Russia. The seventeenth century saw decades of conflict between the two empires as they competed for natural resources and human allegiances along the Amur. For the Manchus, the conflict was conceptualized as one of survival, as Russians threatened the northern door of their ancestral power base. The Manchus gradually achieved military dominance in the region by establishing a system of forts along the river's southern tributaries and recruiting (sometimes forcibly) the local tribesmen into Qing military institutions. Qing-Romanov peace was finally established with the signing of the Treaty of Nerchinsk in 1689—Nerchinsk itself being a town in the westernmost fringes of the Amur basin, east of Lake Baikal on the Nercha River just before it merges with the Shilka.[10] As a result of this treaty, by the late seventeenth century, the Amur basin's valuable natural resources—furs, ginseng, freshwater pearls, fish, falcons, and even wild honey—were formally under the dominion of the Qing empire.

Qing imperial knowledge of the region depended on the knowledge and labor of its local inhabitants. Qing frontier-mapping projects, and thus knowledge of the very extent of the empire itself, relied on indigenous guides' knowledge of the northern Amur basin terrain. In spite of Kangxi's written admonition to imperiled surveying teams ("Forget about remote places that cannot be reached!"), Orochen, Solon, and Dagur trackers led Manchu explorers in the late seventeenth century to establish the empire's northern border dozens of miles above the mouth of the Amur, to inland points along the same northern parallel as the Kamchatka Peninsula and the Aleutian Islands.[11] As it mapped these territories and their resources into the empire, the Qing court also drew lines around the region's human beings. The Qing organized some indigenous families into "New" Manchu military banners and stationed them in forts along the Amur's tributaries to guard the Manchu homeland, or relocated them thousands of miles away to fight the Qing's most stubborn enemies.[12] Most importantly, the Qing created a complex system that harnessed indigenous knowledge and labor to harvest the valuable fauna of the region. As vividly analyzed by David Bello, some were organized into "Hunting Banners," professional mountain men who scoured the forests to fulfill pelt and game quotas set by the Beijing court—a system that Bello has termed "imperial foraging."[13] Others remained embedded in "civilian" village life but were still required to present furs to Qing agents as formalized tax or tribute. Some tribal leaders at the furthest reaches of the Amur even made the perilous 2,500-mile round-trip to Beijing to present their fur tribute to the court in person and receive the hand of Manchu noblewomen in return.[14] The labor and knowledge of the Amur's indigenous peoples provided Qing society south of the Great Wall with its most coveted luxury products, creating what Jonathan Schlesinger has evocatively called "a world trimmed in fur."[15]

Loretta Kim has suggested that we think of the Black River basin not as a remote periphery defined by empires but as "Amuria," an environmentally shaped "zomia" where inhabitants maintained complex transborder connections to cultures and markets, and strategized to cooperate with and resist the organizing efforts of states.[16] Its importance was not peripheral: the biodiversity of the region was central both to Qing identity and to global consumption, and powerful empires quite literally relied on the guidance provided by the diverse knowledge and skills of its indigenous inhabitants. Thinking of Amuria as a whole helps us avoid the territorial splits and reductions caused by modern nation-state boundaries: it reminds us to recognize the remarkable biological and human diversity of the region and not to think in binary terms of "Russia" and "China." It also challenges us to hold in our minds the multiple environments of a landscape defined not only by land but also by the water that flowed through the land.

The Amur and the Global Competition of Empires

By the nineteenth century, what was extensively mapped and known territory for the Qing empire had become virgin frontier awaiting colonization in the eyes of some in the Russian empire. Russia had been expanding eastward for decades but had for the most part, due to the prohibitions of the 1689 Treaty of Nerchinsk, remained hundreds of miles north of the Amur in "Siberia proper." Russian fur traders and adventurers established settlements along the Lena River and had traveled to the Pacific coast at Okhotsk. In the early 1700s, imperial explorers pushed Russian horizons even further: Vitus Bering (1681–1741) explored the Kamchatka Peninsula, probed the Bering Strait, and made landfall in Alaska, and other Russian explorers continued southward along Canada and the California coast.[17] While British and French explorers sailed the South Pacific, Russians had opened up a northern Pacific world that spanned the Eastern and Western Hemispheres. But the trip from Europe-facing St. Petersburg to the far reaches of Siberia was a grueling four-thousand-mile, cross-continental journey on the Great Post Road or an around-the-world ocean journey. Enthusiasm for a Russian Far East was difficult to support, financially and politically.[18]

Heated global competition between empires in the mid-nineteenth century changed this situation. Russia had once competed for control of the Amur solely with the land-based Qing empire, but this newer global competition of the nineteenth century—spanning the Eurasian continent, the Atlantic and Pacific Oceans, the Old World and the New—made the Amur region a more urgent concern for Russia. The British victory over China in the First Opium War (1839–1842) revealed the military weakness of the Qing court and gave Russia's greatest European rival a disconcerting advantage in East Asia. The arrival of the United States on the Pacific Coast with the US victory in the Mexican-American War (1846–1848) and the subsequent discovery of gold in California inspired Russians to emulate America's push to the Pacific. The increased presence of American whalers in Pacific waters also fueled calls to beat back the specter of American military and economic competition in Japan, Russia's neighbor to the south.[19]

Russia's rising imperial expansion was what brought Western botany to the shores of the Amur River. In the autumn of 1853, the Russian frigate *Diana* left St. Petersburg on an official voyage to Japan, part of Russia's attempt to preempt America from "opening" the shogun's realm.[20] One of the passengers aboard the *Diana* was the twenty-five-year-old botanist Karl Maximowicz (1827–1891). In St. Petersburg, Maximowicz was a junior member of the German-speaking intelligentsia who for decades had formed the backbone of Russian scientific institutions, including the Russian Academy of Science and the Russian Geographical Society.[21] Through connections in these circles, Maximowicz was hired to manage the herbarium at St. Petersburg's Imperial Botanical Gardens, then under the leadership of the Austria-born Franz Josef Ruprecht (1814–1870). When the call for an official botanist for the *Diana* came to Ruprecht, he recommended his young protégé for the position. The *Diana*'s west-to-east journey from St. Petersburg to Japan was scheduled to include stops in the Azores, Brazil, Argentina, Chile, Peru, and the Hawaiian Islands—locations that had been the object of legendary imperial exploration from Cook to the *Beagle*. The young Maximowicz was charged with collecting samples along this oceanic route to augment the Russian herbarium's Eurasia-focused collection with plants from South America and Japan.[22]

The Crimean War (1853–1856) dramatically altered this plan. While the

Diana was anchored on the Chilean coast at Valparaíso, word came that France and Great Britain had declared war on Russia. Even though the fighting centered on the Black Sea and the Crimean Peninsula, the French and British navies sought to attack Russian forces wherever they could be encountered— even in the faraway South Pacific.[23] With a British frigate giving chase, *Diana* fled northwest across the Pacific for over ten thousand miles, and in July 1854 arrived at the Bay of de Castries, the narrow inlet between the Asian mainland and the northern coast of Sakhalin Island, a region of "Southern Siberia" newly (and sparsely) occupied by Russian forces.[24] All nonmilitary crew, including the *Diana*'s resident botanist, were ordered to disembark at this point. Thus, due to Russia's East Asian ambitions and the global hostilities of a European war, Karl Maximowicz accidentally found himself deposited at the mouth of the great Amur River.[25]

Russian botany stumbled into Amuria at the same time as the very intentional arrival of the modern Russian navy. General Nikolai Muravev (1809–1881), the former governor of Eastern Siberia, had long argued that the Amur needed to be open to Russian colonization, and in 1854, he finally gained the czar's authorization to realize his vision. The need to aid Russian ships stranded at the Amur's mouth added urgency to the mission. That summer, Muravev assembled a flotilla of rafts led by the steamship *Argun* and traveled over two thousand miles downriver from Chita, a town east of Lake Baikal, to the Bay of de Castries, where the Amur empties into the sea. Most of the sacks of grain they brought to provision the Russian outposts had fallen into the water and were covered in mold, but this mattered little: a modernizing Russian navy had successfully navigated the entire length of the Amur River without the permission of the Qing. Four years later, Muravev negotiated possession of the Amur's left bank from a Qing empire hobbled by internal rebellion and the Opium Wars with Europe. With this 1858 Treaty of Aigun and the subsequent 1860 Convention of Peking, Russia gained over two hundred thousand square miles of Qing territory.[26]

The map in figure 4.1 shows the geographical extent of the Qing's territorial loss. The more northerly light-colored line indicates the old Qing-Russian border established by the Treaty of Nerchinsk in 1689. The darker line along the Amur indicates the new border. Russia gained the land defined by the northern tributaries of the Amur, along with Sakhalin Island and the Sikhote-Alin Mountain region all the way to the Pacific Coast. In terms of North American history, the territory Russia gained is similar to what the United States gained in the Pacific Northwest from the 1846 Oregon Treaty, or

roughly the equivalent of half of the territory the US gained in the 1848 Mexi-
can Cession. Without firing a shot, Murav'ev's maneuvers on the Amur (and at
the negotiating table in Beijing) had brought the borders of Russia to within a
few dozen miles of Japan and solidified his empire's identity as a Pacific power.

Maximowicz may have stumbled into the Amur, but his presence there as
a scientist had everything to do with empire. As the historian of science Janet
Browne has pointed out, naval and military institutions and networks formed
the backbone of European study of animal and plant life in the nineteenth
century, and routes that scientists took around the globe were "determined by
geopolitical and national factors rather than any strictly scientific questions."[27]
Such was certainly the case with Maximowicz. Furthermore, the Russian botanists' scientific work—the determination of regional species in order to map
a global biogeography—was also distinctly tied to imperial forms; indeed, botanical geography was "one of the most obviously imperial sciences in an age
of increasing imperialism."[28] While in Amuria, Maximowicz seemed unconcerned with the glories of the Russian empire—his attention was quite literally focused on the "lilies of the field." Nevertheless, a consideration of how
the botanist gathered, dissected, and scrutinized the Amur's flowers reveals a
single-minded pursuit of empire's universal order.

Flowers along the Amur and the "Universal Order"

After disembarking the *Diana* in the late summer of 1854, Maximowicz had
little time to explore the flora around the Russian settlements before the harsh
Siberian winter set in. The following year, after snows had melted and spring
returned, Maximowicz reemerged to explore the riverbanks.[29] Maximowicz
confessed that the conditions for conducting botanical research along the
Amur were less than ideal: he and his colleagues had to "negotiate immense
distances" while rowing in a small rowboat, "usually upstream," and typically
could spend only a few hours in the field at a time, never venturing far from
the river.[30] In spite of these restrictions of time and resources, within just four
years, Maximowicz was able to publish the *Primitiae florae amurensis* (First
harvest of the flora of the Amur), a five-hundred-page tome that claimed to
categorize all the plants of the Amur into a universal order. An examination
of the *Primitiae*'s sections reveals how this order was accomplished through
different forms of vision and language.

The *Primitiae* begins with an attempt to reproduce, through movement
and words, the length of the river in all its diversity. From his fragmented

field journeys and the observations of others, Maximowicz reconstructed a comprehensive temporal-spatial narrative, descriptions of the landscape viewed from the perspective of an omniscient observer scanning the banks while floating slowly down the Amur from west to east.[31] The pages of *Primitiae* describe in detail the steep embankments, sandbanks, flat prairies, rolling hills, wetlands, and forests along the entire river, and provide detailed spatial groupings of typical plant species specific to each microenvironment. The astonishing detail of these passages reads like a prose equivalent of Dürer's *Great Piece of Turf* (*Das große Rasenstück*), as if the botanist were trying to generate an exact one-to-one correspondence between vision and words. Maximowicz admits that nature along the Amur appeared "magnificent but monotonous." His descriptions present the reader with a two-thousand-mile (or two-hundred-page) tapestry woven from a baroque tangle of species, a weaving that is at once monotonous and masterful.

Occasionally there are breaks in the monotony: Maximowicz is particularly sensitive to spotting boundaries in nature, places where "the country changes significantly." One of the first dramatic breaks Maximowicz notes occurs where the Shilka meets the Argun—here the vast grasslands of the Mongolian steppe quickly give way to dense birch forests. Another spot where the scene changes suddenly is at the influx of the Zeya River, a point where the Manchu town of Aihui and the Russian town of Blagoveshchensk faced each other across the Amur. Here Maximowicz sees the forest suddenly turn once again to prairie, a transition so great that it "forces the beholder to think that he has passed a great natural border of land and nations."[32] A final dramatic transition takes place at the Amur's convergence with the Songhua River. At this point Maximowicz notes a tremendous diversity of flora rising from the surrounding wetlands, with vegetation so dense that the "soil is clothed with an uninterrupted plant carpet."[33]

It is here that Maximowicz finds himself walking through fields of flowers so stunning that the botanist's flat prose becomes painterly and his dense narrative becomes almost cinematic. He invites the reader to "walk beyond the sand dunes" along the riverbank to discover "a perfectly flat surface in front of us, covered in a carpet of grass as tall as a man." With surprisingly vivid adjectives interwoven with scientific plant names, Maximowicz produces the impressionist-like portrait of the prairie landscape that began our chapter: blue iridescent *Vicia pallida* setting off the red sheen of the *Imperata*, *Polygonum divaricum* dyeing the meadow white, luxuriant red clusters of *Aster tataricus* shining "like great jewels of the prairie"—along with a breathless invocation of *Galatella dahurica, Serratula coronata, Bupleurum scorzontrae-folium, Gentiana triflora,* and *Paeonia albiflora.*[34] With its constant use of Linnaean terminology, perhaps this beautiful scene could be appreciated only by a fellow botanist, but in essence Maximowicz has conjured for the reader a field of tall grasses that shimmer with the colors of peacock plumage, accented by pink daisies, brilliant blue bellflowers, delicate yellow Queen Anne's lace, purple thistles, lavender larkspur, indigo crocuses, and creamy-white wild peonies. Maximowicz's detailed description of the terrain along the entire course of the Amur can be numbingly dry, but when he discovers flowers, he is transported.

Maximowicz undoubtedly appreciated the flowers' beauty, but his obsession had another reason: flowers were the key to identifying and classifying plants and thus formed the cornerstone of his life's work. While the flexible standards of scientific writing of the time still allowed Maximowicz to indulge in florid prose descriptions of flowers, his main duty was to probe and cut their sexual parts in order to reveal the category to which the plant should belong. By generating these flower-based identifications, he not only furthered the mission of the Russian imperial collection but also augmented his own reputation as the man who ordered the flora of one of the world's longest rivers according to Euro-American standards of science.

Maximowicz's *First Harvest* is a prodigious work of classification, enumerating "all known plants of the Amur-land" in 904 species of plants divided in over 100 orders, classified according to the "natural system of Candolle's *Prodromus*."[35] Published over the course of decades beginning in the 1820s, the *Prodromus systemalis naturalis regni vegetabilis* (Systematic natural history of the vegetable kingdom) of Augustine de Candolle (1778–1841) was widely influential, followed by the British Joseph Hooker, the American Asa Gray, and other well-known botanists in the mid-nineteenth century. Candolle's system emphasized the importance of flower morphology, and demanded descriptions of the plants' organs of generation in far more detail than Linnaeus could have imagined. The system divides all plants first into dicotyledons and monocotyledons, based on the presence of one or two initial leaves in the seed embryo, and then divides these categories into subclasses, with multiple families under each subclass.[36] Following Candolle, Maximowicz began his enumeration of the Amur's plants with the first order of dicotyledons, the Ranunculaceae, a Latin term translated into vernacular English as "buttercups." This broad category includes flowering bushes such as *Clematis,* herbaceous wildflowers such as meadow rue (*Thalictrum*), and even peonies (*Paeonia*). He then moved from Buttercups to Menispermacaea and on to Berberis (barberries), sorting all the known plants of the Amur into their proper categories.[37]

Using this system, Russian botanists identified 112 new species and 11 new genera, and to this day, as a result, many of the plants of the Amur basin bear some form of the name "Maximowicz."[38] Maximowicz himself identified many of these species, ensuring that an abbreviation of his name, "Maxim.," would formally be appended to the names of dozens of plants, including *Paeonia obovate* Maxim. (the woodland peony) *Juglans mandshurica* Maxim. (the Manchurian walnut), *Pinus mandshurica* Maxim. (the Korean pine), and *Berberis amurensis* Maxim. (the Amur barberry). Some plants were "discovered" by Maximowicz in the field, but the true nature of the specimen plants was identified by senior directors back at the St. Peterburg botanical garden, including Franz Josef Ruprecht and Eduard von Regel (1815–1892).[39] These men bestowed the name of their young employee on many species and even genus designations. These plants include *Prunus maximowiczii* Rupr. (the Korean cherry), *Allium maximowiczii* Regel (the oriental chive), *Geranium maximowiczii* Regel (a plant with an adorable purple, pansylike flower), and *Maximowiczia amurensis* Rupr.

It is with this last plant, with a name that highlights both the scientist and the river, that the young botanist could claim one of his greatest discoveries: an entirely new genus and species of the order Schisandraceae. An examination of how *Maximowiczia amurensis* was produced—through the application of exquisite manual dexterity, intensive vision, and Latin phrasings—provides insights into how human senses and language established the place of a plant within a universal order. This process required a multistep disentanglement: separating the plant from its environment, preserving it in isolation and death, then separating and autopsying its parts. It also required disentangling the plant from the stories, knowledge, and intimacies of local human use and replacing them with a system of order that claimed universal authority.

The Manchurian plant that the Russians called *Maximowiczia amurensis* Rupr. is known today as *Schisandra chinensis* and is commonly called the "five-flavor berry" in English. This perennial climbing plant can spread to a length and height of many feet, its thick vines and masses of light-green leaves capable of encasing tall trees.[40] In the fall, hundreds of plump, brilliant-red berries hang from the vines like clusters of miniature grapes (plate 7). The flowers, in comparison to the massive plant and its showy berries, are modest and almost unnoticeable: each blossom composed of six delicate white petals about a centimeter across, clustered around a small green center. In spite of their visual insignificance, it is the tiny flower that garnered the botanists' attention.

Samples of *Maximowiczia amurensis* gathered in the Amur basin can still be found in the collections of botanical gardens in Europe and the United States. One such specimen is preserved in the Harvard University Herbaria, an institution founded by the American botanist Asa Gray (1810–1888) just thirteen years before Maximowicz took his trip down the Amur. The specimen is approximately ten inches in length, comprised of two thin woody vines to which cling a dozen or so brownish leaves and a few brownish, withered flowers (plate 8). Another sample, from the Botanical Gardens of the City of Geneva (fittingly, in an institution originally established by that great classifier of plants, Augustin de Candolle) was most likely gathered in the fall.[41] This specimen sports a few tendrils to which scatterings of shriveled blackish berries are attached. A small envelope included with the specimen contains more samples of the berries, the white of the paper slightly stained by the dark dried fruits. These specimens in no way conjure the full plant in the field. There is no sense of the living plant's massive expanse, the density of the vines, the intense green of the leaves, or the clustered brilliant red of the fruits. Indeed, like all botanical specimens, these isotypes are small frozen gestures extracted from the vibrant dance of nature, intentionally stripped of their environments and entanglements, whether with other plants or with the humans who gathered them.

The written description of *Maximowiczia amurensis* contained in the *Primitiae* takes us even farther from the plant and its environment, and draws our attention instead to the acute exercise of vision and the precise deployment of language required to classify a plant. These descriptions present a tight, analytical focus on the innermost structures of the tiny flower, including structures that cannot readily be seen but have to be deduced. The Latin text begins with the most important information: *Flores dioici*, shorthand that denotes that the flowers of the plant come in two separate sexes, male and female, and that male and female flowers are on separate plants. The most important secrets to the plant's type are hidden deep within the female flower. Candolle's natural method of classification was a search for symmetry, for patterns indicating a shared grouping, and these patterns were indicated most importantly by "the attachment of the various organs of fructification."[42] Candolle even encouraged botanists to imagine structures that were not visible but should have been in the flower in order to satisfy the principle of symmetry.

The patterns of the "organs of fructification" of the *Maximowiczia amu-*

rensis are revealed in botanical Latin phrases—a standardized language that was universal to plant science in the West:[43] "Femineis: Staminum vestigium nullum. Torus sybcylindraceus brevissime stipitatus, ovariis numerosis obsessus. Ovaria compressa bilocularia. Stigmata obliqua crassa dilatala, subbiloba vel sublacera, interdum cum ovariis bilocularibus margine interiore mediante lamina cellulosa cum toro subconnata."[44]

To produce this description, the botanist in St. Petersburg would have isolated one flower, removed some petals, sliced open the base, and observed the fleshy interior under magnification. He notes the cylindrical shape of the receptacle and identifies flattened ovules inside the ovary, symmetrically arranged in groups of two. The thick stigma arises at an angle from the ovary, somewhat lobular in shape with irregular edges along its open end.[45] As we can see on one of the Geneva herbarium's specimens, the botanist attached the dismembered parts of the flower, its petals scattered from the central stem, a postmortem reminder of the literal "autopsy" required to identify a species.[46] As revealed in this intimate probing of *Maximowiczia amurensis*, minute structures and even principles hidden to the human eye were the basis of determining the order of things.

In the multistep process of Maximowicz's ordering work, the last step was to take this acute vision and explode it into a vision that encompassed the entire world. While Maximowicz and his colleagues spent a tremendous amount of effort peering into tiny flowers, the final pages of *Primitiae* reveal that the ultimate goal of the work was to turn these flowers into numbers and maps—a mathematical and cartographic exercise that could place the Amur's plant life into an emerging global order generated by Euro-American botanists. Maximowicz produced *Floren-Statistik* to further "reveal conditions that escape the eye of the beholder" and illuminate truths about the distribution of plants not only in the Amur region but across the entire globe.[47] Following Candolle, he tabulated the number of species within a variety of classifications, and thus generated ratios that expressed the diversity of species within each grouping. Table upon table of "botanical arithmetic" allowed Maximowicz to compare the Amur flora with that of Scandinavia, Siberia, and, most interestingly, through the work of Asa Gray's *Manual of the Botany of the Northern United States*, that of North America.[48] With these ratios, Maximowicz joined in the work of other botanists of his time who puzzled over the relationship between life-forms and the earth's terrain: the science of biogeography.[49] Men from Humboldt to Candolle held that "the study of geographical distribution was scientific inquiry of the highest order that could lead to the disclosure of fundamental laws of nature."[50] The ratios of *Floren-Statistik* enabled Maximowicz to declare that the Amur was indeed a kingdom of flowers, "distinguished from all neighboring regions by its wealth in Buttercups and Lilies."[51] He found that compared to other northern locales at the same parallel such as Labrador and Scandinavia, the Amur possessed not only a much larger number of plant species but also a greater diversity of types within different categories, thus enabling him to theorize that latitude was not as much a determining factor of environment as had previously been suggested.[52]

Maximowicz's sensitivity to border breaks in plant patterns along the Amur, combined with his mathematical groupings of Amur plants, found cartographic expression in a remarkable colorized map of "plant borders" inserted at the end of his *Primitiae*. This "Karte des Amur-Landes" (plate 9) denotes the Amur's main flow and tributaries, along with the region's mountains and human settlements. The map includes terrain from Chita in the west to Sakhalin Island in the east, from the Stanovoy Mountains in the north to the Changbai Mountains and Liaodong region in the south. Upon this recognizable geography rendered in brown ink rests a tangle of nine dotted blue lines that wind across the map, with roman numerals anchoring the beginning and ends of their complex arcs. These blue lines are the *Florengrenze*, or "botanical borders" of the Amur, delineating the spread and limits of specific tree species and thus constituting what Maximowicz determined to be the "natural borders" of the region. Their seemingly random arcs dive, diverge, and intersect across the landscape, rendering no particularly discernable pattern. Significant for a time when the Amur was used as the ultimate dividing line between China on the south and Russia to the north, almost all of Maximowicz's *Florengrenze* range across both sides of the river. The realm of *Quercus mongolica* (Mongolian oak) goes from the mouth of the Amur, across the Bureya and Lesser Khingan Mountains, west to Nerchinsk, and then south across the Nen River to just north of Qiqihar. The Asian hazel tree (*Corylus heterophylla*) similarly can be found as far west as the Argun and hundreds of miles north of the Amur, and its eastern domain extends as far south as Ningguta. The Asian black birch (*Betula dahurica*) does not penetrate further west than Nerchinsk but can be found in mountain forests both above and below the Amur. Some species seem to prefer Russia: The Manchurian walnut (*Junglans mandshurica*) does not descend much farther south than the area around Aihui; the Manchurian pine and the Amur mackia (*Maackia amurensis*) similarly are found north of the Bureya Mountains. The Ussuri pear (*Pyrus ussuriensis*) clings to the eastern coastlands of Manchuria but can be found south of Lake Khanka and as far north as the Dondon River north of Khabarovsk. It is ironic that at a time when Russian expansionists were fixing the Amur River as the static political boundary between the Romanov and Qing empires, Maximowicz's "imperial botany" created boundaries that had no visible relationship at all with the political project of Russian imperial expansion. Maximowicz's groupings crisscross the Amur, linking the land on both sides in a riotous swirl of trees and flowers.

Maximowicz's understanding of "natural borders" may have been at odds with his empire's scheme to fix the Amur and the Ussuri as the border between China and Russia, but *Primitiae* clearly aspires to a kind of imperial mastery. By following the Russian's activities in the field and in the laboratory, by examining the intersection between perception, manipulation, and the environment, we illuminate the process of making this imperial knowledge. Maximowicz and other St. Petersburg–based scientists arrived on the Amur together with the Russian military, and their activities were facilitated by the Russian military presence. In spite of the constraints of time and logistics that they faced in the field, these men took their limited scope and claimed to comprehensively observe and replicate the entire river and all its life-forms. The process of identifying a species took the plant from the field to the laboratory, removed it from all its in situ entanglements, and made it the object of intense visual scrutiny of its innermost sexual recesses. The specimens, once ordered, generated numbers that could then generate biogeographical maps with a global reach. Maximowicz's view from the tiniest Manchurian flower extended around the globe—from Sakhalin to Newfoundland, from Irkutsk to California, from Finland to Alaska. Maximowicz used Manchuria to reveal, in the words of his botanical hero Candolle, how the smallest blade of grass (or the most modest flower) achieved significance when relating it to a "universal order."

Visceral Knowledge: Qing Perspectives on Heilongjiang Plant Life

The flowers of the Amur were not very important to the Manchu official Xiqing, active in the late eighteenth and early nineteenth century. In his *Heilongjiang wai ji* (Unofficial record of Heilongjiang), he noted that the rustic local inhabitants liked to pick brightly colored wild poppies, impatiens, morning glories, and hollyhocks from the fields, but he lamented that the more sophisticated chrysanthemums and lilacs brought up from the south of China would wither away or stop bearing flowers because of the winter cold, leaving the scholar with nothing to decorate his studio. Without the appropriate climate, it was difficult to cultivate the flowers that really mattered.[53] Another Manchu observer, the exiled imperial courtier Yinghe (1771–1839), tended to notice wildflowers a bit more. In his *Bukui jilüe* (Overview of Qiqihar), he described the fields of the Heilongjiang region as full of purple Manchurian asters, pink and blue periwinkle, wild roses, and fragrant jasmine. He noted the evocative names the local inhabitants gave to flowers: "The Red That Arrives with the Migrating Geese," "Flowers of Eternal Spring," "Autumn in the Han Palace." With the poignant nostalgia of an exile, Yinghe also noted that the flowers of the region were spectacular but short-lived: "The myriad flowers all bloom at once in glorious profusion at the beginning of summer, but by summer's end they all wither away."[54] Ultimately, however, flowers were but a small part of these Qing elites' encyclopedic understanding of the flora along the river they called the Black Dragon.

These two observers, the official Xiqing and the exile Yinghe, can help us make sense of Qing perspectives on the flora of the Heilongjiang region in the first half of the nineteenth century. Xiqing was a Manchu aristocrat of the Blue Bordered Banner, a descendant of Ortai, one of the most trusted officials of both the Yongzheng and the Qianlong emperors.[55] Xiqing had a knack for numbers and languages, and the Qing bureaucracy put those skills to good use: at the beginning of the nineteenth century, Xiqing was appointed to oversee the treasury at Qiqihar, the government seat for the linguistically diverse region of Heilongjiang.[56] Another Manchu aristocrat, Yinghe, was also sent from Beijing to Qiqihar in the early nineteenth century, but he arrived for different reasons. He had once been at the empire's very center of power as a scholar in the Hanlin Academy who served as personal secretary to the emperor and as governor of multiple provinces south of the Great Wall. Because of a drainage snafu in an imperial tomb construction project that he was overseeing, Yinghe was banished to Qiqihar.[57] Neither of these men was an expert in the region's plant life per se—one was there to work as a government accountant, another to suffer a humiliating punishment—but as highly educated scholars, both combined detailed firsthand observations with references to Chinese histories and geographies to create knowledge about Manchuria's environment.

The writings of both men reveal their focus on a landscape of human use: a more visceral, body-based way of knowing. A plant was worthy of investigation only if people ate it, made things out of it, or used it to cure their ills. Plants were known because of the force of human labor needed to bring them into being; they were known because of the way they altered human bodies when consumed. As they were intertwined with human use, Qing observers' descriptions of plants were also full of stories: stories about where the plants were found, stories about people's encounters with plants—knowledge that embedded the plants in a human world.

Xiqing's and Yinghe's observations about plants were anchored in descriptions of Heilongjiang's peopled landscape. The two men both lived in Qiqihar (Manchu Cicigar, also known in Chinese as Bukui), located on the Nen River approximately two hundred miles south of the Amur.[58] In the first half of the nineteenth century, Qiqihar was a fairly major town for the northern frontier—Xiqing put its population at fifty thousand people. The town hosted a remarkable mix of people, including Manchu, Han, and Mongol bannermen; Solon, Orochen, Dagur, and other indigenous peoples; and a large number of Han Chinese merchants and exiles from south of the Great Wall. In his *Unofficial Record*, Xiqing enumerated Heilongjiang's rivers, lakes, mountains, and forests and noted its unique natural features, including sandy deserts, vast marshes, and sulfurous volcanic fields. But he spent far more time describing the human landscape, providing information on the locations and populations of major towns. These included Mergen and Buteha, outposts further north on the Nen River; Heilongjiang town (also known as Aihui), to the far north on the right bank of the Heilongjiang River; Hulun biur, a bustling western town on the edges of the Mongolian steppe; and Hulan, a small but growing settlement on the banks of the Songhua River near what would later become Harbin. Xiqing described the rustic wooden palisades that surrounded the towns and the temples and government offices within, along with the relay stations, post roads, and frontier guard outposts scattered across the hundreds of miles that lay between each settlement.

In Xiqing's eyes, Heilongjiang was a peopled landscape, and since people had to eat, Heilongjiang was first and foremost an agricultural landscape. Heilongjiang's wilderness could provide food, but it only offered small pleasures such as the wild cherries and strawberries that could be made into sweet wine and jams, or the fire-roasted hazelnuts that vendors sometimes sold on the streets of town. There were a few vegetable crops—including bok choi, Chinese broccoli, cucumbers, and squash—but these were plants grown exclusively in the gardens kept by Han Chinese exiles. There was one leafy green that was cultivated widely, but it was not for eating: indigenous Dagur women planted, harvested, and cured tobacco for their own enjoyment.[59] According to Xiqing, the soil (literally, the "pulse of the earth," *tumai*) along the Black

Dragon River was, first and foremost, fit for the growing of grains. A true portrait of the Heilongjiang landscape featured cultivated fields of green and gold, the product of human labor designed for basic human consumption.

Xiqing's descriptions of plants could be vivid and detailed but always linked each plant to the benefits it provided to humans. With this principle in mind, "staff of life" grains were listed first. The most important plant was millet, common but exceedingly complex. There were many different types of millet grown in the Amur basin, and no one term unified them all. There was *mizi*, with its large red or yellow grains loosely fanning out from the tops of its stalks like straws on a broom. There was *guzi*, with its fat spikes full of thousands of little seeds drooping from stalks like bushy fox tails. *Baizi*, with its purple grains standing upright like cattails, could easily grow on marshy wasteland or in the unattended corners of barnyards. Some millets produced sticky sweet grains (*shu*) good for making wine or rich puddings. Others were less starchy, best when roasted, husked, and made into gruel for everyday meals.[60]

Farther north, especially around Aihui on the banks of the Heilongjiang River, the land was perfect for different types of *mai*, grasslike plants that produced sweet nutty grains. The "little *mai*" (*xiaomai*) or wheat of Aihui was the sweetest and most fragrant wheat in the empire. In the late fall, dozens of heavy sacks filled with snow-white flour were sent south as tribute to the imperial court. There was *qiaomai*, or buckwheat, shorter plants with sprays of white flowers that produced deep brown kernels in the fall. The buckwheat from Aihui was like none other—sweet and moist—good for making fried buckwheat cakes and buckwheat noodles. The *lingdang mai* (naked oats) were used in the northern outposts of Mergen and Heilongjiang town to fatten the bannermen's horses when pastures were thin. People also grew other types of grains, like sorghum and amaranth, and even used tiny black beans, green mung beans, or mottled black-eyed peas for sustenance. As for rice, it was imported from the south and was far too expensive for most locals to eat.[61]

But even though the "pulse of the earth" was good for growing these diverse grains, Xiqing emphasized that farming in the north was incredibly difficult: the environment was characterized by the need for backbreaking human labor. Not only was the growing season short and the winters brutally cold, but the thick black soil was extremely difficult to plow. The plows of the region were longer, heavier, and more expensive than the plows of the southlands, and each plow needed to be dragged by two or three oxen, yet another major investment. The problem with Heilongjiang, Xiqing explained, was there was lots of land but not enough labor. The land was supposed to be reserved for the Manchus, but when the Qing state tried relocating idle urban bannermen from Beijing to grow their own food in the northeast, most of them found the harsh work to be unbearable.[62] Much of the farming work was therefore done by unfree labor: Chinese criminals who worked on government-run plantations to produce required grain quotas for the state. These laborers often ran away, and when they did, the government issued detailed descriptions of the escapees' clothing, facial hair, and even the hue of their skin in order to track them down and return them to the farms.[63] Xiqing defended the local bannermen: it was not that they were lazy (as some said), but they were busy with their military duties, and since most lived near major towns, they didn't have the resources to farm the large tracks of vacant land that were spread out over dozens of miles. For lack of labor and lack of means, many bannermen barely got by, gathering firewood and herding a few sheep while tens of thousands of fertile acres lay fallow. Plants were fundamental to man's existence in Heilongjiang, but Xiqing clearly envisioned man's relationship with the region's plants as one of endless, overwhelming physical struggle. This struggle formed one way of knowing Manchuria.[64]

Even as he lamented their uncultivated state, Xiqing admitted that the many fallow fields and marshes still provided useful plants for the people of Heilongjiang. Locals recognized dozens of different wild reeds and tall grasses, each of which had its own important use: the "sheep grass" (*yangcao*, or false rye grass) with its seeded tufts like thin wheat was a perfect feed for banner horses; *zhangmao* and *huangmao* grasses were used for roof thatch; sagebrush and wormwood were abundant and sold as a cheap fuel; while the famous *ula* grass could be used as insulation in clothes and shoes to ward off the bitter Manchurian cold. Moisture-loving willows grew in profusion at meadows' edges: their branches made the best fuel in winter, clean and long burning, and could be used to make shelters for shade in the heat of the summer. Yinghe also observed the abundance of useful reeds in the Heilongjiang region and noted that these grasses grew in tremendous profusion in a vast wetland along the Uyur River thirty miles to the southeast of Qiqihar. It was a spot where thousands of cranes would gather in the spring, but local people who traveled there went mostly to glean marsh plants for fuel.[65]

The fields and forests also held many valuable medicinal plants. Xiqing heard there was ginseng to be had in the Khingan Mountains, but to get there required crossing a vast poison swamp and few risked the journey.[66] While ginseng proved elusive, the local fields were full of other useful but more common medicinal plants: peony (*chishao*), lily (*baihe*), skullcap (*huangcen*), mugwort (*yinchen*), and motherwort (*yimucao*), roots and leaves that could relieve stomach pains, reduce fever, boost eyesight, and even stop hemorrhages. While Xiqing dutifully recorded the existence of these plants in the wild, he confessed that most urban dwellers he knew simply bought processed herbs from the south at local pharmacies. While this was convenient, Xiqing warned that the consumer ran the risk of encountering fakes, expensive dried roots and leaves passed off as potent medicine but actually comprised of useless weeds. The translation of plants into commodities came with its own risks.[67]

The main challenge in identifying the useful plants of Heilongjiang was aligning (or "rectifying") their local names with accepted standard names from the textual tradition. As described by the historian of science Georges Métailié, the rectification of plant names was "a firm belief that the importance of correct names really does constitute the basis upon which . . . all thought on natural living objects was founded."[68] Texts dictated the way that these Manchu observers ordered their environment. In their writings, both Xiqing and Yinghe followed the same precepts and the same order when enumerating the useful plants of the Black Dragon River: beginning with grains, moving on to vegetables, then medicinal herbs, flowers, and fruits. Then came consideration of other useful plants—grasses and reeds for fuel and roofing, trees used for building houses and boats or for fashioning tools. This "useful order" was determined by conventions embedded in millennia of writing about nature, shaped by poetry, dynastic histories, agricultural treatises, pharmaceutical manuals, and local history gazetteers.[69]

In spite of their reliance on texts, Qing observers knew that the rich diversity of Heilongjiang plant life went far beyond the ken of Chinese texts from south of the Great Wall. This diversity was hard to manage linguistically, and both noticed discrepancies between text and local usage that bothered them. As Manchus under the Qianlong emperor, they were both proud of their abilities in the Manchu language, and as elite scholars of the imperial court, both were also deeply steeped in the latest trends in classical Chinese philology. For Xiqing and Yinghe, the best way to solve problems of classification was to triangulate observations of the environment, living local languages including Manchu, and ancient Chinese texts.

Setting names was not an easy task, and nomenclature was a subject of debate and controversy. For example, Xiqing was dismayed that the local bannermen used the wrong names for their millet. As anyone who has eaten millet gruel in the company of friends from northeast China knows, the names and differences among the different varieties of millet grains can be a subject of intense disagreement even today. The historian of technology Francesca Bray has observed that debates about millet terminology have been "rampant ever since the Han," and this confusion has "persisted throughout the historical period and has survived triumphant."[70] From Xiqing's perspective in nineteenth-century Manchuria, this problem was complicated by the inaccurate use of Manchu terminology by Heilongjiang's bannermen. Xiqing was confident that what was called *mizi* by Chinese speakers was the same as the *ji* of classical texts. However, the people in Heilongjiang used the Manchu word *ila* to refer to *mizi*, when the appropriate standard Manchu term for the grain identified as *mizi* should have been *feishihe* (Manchu *fisihe*).[71]

To establish the differences among types of millets, Xiqing cross-referenced ancient Chinese texts with his knowledge of the Manchu language along with the agricultural realities he observed in the fields around him. The *Shuowen* dictionary (second century CE) and the *Annals of Master Lu* (third century BCE) all provided information on harvest times for different grains, thus allowing Xiqing to match the fast-ripening *ji* of ancient texts with the short growing season needed by the *mizi* of Heilongjiang's fields. The clearest information on the morphology of plants could be extracted from *materia medica* compilations such as the *Shennong bencao jing* (The divine farmer's materia medica, third century BCE), Li Shizhen's late sixteenth-century masterpiece, *Bencao gangmu* (Systematic materia medica), or, in Xiqing's case, from the more recent and quite popular *Bencao beiyao* (Essentials of the materia medica) by the early Qing scholar Wang Ang (1615–1695).[72] Xiqing's careful parsing of the *materia medica* descriptions of plants established that the grains of *ji* were larger and less densely packed than other forms of millet. From these descriptions in ancient texts, combined with his own careful firsthand observations, it was clear to Xiqing that what the locals mistakenly called *ila* was actually *mizi* (panicum or broomcorn millet), to wit, *fisihe*—a conclusion that seemed to give the Manchu scholar great pride in his philological and empirical skills.

Pharmaceutical texts were also central to Xiqing's understanding of the categories, natures, and uses of plants that abounded in Heilongjiang. Indeed, if the Russian botanist Maximowicz had only looked within texts such as Li Shizhen's *Bencao gangmu*, he could have found descriptions of many of the plants he so painstakingly described in the field, including several of the species he claimed to have "discovered." When Maximowicz rapturously walked into a stunning field of Amur flowers, with its deep blue buds of *Gentiana triflora*, the delicate yellow sprays of *Bupleurum scorzonerifolium*, the supple white petals of *Paeonia lactiflora*, and the showy purple spikes of the *Veronica sibiricum*, he was, in essence, walking into a field of Chinese medicine. The *Gentiana triflora* was *sanhua longdan* (three-flower dragon bladder), used to relieve the symptoms of the flu and treat hemorrhoids; *Bupleurum* root, or *chaihu*, eased pain and calmed the emotions; the root of the peony, *shaoyao*, vitalized the blood and promoted circulation, while the leaves of the *Veronica sibiricum* (*fushuicao*, literally "belly-water grass") promoted urination and re-lieved swelling. Maximowicz was certainly right that the flowers of the Amur were important, but here we find an entirely different reasoning for their im-portance. A knowledge of flowers would help the herb gatherer identify which plants were the most effective, but the parts that were the focus of concern—indeed, the parts that could mean the difference between life and death—were typically not the flowers that decorated the landscape but the medicinally active roots that lay beneath the ground.

The most casual glimpse into *materia medica* actually reveals that Maxi-mowicz's most important discoveries had been part of the Chinese pharmacy for centuries. Perhaps the best example of this is the new genus and species of the Schisandraceae family that was named after the botanist himself: *Maxi-mowiczia amurensis*. This plant, the tiny flowers of which were painstakingly dissected by Russian botanists in order to establish the new genus, was in fact the well-known five-flavor berry plant of Chinese *materia media*, the *wuweizi*. One could easily identify the plant from descriptions collected in Li Shizhen's *Bencao gangmu*—indeed, the descriptions evoke the plant far more readily than the exhaustive Latin descriptions penned by Maximowicz. The five-flavor berry plant was classified within the category of creeping vines (*man cao*).[73] In the wild it could be found climbing up tall trees, and thus it required trellises to be successfully cultivated in gardens. The leaves are round shaped and come to a point, similar to the heart-shaped serrated leaves of the apricot tree but larger. The woody stems have a dark reddish color and bear small cream-colored flowers that open in the fourth lunar month. In the seventh lunar month, berries the size of peas appear in clusters at the extremities of the stems. The berries are green at first, and then ripen to a distinctive reddish purple in the fall.[74]

While Maximowicz saw his new genus as unique to the environment of the Amur River, the Chinese texts held that the plant could be found in many places, although there was acknowledgment that different regions gave rise to different varieties. Five-flavor berry plants grew in the mountain valleys and ravines of Shandong, Shanxi, and Shaanxi, all northern China locations south of the Great Wall. A famous scholar of the fifth century mentioned that the highest-quality five-flavor berry came from Gaoli, or Koguryŏ, then a kingdom that encompassed territory from the banks of the Songhua River to the mountains outside of Seoul—regions above the Great Wall clearly associated with Manchuria. The sixteenth-century naturalist Li Shizhen noted that his contemporaries typically distinguished between a "northern" five-flavor berry plant and a "southern" one. The northern plant, with its dark red berries, was the superior variety.

The role of pharmaceutical knowledge in understanding plants points to a different, more visceral sensorial base for knowledge. What made varieties *varieties* was their impact on the human body when used as medicine: knowing a plant meant knowing the effect it had on the human body when consumed. The *wuweizi* plant gives a particularly complex example of this mode of knowing. While texts described the overall appearance of the plant in the field, the small, medicinally active part of the plant—the berries—garnered the most attention and required the most painstaking analysis. This analysis, however, was not visual. The *wuweizi* plant was unique because its berries possessed all of the five flavors (or "tastes," *wei*) that a drug could possess: the skin and meat of the berry were both sweet and sour, the kernel was pungent and bitter, and all of the parts of the berry had the property of saltiness. This did not mean that the berry *tasted* like all of these different flavors on the tongue, although taste could certainly figure into this property. The "flavor" of a drug is closely linked to the action that the drug is thought to have on the body. Sweet drugs boost weakened energies and bodily constituents, especially blood. Sour drugs restrain the leakage of fluids and are used to stem diarrhea or reduce bleeding. Pungent drugs are used to eliminate heat and toxins from the body, while bitter (or acrid) drugs reduce coughing and stimulate the circulation. Salty drugs are useful for reducing swelling, promoting bowel movements, and breaking up obstructions and masses. The five-flavor berry, as the name suggested, encompassed all these visceral impacts.

Given all the functions attributed to it, it is not surprising, then, that Chinese physicians and scholars had debated the true qualities of the drug for almost two millennia. In the sixteenth century, Li Shizhen sampled texts from across the ages and found dozens of uses. *Wuweizi* could tonify *qi*, alleviate weakness, expel heat, stop indigestion, brighten eyesight, improve mood, strengthen bones and joints, assist digestion, reduce swelling, relieve inordinate thirst, dispel anxiety, cure hangovers, relieve dry coughs, and even boost flagging male virility. Some physicians emphasized the berry's affinity for the lungs; others focused on its action on the kidney system. Some posited clinical differences between the northern variety and the southern variety. Even later nineteenth-century texts that have been hailed as advances in Chinese botanical knowledge, such as Wu Qijun's 1848 *Zhiwu mingshi tukao* (Illustrated explorations of the names and realities of plants), still took up this same debate about "flavor." In his entry on the five-flavor plant, Wu did not provide new descriptions of the plant's morphology or environment but sought to solve the mystery of how one berry could contain all five *wei*. While illustrations that accompany Wu's discussion may have more closely resembled the plant in the wild—seen as an indication by some that Chinese botany was becoming more "modern"—as late as the mid-nineteenth century, the most important distinguishing characteristic of a wild plant for many Qing scholars remained its impact on the body.[75] Ultimately, plants in the Qing were known through their deep and sustained entanglement with the human: this was, quite literally, a visceral knowledge. The standardizations of language could only go so far. As bodies varied, so too did knowledge of the things entangled with them.

Indigenous Plantways

Although their approaches to categorizing nature were remarkably different, representatives of the Russian and Qing empires in the nineteenth century had one thing in common: they both relied on the native people of the Amur basin to provide knowledge about the region's environment. The most impressive manifestation of this knowledge can be found in an extensive list of plants on the last pages of Maximowicz's *Primitiae*. The list is a compilation of the names of over four hundred different plants as expressed in ten different languages of peoples of the Amur region, including Solon, Ulch, Oroqen, Gorin, Manchu, Chinese, and two dialects spoken by the Nanai people (whom Maximowicz refers to using the Russian term, "Goldi"). Of these four hundred names, fewer than ten are Chinese or Manchu. Names in "Goldisch" clearly predominate the botanist's list. Given the predominance of "Goldisch" names, the Nanai people living near the intersection of the Amur and the Songhua clearly had an important role in helping Maximowicz identify plants.[76]

Maximowicz named plants but did not name his native informants. In the more than five hundred pages of *Primitiae*, Maximowicz only mentions the *Eingeboren* ("the natives") a handful of times, and always obliquely. They give him information about the lay of the land in places he himself cannot reach, occasionally mention the geographical extent of certain plants, and offer advice for how to avoid mosquitoes. The list of indigenous plant names at the back of the book seems to have been an afterthought, added out of purely linguistic or anthropological interest. Nevertheless, a perusal of Maximowicz's huge Candollian species listings (which comprise the bulk of the book) reveals that almost half of all the species identified in Latin also include native names for plants. Significantly, this included all of the plants that Maximowicz identified as new species.[77] The role of indigenous guides was clearly essential to the success of Maximowicz's work. Even though Maximowicz includes no direct mention of the participation of indigenous people in his narrative, one can easily imagine local guides navigating his rowboat up the Amur, leading the botanist to the location of rare plants or patiently describing a pile of picked samples as the botanist dutifully recorded their names. While he preserved the names of the plants, Maximowicz erased the people who identified those plants from his narrative. Indeed, such attempts at erasing the human is exactly what defines Maximowicz's work. Maximowicz made the minutest details of plant anatomy visible, but to generate "universal knowledge," he rendered the entangled human-plant context of the local environment all but invisible.

✳

In their texts, Qing observers like Xiqing preserved mention of indigenous informants and attributed specific knowledge to them. It is possible to imagine that Xiqing would have had ample opportunity to speak with indigenous men from the Hunting Banners who delivered their pelt quotas to the great tribute-collection depot at Qiqihar. These men had knowledge of the Greater Khingan forest and even the lands and tributaries far to the north of the Amur—lands they had to scour in ever-widening circles as hunting for the Qing court (and for the global fur market) continued to drain the Amur forests of sable. As a Qing official sent to rectify accounting corruption in Heilongjiang, Xiqing would have been focused on accurate evaluation and tabulation of pelts, but pelts came from people.

Indeed, in some Qing records, indigenous men appear as fully recognized individuals through a one-to-one correspondence with the furs they contributed for imperial tribute. For example, fur-collection records from the early to mid-nineteenth century generated from the Manchu outpost at Yilan on the upper Songhua River note the village location, clan associations, and even the given names of specific Nanai men who submitted furs for tribute. Records could be remarkably detailed: in 1804, from the village Malin, a man named Gaohe from the Fisi clan delivered ten skins, while another man from the same village and clan named Chikesi presented five. From the Hechikeli clan, men including Niyafeyinga, Wuke, Biyaketa, and Madake all submitted tribute. And from the large Ujala clan, the men Bumubu, Molenta, Keba, Agakulo, Mase, Shinga, Dangge, and many others brought a total of 127 sable skins.[78] While scholars have noted discrepancies between what collecting agents recorded at Yilan and Qiqihar and which furs were actually received by imperial warehouses in Beijing, it is clear that the people who hunted the animals were known as individuals to agents of the Qing state. A Japanese observer of tribute ceremonies at a village on the upper Amur in the early nineteenth century even noted that Manchu tribute-collection officials "fraternized freely with the natives," eating and drinking with them, "romping with the native children," and even hugging them goodbye as they prepared to leave.[79] If Xiqing had such warm encounters with indigenous peoples, he does not tell us, but he certainly learned that indigenous people had knowledge of nature that went far beyond a simple association with fur-bearing mammals.

In particular, Xiqing recorded that indigenous hunters had knowledge of the medicinal (and even magical) properties of many different plants that tribal peoples used for survival in the forests and plains of Heilongjiang. There was a type of tree bark that could quickly cure wounds and remarkable herbs that could repel and kill insects. Xiqing lamented, once again, that the native people did not know the proper names for these marvelous plants. For Xiqing, the "rectification" of these names with written texts was the most important fact about knowing a plant. He searched his books to identify native plants and only found a possible reference for the bug-killing herb: perhaps it was *mobei yabulu*, the *yabrūh*, or mandrake, of the Gobi Desert, a potent poison that could be used as a pesticide.[80]

Xiqing reported the presence of another plant whose leaves, if chewed in large quantities, could bring on strange visions and perhaps even impart longevity. The Qing official offered a delightful story about this particular plant's discovery: an indigenous Dagur hunter, on a trip far into the mountains, encountered a storm and wanted to protect his gun from the rain. Knowing that wild grasses had insulating properties, the hunter picked some nearby weeds, chewed them to soften them up, then stuffed them into his gun barrel. This kept the gun dry, but the hunter suddenly felt on the verge of passing out, experienced "uncontrollable thoughts," and had to return immediately to his cabin to rest for several days. Ironically, this Dagur hunter lived to a ripe old

age in hale and hearty health, an outcome that many attributed to his hallucinogenic encounter with the magical herb.[81]

While neither Xiqing nor Yinghe ventured a guess as to the identity of
this mystery plant, both observed that Heilongjiang's marshes, meadows,
and riverbanks abounded in a weedy plant called *yinchen*—a common plant
found all across China, and one of the most common herbs of the Chinese
traditional *materia medica*, today typically identified as *Artemisia capillaris*,
or capillary wormwood. According to Li Shizhen, *yinchen* was a type of *hao*,
a term today generally translated as an approximation of the genus *Artemisia*.
This is a broad category that includes over two hundred species, including
many commonly known in English as wormwood, sagebrush, sagewort, and
mugwort. These plants literally grow like weeds, their bushy, woody stems
rising three to six feet tall, poking out of rocky soils in wild fields and on the
edges of forests. The leaves typically form fernlike fronds, with colors ranging
from deep green to a dusty silver white (the latter color is called "old woman"
artemisia in vernacular English). Many artemisia plants produce volatile oils
and are used in the West as flavoring herbs, including sage, tarragon, and, perhaps most famously, the herb used to flavor absinthe—*Artemisia absinthium*
being the ingredient that imparted that drink's fabled hallucinatory effect.[82]

In the Chinese *materia medica*, dozens of herbs are recognized today as
part of the *Artemisia* genus, ranging from *ai*, Chinese wormwood, the ubiquitous herb burned in moxa treatment (and also burned to repel insects), to
qinghao, or *Artemisia annua*, the plant that produces the active chemical in
artemisinin, one of the most prominent treatments for malaria today. It is not
difficult to imagine that the "miracle plants" of the Amur peoples described
by Xiqing—with their fever-reducing, insect repelling, and dream-inducing
effects—were indeed different types of artemisia.[83]

With their remarkable diversity, it is no wonder that observers from the
metropole saw linguistic confusion surrounding the identity of different artemisia plants in Heilongjiang. Xiqing and Yinghe both noted disjunctures
between the names of plants in Chinese, their names in Manchu, and their
uses. For example, in the spring, bannermen took to the fields to pick a sort
of wild vegetable that some called (in Chinese) *liuhao*: willow (or willow-leaf)
hao. In Manchu, the bannermen called it *empi*, and with its subtle fresh taste,
it was considered a local delicacy. But the Qing court officials saw a problem:
empi, as any educated Manchu who had studied the Chinese classics knew,
was not necessarily something that you'd want to eat. *Empi* was the Manchu
translation for *fan*, a botanical term dating back to the *Book of Odes*—China's

oldest existing collection of poetry and the locus classicus for many plant and animal names from antiquity. According to classical *materia medica*, this *fan* was the same as silvery (or white) wormwood—*baihao*, or "old woman" artemisia—something quite different from the tender green shoots used in Heilongjiang's springtime stir-fries.[84]

The indigenous guides who accompanied Karl Maximowicz as he picked reeds and flowers from fields along the Amur in the mid-nineteenth century were much clearer about their *Artemisia*. From samples collected in 1855 and 1856, Maximowicz identified thirteen different species of the genus, and for many of these species, he noted Nanai names. Maximowicz provides a short entry for *Artemisia vulgaris*, the common "generic" mugwort that was found "frequently and everywhere *im ganzen Amurlande*." He noted that the Nanai simply called this plant *ssóachta*—in their conception, this was the generic or base variety of *Artemisia*. While they recognized an overarching unity to the genus, the Nanai differentiated different types primarily according to the environment in which they grew, and Maximowicz's observations seemed to follow these groupings. *Artemisia campestris* (artemisia of the plain) with its panicled stalks, was found in flat fields along the Amur—the Nanai named it *ssjússun ssóachta*, or "sorghum artemisia," after the heavily panicled grain that was grown in the Amur and Songhua plain. Maximowicz found *Artemisia sacrorum* growing in steep rocky slopes above the river—the Nanai called it *churren ssóachta*, or mountain artemisia. *Artemisia selengensis* could be found on sandy shores and island outcroppings in the lower Amur—the Nanai called it *audan ssóachta*, literally "inlet artemisia." Maximowicz, after describing its vibrant green leaves with deeply incised lobes, mentions that the young shoots of this plant are eaten and notes that in Manchu, some called it *ymbi*. Where Manchu observers noted confusion and a potential for poisoning, the native peoples whose survival depended on recognizing difference provided classificatory clarity.[85] It is clear that Nanai "ethnobotanic" categories helped Maximowicz differentiate plants in the field, but the value of this knowledge went far beyond its role in producing European science.

There is one section of the *Primitiae* that inadvertently pulls back the curtain on the immense store of embodied botanical knowledge possessed by the peoples of the Amur. In a separate and, for Maximowicz, insignificant section on *Kultur Pflanzen*, Maximowicz offers notes on cultivated crops and other human-plant interaction in the Amur basin. Like the Manchu writers of local histories, Maximowicz in this section prioritizes grain production. He notes that agriculture in the Amur basin is mostly confined to the areas of significant town settlement, the Nen River basin and the area around Aihui, and that sorghum, millet, wheat, and barley are the major crops. He dismissively notes that "those who farm are almost always Chinese or Manchu, and not the *Eingeborener*." The Chinese "criminals," where they have settled, have abundant, well-maintained gardens with spinach, eggplants, melons, chilies, cucumbers, pumpkins, corn, and a variety of greens. By contrast, the less industrious natives around the Songhua and middle Amur have small gardens, "insignificant and carelessly kept," where they grow tobacco, squash, and occasionally beans. The Russian settlers at the Bay of de Castries, Maximowicz notes with some pride, have produced bumper crops of tasty beets, cabbage, and potatoes, but he reports that the natives in the upper Amur "know nothing about agriculture at all," leading the reader to believe that the poor native peoples eat nothing but fish, with any plant-based food provided by Chinese merchants or offered through tribute exchanges with Manchu officials.[86]

After this belittling observation, Maximowicz proceeds to provide a listing of wild plants used by indigenous people that is astonishing, even awe-inspiring in its complexity and diversity. He notes, with some disdain, that the natives eat "indifferent-tasting" parts of many wild plants, items that would "not meet with the approval of a European palate." Maximowicz considers these to be primitive foods that required neither effort nor knowledge to collect—they could be "gathered even on the most casual walk." A glance at the list, however, reveals the deep knowledge that was necessary to identify and process wild plants for human consumption. According to Maximowicz, the Nanai dry the bulbs and shoots of wild onions in large quantities, particularly *Fritillaria kamtschalcensis* (Kamchatka lily), along with the leaves of the *Allium shoenoprasum* (a type of chive) and *Cetraria islandica* (a type of lichen). The native people also eat a variety of roots and rhizomes as starches, including the roots of the peony (Nanai, *hulleto*) and the thick white roots of *Platycodon* and *Adenophora* (Chinese bellflower, ladybells) (in Nanai, *la-ke*). He also lists multiple different rhizomes and roots eaten by the peoples of the upper Amur, describing their appearance and taste with adjectives ranging from "sweetly disgusting" to "disgusting," but he admits there are so many that he cannot identify all the species.[87]

Among what he allows are native "vegetables," Maximowicz identifies a dozen plants, including the leaves of *Allium victoriaiis* (Alpine leek); *Cacalia hastata* (a type of ivy), *Sambucus racemosa* (red elderberry), and *Senecio pseudarnica* (ragwort, similar to dandelion); also the young stems of *Polygonum Hydropiper* (smartweed), *Chenopodium album* (pigweed), *Limnanthe-*

mum nymphoides (water lily), *Pleurospermum austriacum* (garden angelica, or wild celery) and *Heracleum* (cow parsnip); along with the young shoots of *Epitobium angustifolium* (fireweed) and *Artemisia selengensis* (this is the *empi* of Xiqing and Yinghe that Maximowicz allows "should be quite tasty"). He also lists other edible plants that he had not encountered personally but had recorded names from native informants: *aughoch, jagdsha, chira, tachssolyé, adirgan ssolgé, kanje ssilekla, arsj,* and *kuchj*.[88]

Maximowicz chides the natives for not being interested in smaller fruits such as wild strawberries—the suggestion being that such fruits, while tasty, would be too much effort for the indolent natives to gather.[89] The list of fruits enjoyed by natives that Maximowicz provides, however, indicates a varied diet remarkably rich in fruits. The Nanai collect wild plums, dry them, crush them into a pulp, and use the pulp to make portable flat cakes (perhaps like a cross between an energy bar and fruit leather). They collect lingonberries, rose hips, hawthorne fruits, crowberries, gooseberries, raspberries, red currants, blueberries, and the red fruits of the Sakhalin honeysuckle (which the botanist dubbed *Lonicera maximowiczii*). The lingonberries, in particular, were gathered in large quantities in the fall, frozen with the first snows, and eaten throughout the winter.

While Maximowicz observed the native enjoyment of tobacco acquired from Chinese traders, he also reports that the Nanai were confident they could find tobacco substitutes for free in the wild, including a species of *Viscum* (a type of mistletoe), *Bupleurum* (known in Chinese as *chaihu*, a major ingredient in herbal formulas used today to relieve anxiety), and the leaves of shield ferns (*Polystichum*). Maximowicz also notes that native villages were frequently surrounded by clusters of *Artemesia vulgaris* (mugwort), *Urtica dioica* (stinning nettle), *Leonurus sibiriens* (Siberian motherwort), and cannabis plants—suggesting an intentional cultivation of medicinal and mood-changing herbs. Tantalizingly, Maximowicz notes that the native peoples had deep knowledge of the preparation of medicinal plants, including decoctions of wild rosemary, chrysanthemum, burdock, Solomon seal, and cliff ferns, but the true knowledge of these remedies, "of which there are as many as herbs in the forest," were primarily the "concern of women," and thus, apparently, not of particular concern to the botanist.[90]

In-depth knowledge of plants also created the remarkable material culture of indigenous peoples. While Maximowicz evinced disdain and disgust for native plant foods, his description of the Nanai uses of plants to create shelter, clothing, and even art contain a grudging admiration of native ingenuity. Willows were used not only for kindling but also for insulation, scaffolding, and rope. Through steaming and boiling, birch bark was ingeniously transformed into a fabriclike material to make everything from watertight canoes to rainproof coats. Plants gave dyes that were expertly used by women to impart brilliantly colored decorations on bags and clothing. The soft wood from small pine trees and young larch trees was used to make everything from eating utensils and elaborately carved boxes to the wooden cabins the Nanai called home. Different types of reeds were used as cover for roofs and insulation for walls, while mosses were used as caulk. The use of reed insulation in particular made the native homes relatively comfortable even through the cold Siberian winters. While Maximowicz allowed that plants seemed to provide the indigenous people with everything that they needed, he noted that the natives were unable to take advantage of the lucrative market for hardwood trees since they lacked the large metal implements needed to fell them. The harvesting of these commodities, Maximowicz noted, awaited the axes of Russian colonists, who were already hard at work cutting down the Amur's old-growth forest.[91]

Soon, Maximowicz optimistically predicted, the fertile soil of the Amur would attract energetic Russian settlers who would clear the forests and drain the marshes. In his scientific opinion, these actions would actually result in a felicitous increase in the diversity of species, since they would eliminate the old grasses and light-blocking tree canopies that inhibited growth. Once this civilized "parklike" landscape along the Amur was established, the botanist concluded, his "initial harvest" of the Amur flora might serve as a reminder of the landscape's primitive past. In a passage that points to the use of science to further colonialism, Maximowicz concludes his botanical tome with a prediction of the Amur's bright environmental future: "We therefore considered it our duty to avail ourselves of this rare opportunity to describe, to the best of our powers, a country in its original state, in the hope that the land of the Amur will soon become a cultivated land. When that time comes, this description of the flora-physiognomy (which by then would no longer correspond to reality) can perhaps offer an instructive comparison between then and now—between what once was and what will be."[92] In Maximowicz's view, there would be no lament for the demise of either the primitive local environment or primitive local knowledge.

Maximowicz was quite right about the future of the Amur basin landscape, although he could not have predicted the mechanisms through which such change would take place. The Russian presence certainly provoked a great environmental transformation, but this transformation did not take place at the hands of Russians. The Romanov assumption of the land to the north of the Amur caused the Qing government to radically change its policy toward Han emigration to the river's south. The Han presence in the area had been steadily increasing (in spite of government prohibitions) during the eighteenth and early nineteenth centuries as Chinese convict laborers escaped plantations and as Chinese migrants quietly established themselves as long-term cultivators on lands supposedly reserved for bannermen. But after 1860, the Qing government abruptly ended the centuries-long official Manchu monopoly of the land and actively sought Han settlers for millions of "decommissioned" acres. The hope was that populating the land with Qing subjects—even if they were Chinese—would bolster the Qing's claim to dominion.[93]

This shift in policy led to a profound and rapid shift in the landscape of Amuria. For example, in the 1880s alone, Jilin opened up a total of over 1.6 million acres to Han farmers, while Heilongjiang opened about 1.1 million acres—mostly fertile river-valley land in the area where the Nen River and the Songhua River meet, and farther east in the area around present-day Harbin. By the turn of the twentieth century, cultivation extended east toward the vast prairies and wetlands located at the confluences of the Songhua, the Amur, and the Ussuri Rivers—the area that would come to be known as the Three Rivers Plain (Sanjiang pingyuan). In the last years of the Qing, over 750,000 acres were opened in the Mishan area around Lake Khanka (Xingkai hu). Yilan, the Manchu administrative outpost on the Songhua River where Qing officials would meet with indigenous fur-trappers, opened 250,000 acres to Han settlement, while Fujin, a region close to Maximowicz's flower-strewn prairies, opened a similar amount. By the end of the Qing, Jilin recorded over 4.7 million acres under cultivation, and Heilongjiang 2.7 million acres. These figures are small in comparison to the overall size of these regions, but they signaled a tremendous impact on the knowledge and lifeways of those who had called the region their home.[94]

✳

In the spring of 1902, the Russian surveyor Vladimir Arsenyev (1872–1930), on assignment in the Sikothe-Alin Mountains near the Ussuri River, encountered a solitary indigenous Goldi hunter named Dersu Uzala.[95] In Arsenyev's telling, the hunter simply showed up at the surveying team's campfire one evening, laid down his ancient rifle, unceremoniously sat down, and ate the food offered to him by the surprised Russians. Arsenyev's initial description of Dersu is drawn in physical-anthropology terms—he details the shape of his skull, the presence of an epicanthic fold, and the color of his hair, beard, and eyes[96]—but as the two men communicate (Dersu in broken Russian) and experience the forest together, a relationship grows between them. For months, Dersu guides the expedition through the Sikothe-Alin's dense forests, through river gorges, over mountains, and across the vast expanse of wetlands surrounding Lake Khanka. In Arsenyev's eyes, Dersu is possessed of an uncanny knowledge about nature born of the need to survive. His ability to read the clouds and sense the air allows him to predict accurately the region's extreme and changeable weather. Subtle changes in the forest environment allow Dersu to know the comings and goings of potentially dangerous inhabitants, whether Amur tigers or Chinese brigands.[97] With each revelation of Dersu's skill, Arsenyev's admiration for "the Goldi" increases, but it is an admiration tinged with a poignant awareness of the decline of Dersu's people in the face of Russian development. As their friendship deepens, Arsenyev learns that Dersu's entire village, including the hunter's wife and children, had been decimated by disease, leaving him the sole survivor.

Arsenyev seems unaware that the challenges faced by Dersu's people could be traced back centuries, challenges created by their position as inhabitants of Amuria. Dersu's ancestors were among the thousands of indigenous peoples who were, as David Bello ironically but aptly describes, "hunted and gathered by both the multiethnic Romanov and Qing empires"[98] His clan, Uzala (rendered in Manchu as Ujala, in Chinese as Wuzhala, and pronounced, according to Arsenyev, something like "Ochzhal"), shows up in the long list of northern clans subjugated by Manchu armies as early as the 1630s. Harassed by Cossacks to pay fur tribute in the 1650s, they were also on the roster of people conscripted to form the "New Manchu" banners in the 1670s.[99] Integrated into the Qing fur-tribute system, individual members of the Uzala (Ujala) clan—Bumubu, Molenta, Geba, Agakulo, Mase, Shinga, Dangge—can be spotted in the Qing pelt accounting archive dutifully delivering their sable-skin tribute. Uzala hunters were undoubtedly among those who skimped on sable-fur procurement as the years wore on, substituting squirrel skins for the hard-to-find sable or even buying sable skins as commodities on the open market to maintain the facade of the culture-laden tribute system.[100] Even so, hunters such as the Uzala continue to appear in fur-tribute records through the nineteenth century. Perhaps participation in the festivities surrounding tribute exchange was still enough of an honor, and the days of feasting in Manchu towns on the Qing tab was worth the trouble.

Participation in tribute exchange was, after all, a sure way of acquiring part of the Qing agricultural harvest. Qing receipts for the expense of feeding the gathered indigenous clans at Yilan show that these feasts were grain-centered events. The Qing used the bounty of their farming in the Amur basin (provided in part by the unfree labor of Han convicts working on government plantations) combined with expensive imported grains to fete the native hunters. Food served to the indigenous people during tribute feasting in 1804 included 98 *dan* (almost 300 bushels) of cooked rice along with thousands of *wowo* cakes (a kind of conical steamed croquette) made from 49 *dan* (150 bushels) of yellow millet. The clansmen also received grain supplies to tide them over on the long trip back to their settlements. Over one thousand members of the gathered clans—the Biladachili, Hechikeli, the Jakesulu, the Ujala—each got from one to three *dou* (pecks) of grain, and the farther they had to travel, the more grain they got. The total amounts distributed as "grain for the road" were not insignificant: reports show the direct distribution of over 628 *dan* (almost 2,000 bushels) of rice and 580 *dan* (1,700 bushels) of millet. Notably, the largest single amount of grain used for the feast itself was not consumed directly but went into the making of alcohol: over 250 *dan* (750 bushels) of grain to make the spirits that were consumed during the feasts each year.[101]

After the emergence of the Russian threat at the middle of the century, the Qing government ended the era of the "free lunch." Assuming that Nanai would make good soldiers because of their mobile lifestyle and "proficient use of firearms," the Qing conscripted them into fighting units in order to bolster Qing defenses south of the Amur.[102] In return for their military service, the Nanai were given parcels of land in the Sanjiang region in and around what is now Fujin county along the upper reaches of the Songhua River. Some joined the trend toward commercial agriculture in the region and took up farming cash crops, especially opium.[103] Most indigenous families rented the land out or sold it outright to Han settlers. By the late nineteenth century, local officials were sending urgent appeals to the governor for more grain to make direct payments to their Nanai recruits.[104] Qing authorities feared that if they failed to provide these men with grain, they would turn their allegiance to Russians, who could supply what had become necessary staples in the lives of indigenous people: millet and alcohol. At the beginning of the twentieth century, then, Arsenyev encounters Dersu Uzala in this context of the waning years of the Qing. The hunter's people have been conscripted, relocated, proletarianized, and "grainified." Dersu's solitary predicament had been long in the making.

Nevertheless, Dersu's knowledge of the plant life of the Amur basin seems to have remained intact, and at times even meant the difference between life and death for his Russian charge. During one trek, Arsenyev encountered a nest of massive hornets and was stung multiple times. Dersu wrapped the afflicted areas with an herb from the forest, and within minutes Arsenyev found his pain relieved and infection avoided. Arsenyev identified the herb as *Clematis mandshurica* Maxim., a flowering bush that Maximowicz had once noted "decorating" the Amur forest with its white blossoms.[105] During another expedition, Arsenyev came down with severe dysentery. Dersu "rushed into the forest" and returned with an herb that stopped Arsenyev's suffering. The herb, Arsenyev later noted, was Solomon's seal, or what Dersu (if Maximowicz's record is correct) would have called *kóhgliachta*.[106] In the most dramatic manifestation of Dersu's plant knowledge, the two men found themselves lost on a frozen marsh near Lake Khanka with night approaching and a ferocious blizzard closing in. Dersu urged the "Kapitan" to start chopping down and piling the tall reeds that surrounded them. Arsenyev was not entirely sure of the value of this exercise but joined Dersu in frantically collecting reeds until his hands were bleeding and he fainted from exhaustion. He awoke the next morning to find himself safe and warm inside a shelter that Dersu had constructed using mounds of weather-resistant reeds, with Arsenyev's surveying tripod serving as an internal frame. Dersu's knowledge of the plants of Amuria had once again saved Arsenyev's life.[107]

After years of such adventures, Arsenyev's tale ended on a sad note: Dersu Uzala died in Khabarovsk in 1908 after the aging Nanai hunter was unable to adjust to life in the Russian city. Arsenyev used the life of Dersu the Hunter as an extended metaphor for the embodied beauty and diversity of nature in the Amur basin. With Dersu's passing, Arsenyev signaled the inevitable demise of Amuria's wilderness in the face of modern civilization.

✳

The diversity of Amuria plant life that astonished imperial observers in the mid-nineteenth century began to decline and disappear soon after Dersu's death. The search for productive, large-scale, grain-based agriculture that began in the late Qing continued into the twentieth century. Today, the Amur prairies where observers saw fields teeming with *Clematis*, five-flavor berry (*wuweizi*), *Aster tataricus*, and a dozen types of *Artemisia, empi, ssóachta*, or *hao* are now the home of an industrialized straight-line monoculture, with vast fields of crops that are planted, maintained, and harvested with the use of machines (a process examined in chapter 8 of this book). Xiqing would be amazed that the wetlands at the juncture of the Songhua, Heilongjiang, and Ussuri Rivers (the Sanjiang region), once the marshy homelands of the Nanai people, have become the center of production for China's most coveted varieties of rice and help to feed a nation of over a billion. Along with the midwestern region of the United States, the Amur basin exemplifies what the environmental historian Donald Worster has called the total "transformation of the earth," a landscape radically changed by the transition from subsistence farming to [socialist and then] capitalist modes of food production."[108] The diversity that once characterized the plant life of the Amur has literally been plowed under.

With this transformation of the earth came an erasure of people and of knowledge. Scholars such as Philip Deloria have pointed out the danger of negating the lives and lifeways of indigenous peoples through assumed narratives of decline and disappearance.[109] However, for the native peoples of Heilongjiang, such a narrative is justified. The Nanai, categorized within the PRC "national minorities" scheme as the Hezhe, have become the least populous of all the officially recognized indigenous ethnic groups of China. Historical demographers suggest that their population fell by 50 percent during the nineteenth century, and again fell precipitously in the beginning of the twentieth. Part of this can be explained by their absorption into the ranks of the Manchu banners. But other factors—including disease, agricultural pressure on the land, opium and alcohol addiction, and violence—resulted in a drastic decline in the number of people living as Hezhe. By 1949, the total population of Hezhe in the PRC was estimated to be only three hundred.[110] Dersu Uzala's people– and their astonishingly complex, deeply embodied knowledge of the Amur environment—were rendered invisible by the universalizing impulses of empire.

Conclusion

The Amur River offers a unique and revealing window into Manchuria's global changes during the nineteenth century. For three hundred years, the region of the "black water" comprised the northeasternmost expanse of the Qing empire. The river flowed from Central Asia to the Pacific Ocean, fostering a diverse abundance of plant life and human societies that thrived on either side. New global competition of empires changed that status. In 1858, the Amur River went from being the watery network that connected life across a vast region to being a border object: a line dividing two of the world's largest empires.

Western science came to the Amur as an active part of global imperial competition. Chased halfway across the world by an enemy British frigate, on a mission designed to preempt American power in the Pacific, the botanist Karl Maximowicz was deposited on the banks of the Amur just as Russia sought to expand its imperial reach vis-à-vis China. Within this imperial context, Maximowicz sought to bring the nature of Manchuria into a new universal order of knowledge. With just a few rowboats, native guides, and a network connecting him to St. Petersburg's scientific community, the twenty-five-year old set out to catalog all the plant life of the massive Amur down to "the smallest blade of grass." Maximowicz's biogeographical enterprise aimed to place Manchuria within a vast cartographic order, a "universal system" produced by Western empires as they claimed territory around the globe. The massive themes of science and empire were inextricably woven in Maximowicz's work: indeed, *Primitiae florae amurensis* was published just a few months after Russia forced the Qing to cede hundreds of thousands of square miles of its territory. By placing a lone Western botanist together with certain non-Western observers in the same quiet field of Manchurian flowers, we can unpack the small gestures and modes of attention that went into making both science and empire.

An examination of the senses in making natural knowledge shows how Maximowicz relied on two exhaustive ways of seeing. In the field, when encountering entire landscapes, he aspired to a comprehensive, omniscient overview of all plant forms. Within the confines of the botanical study, however, as he encountered individual samples, Maximowicz trained his vision on tiny, invisible things. The minute internal anatomy of flowers dictated the place each plant would occupy within an imagined universal order that transcended the local. This process required the mobilization of a vast network of people wielding multiple forms of expertise. In spite of this, Maximowicz's goal was to de-people natural objects, to disentangle them from local meanings and redefine them in a precise but dead language.

Maximowicz's work reflected science's attempt to subjugate the local to the universal. While on the lookout for useful commodities in the New World, Europeans in the seventeenth and eighteenth centuries encountered plants that were deeply embedded in indigenous knowledge and sought to understand their nature from indigenous ways. The result was knowledge that recorded, however incompletely, local uses, local names, and local stories. By the early nineteenth century, however, European study shifted away from use-value to abstract categorization. Plants had to be conceptually disentangled from their native places, detached from their native cultural moorings, and placed within schema comprehensible first and foremost to European botanists. Through a process which Londa Schiebinger has dubbed "onomastic imperialism," plants were "uprooted from their native cultures and acclimatized to colonial rule by being given European names." Through standardized naming, European botany thus "swallowed into itself the diverse cultural identities of the world's flora."[111] This process of universalization required the erasure of local meanings and local people from writing, even if the explorer's experience of the land was entirely structured and mediated through local presence. In this light, Maximowicz's knowledge making was clearly colonial. His work of removing knowledge from its indigenous moorings was part of the larger project of erasure that D. Graham Burnett has characterized as "the first step toward transforming *terra incognita* into *territory* . . . a landscape in which the old adage 'master of all I survey' could be meaningful in its most literal sense."[112] Maximowicz's listings of Nanai plant uses, while they hint at the abundance of indigenous plant knowledge, subsumed the vast embodied knowledge possessed by the denizens of Amuria into his own.[113]

In contrast, people were explicitly central to Qing natural knowledge. Qing observers in the nineteenth century looked out over the diversity of plants in the Amur and made sense of them according to their impacts on the human body: a more visceral knowledge that spoke of human-plant entanglements through labor, consumption, and internal transformations. To organize this array of knowledge, Qing observers combined real-life observations with knowledge of classical Chinese texts and standard Manchu terms: a multilingual "field philology" that signaled the authors' membership in an empire-wide community of scholarship. Within this practice there was room to retain local meanings and local stories. As described by the historian of science He Bian, Qing scholars "expressed strong interest toward individual experience and description of natural particulars," part of constructing an idea of "place-based authenticity" for plants.[114] Knowledge was valued for its ability to preserve local authenticity, attributes that could enhance both the value of the plant itself and the reputation of the scholar who recorded the knowledge. This knowledge maintained, if only in a partial and unsystematic way, the stories, lives, and rituals of people who provided the knowledge. Place-based authenticity in an expanding empire meant expanding through accretion the catalog of ever-diverse items and peoples.

In the expansion of the Qing, natural knowledge did undergo certain forms of standardization: transformation of knowledge into categories that made sense in the terms of the metropole. As Carla Nappi has shown, for example, the incorporation of the *Cordyceps* fungus of Tibet into the Chinese pharmacopeia during the eighteenth century entailed an erasure of its local Tibetan roots, a form of "colonial medicine" that perhaps echoes current attempts to engulf "national minority" medicines into a seamless tradition of Chinese medicine under the PRC.[115] Certainly, the importation of plants as commodities in the vast Qing market resulted in a paring away of some local meanings as they became removed from one use-economy and fixed into another. But the writings of Xiqing and Yinghe from the far northeastern frontier demonstrate that nonhuman things at the edges of the Qing empire remained entangled with the human: even as they were being described and translated for the center, these things resisted standardization and retained their peopled content. Qing multilingual observers "in the field" struggled to fit plants into standard language but did so by triangulating between empirical observation, ancient texts, and persistent local languages and local lore. Snippets of stories, tales of individual hunters, and mention of unique beliefs survived even as Qing observers attempted to find the "true" name—a far cry from the "onomastic imperialism" of nineteenth-century Europeans.

Thus, by following observers in a field of Amur flowers, we can get a sense of the state of natural knowledge at the beginning of the end of an era. Well into the nineteenth century, the Qing remained an empire of diversity, an empire of "ten thousand things" (*wanwu*) grasped through the approach of *gewu zhizhi*—"investigating things and extending knowledge"—where "things" were embedded in and encompassed the affairs of man and the goal of extending knowledge was to guide the self and the polity.[116] As Matthew Mosca has pointed out, the Qing system was not particularly successful at taking the unique knowledge of frontier peoples and transforming it into standardized, unentangled, instrumental knowledge at the metropole.[117] As reflected in the *Primitiae*, nineteenth-century Western science in Manchuria utilized and recorded aspects of indigenous knowledge but stamped over it a standardized "universal" order. The complex knowledge of indigenous actors appears as fragments in the interstices of both Qing and Western writings, discernible only as fleeting glimpses.

The social, economic, and environmental processes unleashed by competition between empires in Manchuria would soon result in the radical transformation of the land and the radical subjugation of local knowledge. But for a time in the late Qing, the profusion of flowers along the Amur remained, until a new century—and a new way of seeing—would erase them from view.

FOSSILS OF EMPIRE

The Jehol Biota and the Age of Coal

"To the true man of science, the earth is a unit, and
cooperation is a sacred duty."
—E. AHNERT, address to inaugural meeting of
the Geological Society of China, 1922

The arthritic American professor peered intently at the fossilized mollusk shells arrayed on his Beijing office table. Embedded in thin gray shale, each tiny shell was distinct, its jointed sections spiraling upward until coming to a delicate point. To the uninitiated, they looked like insignificant snail shells, but subtle differences in minute structures could signal the discovery of a whole new species. Adjusting the magnification of his lens, the scientist counted the number of whorls in each shell and noted patterns of barely discernable growth lines. He then calculated the angle of the whorls' emergence from the central axis and measured the distance between the shell's opening and its pointed top, generating lists of numbers that could reveal the shell's identity.

While his eyes focused on these minute anatomical differences, the professor's mind envisioned far-flung places around the world. The fossils were from a southwestern part of Manchuria known as Jehol, but they looked very much like the fossils he had studied from the Laramie Formation in Colorado, a site well known for its coal deposits and dinosaur bones. The thin brittle rocks in which they were embedded reminded him of the Flinze of Bavaria, a formation that had given rise to the famous *Archaeopteryx* fossil. Were these Manchurian fossils unique, representative of a singular ancient environment, or were they the same as those found on other continents, continents that were once melded together as part of a unified terrestrial whole?

After hours of painstaking measurement and analysis, the professor reached his conclusion: the snail shells were indeed a new species of the genus *Campeloma*, and the environment of northeastern China they emerged from was unique. In a 1923 publication on fossil mollusks, he recorded this discovery of several new species, appending his surname, "Grabau," at the end of each name to signify his role as the taxons' author. One fossil he dubbed *Campeloma tani*, named after H. C. T'an, the Chinese geologist who had discovered the shell in the field. He named another species after the place where it was discovered: *Campeloma jeholense* ("of Jehol province"). He concluded that Jehol contained a wealth of fossils, a "unique horizon" that would one day achieve global importance.[1]

This "unique horizon" suggested by the American professor Amadeus Grabau (1870–1946) has become known as the "Jehol biota," and fossils from this location now represent some of the most spectacular paleontological finds made anywhere in the world. The barren hills three hundred miles northeast of Beijing where *Campeloma jeholense* originated have recently yielded evidence of "feathered dragons": fossils that cemented the evolutionary link between dinosaurs and birds. *Sinosauropteryx prima*, discovered in 1996, was the first definitive evidence of feathers on a dinosaur. *Beipiaosaurus inexpectus* ignited imaginations in 1999 with its *Jurassic Park*–like raptor anatomy combined with birdlike features. *Confuciusornis sanctus*, named after the Chinese sage Confucius, represents the world's earliest known beaked bird.[2] These discoveries from the Jehol biota have revolutionized scientific understanding of past and current life. Ancient dinosaurs are now thought of as brightly colored, limber, social animals, and today's birds are now seen as "modern-day dinosaurs."[3] Yet while the fossils of the Jehol biota have made headlines around the world, little thought has been given to the significant ground in which they were found.

This chapter examines the history of the early Jehol biota discoveries, focusing on this corner of Manchuria as a site that makes visible the inextricable twisting of natural knowledge, energy extraction, and empire. In the first half of the twentieth century, Jehol became a microcosm of the radical military, political, and economic transitions that characterized the birth of modern East Asia on—and in—the ground of Manchuria. The Qing court had once considered Jehol a place "where dragon veins meet," an auspicious location strategically positioned to imbibe the sacred natural energies that flowed from

the Long White Mountain.[4] The significance of Manchuria changed dramatically with the advent of industrialization, modern resource extraction, and
global imperial competition. Dragon veins in the earth meant very little to
the Han Chinese revolutionaries who overthrew the Manchu imperium in
1911 and established a modernizing republic of China. These men sought to
acquire the techniques of the West, particularly science and technology, to
ensure the existence of their new republic in the face of incursions from other
empires. As the historian of technology Victor Seow observed in his pioneering study of mining in Manchuria, the discovery of major coal deposits in
the area around the original Manchu homelands turned Chinese visions of
dragon veins into desires for subterranean mineral deposits that could fuel
the armies and factories of modern-day states.[5] From 1894 to 1945, multiple
entities fought brutal wars for possession of Manchuria's territory both above
and below the ground: wars that ultimately became a fifteen-year long clash
between the republic of China and the empire of Japan. While much of Manchuria's natural resources were exploited by foreign powers, Jehol remained a
sort of buffer zone for decades, a unique site of Chinese sovereignty and Chinese scientific endeavors until the region fell to the Japanese military in 1933.
It is against this dramatic backdrop of modernization, war, and empire that
the mundane excavations, precise measurements, and careful identifications
of Jehol's fossils took place.

In current studies of the Jehol biota, the history of the region's unique fossil finds is usually conveyed with a few brief citations that proceed as follows:
Woodward (1901), Grabau (1923), Endō (1940).[6] These citations reference
obscure articles describing the fossilized remains of a fish, a snail, and a lizard,
respectively—discoveries that seem far removed from the spectacular winged
dinosaurs of the late twentieth century but were nevertheless crucial to the
growing scientific awareness of this "unique horizon." This chapter probes the
relationship of these three fossils and the place that gave rise to them. With
their British, German, and Japanese surnames and succession of dates, the
early discoveries of the Jehol biota seem to represent the seamless progress of
a collaborative international science. However, Jehol was not only a place of
significant paleontological interest; it was also a part of Manchuria, a "cradle
of conflict," ground zero of the twentieth century's violent imperial competition in East Asia. Knowledge of the fossils of the Jehol biota was inseparable
from the political fate of the territory in which they were discovered.

This chapter also explores the different ways this knowledge about fossils
could be deployed. Creatures etched in stone had long fascinated observers

with their intimations of unfamiliar life and inspired musings about ancient landscapes.[7] These musings remained fairly localized until the nineteenth century, with the emergence of the sciences of geology and paleontology in Europe and the United States. Modern scientists looked at fossils through different eyes and communicated about them in radically different ways. As we saw in the work of the botanist Karl Maximowicz in the previous chapter, precise measurement of tiny anatomical differences was required to define individual species in the "universal order" that science aspired to. These minute differences could become crucial ingredients in a cartographic enterprise to map the history of the earth's terrain. Patterns of fossil distribution were used to indicate landmass formations that were radically different from today's configuration of continents. Disjunctures and continuities in the fossil record present the earth as a giant moving puzzle, and paleontologists around the world attempted to understand how minute local pieces—such as the fossils of Jehol—fit within the global whole.

While one goal of paleontology and geology was to create a vision of the earth as a unified whole, the earth sciences emerged during what Jürgen Osterhammel has dubbed "the century of coal."[8] While fossils helped to conjure a vision of the earth's ancient continents, they were also used to gauge the presence of coal deposits. We shall see in this chapter that all the discoveries of the Jehol biota fossils in the early twentieth century were related in one way or another to the extraction of mineral resources: indeed, the fossils of the Jehol biota were discovered exactly because of the competition over the subterranean wealth of Manchuria.

Many of the men responsible for the Jehol biota thought of their work as immune from the military conflicts, economic competition, and political divisions of the day. For them, the fossils of Jehol even inspired transcendent visions of a unified world. Such sentiments were expressed by the Russian geologist (and colleague of Amadeus Grabau) Eduard Anhert (1865–1946) in words from a speech that form the epigraph to this chapter. As Grace Shen has shown in her pioneering study *Unearthing the Nation*, many professionals involved in the earth sciences in China during the interwar years shared an ideal view of science as a community striving together to create a universally shared knowledge of nature. Addressing the cosmopolitan Geological Society of China in 1922, an organization that included Chinese, American, Scandinavian, Russian, and Japanese members, Anhert proclaimed, "To a true man of science, the earth is a unit, and cooperation is a sacred duty."[9] Anhert had found professional fulfillment through science in northeast Asia and celebrated this spirit of collaboration with colleagues of the fledgling Chinese scientific community. Nevertheless, his inspiring speech failed to acknowledge that his ability to explore the geology of Manchuria was predicated on imperial expansion into the region, an expansion characterized by economic exploitation and military violence. The forms of vision involved in identifying ancient life-forms and projecting them onto the earth were embedded in visions of extraction, blood, and conquest that accompanied the world's transition to fossil fuels.

This chapter recapitulates the discovery of the Jehol biota—Woodward (1901), Grabau (1923), Endō (1940)—placing scientific ways of seeing in the context of empire and nation. To bridge the gap between the previous chapters and a world dominated by science, this chapter starts with the intriguing observations that the Qing emperors made about Jehol's fossils, observations that included musings about anatomical structures and the structures of deep time. Modern geology and paleontology, however, viewed these fossils with different eyes and for different purposes, and with a different understanding of the workings of the cosmos. Beginning with the Kangxi emperor's observations about fossils highlights the radical shifts in sensing the nature of Manchuria that took place at the turn of the twentieth century as the region shifted from being a land of dragon veins to a land of fossil fuels.

Jehol in the Eighteenth Century: Land of Stone Fish and Dragon Veins

The Jehol biota may be world famous, but a contemporary map of the PRC will not reveal a place called "Jehol." Maps from the early twentieth century will show a Jehol or "Rehe" province nestled between Beijing and Inner Mongolia. Fig. 5.1 replicates the political geography of this time period to show Jehol's positioning. Its eastern borders begin about sixty miles inland from the Bohai Gulf, its northern and western extent ranges into the Inner Mongolian grasslands, while its southern border runs more or less along the Great Wall. Historically ambiguous, Jehol is a transitional space, a place where the mountains of the Great Wall gradually give way to the Central Asian steppe. Its identity was created from the legacy of the Qing emperors, for whom it was a gateway to the world "beyond the Wall."

Jehol is most clearly associated with the Qing imperial palace complex at Rehe, in today's city of Chengde. Here in the mountains just one hundred miles northeast of Beijing, the Qing emperors established their "Villa

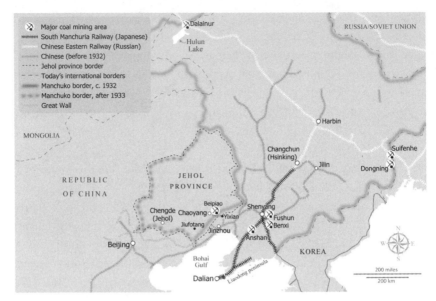

Fig. 5.1 Jehol and Manchuria/Manchukuo, showing major mining areas.
Map by Jeff Blossom. See plate 18 for color version.

for Avoiding the Summer Heat" (*Bishu shanzhuang*), a remarkable highland
retreat complex that included multiple palaces, temples, and vast imperial
gardens and forests. Buildings on the 1,400-acre grounds encompassed architectural styles reflecting the multiethnic polity of the Qing from around
the empire—including even a replica of the Dalai Lama's Tibetan palace, the
Potala. At Jehol, generations of Manchu emperors passed the time living in
hardy Inner Asian "beyond the Great Wall" style, riding, hunting, and relax-
ing in the company of Mongol dignitaries and Tibetan lamas. Some emperors
spent so much time at their summer villa that Jehol served as a de facto capital
of the empire. It was here that the Qianlong emperor had his famous meetings
with the British emissary George Macartney in 1793. While the Irish earl was
impressed by the rugged landscape and the massive tents erected for imperial audiences, he had little understanding of the significance of the ground
beneath his feet.[10]

As the art historian Stephen Whiteman has pointed out, the Qing conceptualized the landscape around Jehol as an extension of the auspicious energies
of the Long White Mountain. As we saw in chapter 3, in Kangxi's thinking
about the geomancy of his realm, the dragon vein of the Manchu's sacred
mountain, Golmin Šanggiyan Alin, or Changbaishan, was the source for the
dragon vein for China's most sacred mountain, Taishan. But Kangxi also dis-

cerned several other important dragon veins that descended from the Long White Mountain. One dragon vein flowed from the Long White Mountain to the south, where it formed the mountain ranges of Korea. Another flowed from the Long White Mountain in a more westerly direction, through the forests around the old Aisin Gioro stronghold of Hetu Ala, across Mukden, and then on to Mount Yiwulü, a mountain north of Jinzhou that was sacred to the Liao and Jin dynasties. Through an examination of Kangxi's essays and poetry, Whiteman has hypothesized that the Qing emperor envisioned that this Yiwulü line extended west for two hundred miles and emerged to inform the stunning scenery of Jehol. But that was not the end of the dragon vein's power. From Jehol, several other dragon veins spread their energies north to the Amur (through the Greater Khingan Mountains) and to the west all the way across Mongolia, where they formed the remote Altai Mountains on the border with today's Kazakhstan. Jehol was thus a "geomantic axis" that encompassed the energies of the whole empire, a pivotal place "where the dragon veins meet."[11]

Today, it is hard to sense the regal energies in this region. The Rehe palace complex in Chengde is a major domestic tourist center that welcomes millions of visitors each year, and hundreds of tourist "villas" and hotels have been built in close proximity to the old imperial pleasure grounds. Once outside the main tourist areas, rural roads open to vistas of compacted clay soil and dry rocky outcroppings. Farmers tend corn and herd sheep and cattle that graze on the denuded hills. Sandy scrub alternates with groves of trees recently planted to beat back desertification. The region is dotted with struggling heavy industry and spent mines. It is difficult to envision the current landscape as enlivened by the presence of multiple auspicious dragon veins.

It is perhaps even harder to envision the lush prehistorical landscape that ultimately gave rise to the Jehol biota. Scientists hold that during the Mesozoic era (251 million to 66 million years ago), this dry, hardscrabble landscape was covered with dense rain forests and swamps teeming with life. A series of volcanic eruptions in the region during the early Cretaceous period (approximately 145 million to 130 million years ago) dramatically altered the environment. In this "mass mortality event" that some scientists have dubbed a "Mesozoic Pompei," the area's copious wetland life-forms were instantaneously killed by poisonous volcanic gases and buried beneath layers of mud and ash. Today, the landscape is characterized by exposed outcroppings of sedimentary rock that hold the remains of the watery landscape that once was. A casual observer might miss the fine remains of snails, ferns, dragonflies,

Fig. 5.2 The "stone fish" of Jehol, a typical *Lycoptera* fossil from the Jehol hills.
JasonS/Shutterstock.com.

and mosquitoes that have been captured in the Jehol stones. But even the smallest slabs of the region's thin shale, when broken open, can reveal the vivid imprints of fish fossils, their delicate fins, tails, and even scales perfectly preserved (fig 5.2).

Well before the advent of the Jehol biota, the Jehol landscape was famous as the home of "stone fish" (*shi yu*). Stone fish was a name given to one type of fossil manifestation, usually a flat stone bearing the clear two-dimensional imprint of fish. In premodern China, there was no one generic name for the objects we would recognize as fossils: the modern Chinese word *huashi* (literally "transformed stone") was a neologism borrowed from Japan in the nineteenth century. A common term used to describe fossils was "dragon bone," although this term was typically applied to three-dimensional objects such as the fossilized teeth and bones of higher-order animals. As Sigrid Schmalzer, Carla Nappi, and others have detailed, during imperial times these bones were often imbued with magical properties or miraculous origins. "Dragon bones" were also used as an ingredient in medical formulas, where they were thought to be a particular aid in illnesses related to water retention and the kidneys.[12] The more commonly encountered imprints of shells or fish skeletons may not have been in the same category of interest as the "dragon bone," but Chinese writers had mentioned their curious existence, particularly in regions where they could be found in abundance, such as the rocky palisades to the northeast of Beijing.

It is not surprising, then, that the region's "stone fish" caught the attention of an important frequent visitor to Jehol, the Kangxi emperor. The Kangxi emperor (as well as his grandson after him, the Qianlong emperor [r. 1735–1796]) collected stone fish as part of his imperial curio collections and recorded his thoughts about the objects for posterity. In his essay "Stone Fish" from the early eighteenth century, the Kangxi emperor begins with a vivid description of the fossil that leaves no doubt as to its identity:[13]

> In the land of the Kaerqin [Mongols] there is an ashen gray stone, which, when cracked open, always contains the shape of fish.[14] The fish are golden brown, as if they were painted with *cihuang* [an arsenic sulfate pigment known in the west as "orpiment"]. Sometimes three or four fish appear, each with the shape of the scales, fins, head, and tail all complete. Each fish is a few inches long. They are no different from the fish which is now called "*makouyu*" [a freshwater fish related to carp]. With raised gills and waiving whiskers, they appear as vivid as if they are still swimming through the billows. I ordered craftsmen to hone one of the stones and set it in the frame of an ink-grinding vessel. Paired with a Songhua River Stone, it makes for an elegant object to place upon my writing desk.

Kangxi then muses on the relationship between textual and physical evidence, and offers thoughts on the process through which the fossils were created:

> I've read in the *Commentary on the Classic of Waters* and the *Miscellaneous Tales from the South Slope of Mount You* that in Hengyang [in Hunan Province] there is a "Stone Fish Mountain," where the stones also contain the bodies of fish, appearing as graceful as if they have been engraved. Also, *Casual Chats from North of the Pond* describes a "Stone Fish Gully" in Yiyang County [in Jiangxi], where there are rocks which, when split in two, reveal the mirror images of two fish, one on each side . . . Is it that these fish all lived among the stones, which pressed down upon the fish, and the fish were thereby transformed . . . ? Thus we see that the underlying principles of things can never fully be exhausted.[15]

Kangxi was a careful observer of the curious stone fish. He described its shape, size, and color so vividly that its form is immediately graspable by today's reader. The stone fish also inspired the emperor to ponder ideas far beyond the physical object itself: the nature of change, the geographical continuity of his empire, and the importance of empirical evidence. He suggested that the fish had existed in ancient seas, were buried in mud, and with the passage of time, the mud had turned to rock.[16] The stone fish from Jehol were similar to stone fish found elsewhere in the empire, in places like Hunan, which was over a thousand miles away from the imperial palaces. Several texts included mention of the fish, but firsthand observation of things was needed to either confirm or reject received wisdom.

For Kangxi, the stone fish of Jehol were one of many strange phenomena from around the empire, but it is clear that the northeasternmost region of his domain was a place that had the greatest concentration of natural wonders. For Kangxi, Manchuria was a land of mirages and miracles. Some of its mountains spewed smoke and fire, and were unlike anything found within China proper. The northeast was a land where the sun did not set, and where there was ice in the ground that never melted. The dense forests of Jilin hosted such a profuse diversity of plants and animals that many remained unknown and unnamed. It was also a place of strange transformations. There was wood that transformed into stone, crustaceans that morphed into deer. In this land where the energy of the dragon coursed under the earth, it was entirely plausible that fish could turn into rock. In Manchuria, things that could not be imagined could turn out to be real.[17]

For Kangxi, the marvels of Manchuria, including the stone fish of Jehol, were evidence that "the underlying principle of things in the universe can never be fully exhausted." Firsthand observation of the myriad things of the world demonstrated that the strange tales of the ancients could be true. One had to keep an open mind to the diversity of phenomena and not assume that knowledge was ever complete. Such open-mindedness kept the curious emperor ever on the lookout for novel discoveries, items that he could add to his collected trove of knowledge, or formed into curios that could remind him of the wonders of his homeland.

Kangxi was an astute observer of the fossils of Jehol, but there were stark differences between the emperor's wide-ranging observations and the hyperfocused, intensely analytical calculations that typified emerging European approaches to the same object. For Kangxi, the stone fish was one of the myriad marvels that emerged from the unique land where the dragon veins met. By the nineteenth century, the emperor's stone fish would become the genus *Lycoptera*, its ribs, fins, and bones counted, probed, measured, and compared with categories of numerous other "stone fish" from around the world. This marvel of Manchuria, once reworked through the networks of other empires, became a datapoint in the universal order.

The Birth of *Lycoptera* (1845–1901)

In the nineteenth century, the stone fish that graced the desks of Qing emperors became *Lycoptera*, a genus with a defined place in a "global" order of life dictated by European scientists. The birth of this most representative (and common) fossil of the Jehol biota—a small fish with a bony body and bulbous eyes—is representative of the effort of Western empires to catalog and exploit the world. Europeans took objects found in the field in far-flung locales and sent them back to laboratories in the metropoles, "centers of calculation" that channeled the networked circulation of personnel, ideas, and nonhuman actants.[18] Discovery and analysis necessitated networks of circulation between the field and the laboratory—the ability to range far and wide on the face of the earth, coupled with a more contained indoor environment, equipped with specialized references and tools designed to facilitate close observation—conditions that were facilitated by networks of empire.[19] The emergence of *Lycoptera* as an official genus is the story of how European scientific networks encompassed Manchuria, manifested through debates over minute differences in the anatomy of an ancient fish.

The first European description of *Lycoptera* was published in the Russian explorer Alexander Middendorff's epic report of his 1845 *Reise in den äusserten Norden und Osten Sibiriens*. Middendorff (1815–1894), an employee of the Imperial Russian Academy of Sciences at St. Petersburg, was one of the earliest European explorers to range from Lake Baikal to the Stanovoy Mountains, an area that skirted Qing territory but which the explorer cryptically referred to as "Eastern Siberia." Middendorff collected hundreds of objects in the eastern end of the Eurasian continent and sent them back to multiple sites in the western end. Among the specimens that he shipped back to Europe for identification was a slab of thick gray shale containing the impressions of numerous small delicate fish, found in a riverbed forty miles west of Nerchinsk. He entrusted the interpretation of these finds to Johannes Müller (1801–1858), a noted physiologist and anatomist at the University of Berlin.[20] With Müller's 1848 report on Middendorff's "Fossile Fische," the "stone fish" of the Qing emperors began their lives within Linnean taxonomy, reborn as the genus *Lycoptera*.[21]

The European scientific approach to rocky fossils was very similar to the European scientific approach to tiny flowers: they both required the application of a very unusual way of seeing in order to identify separate species. The identity of the fossil from Nerchinsk as a species separate from other fossil fish required an intensely detailed focus on the minute structures of the fish's skeletal remains. In his Berlin study, Müller noted the overall size and proportions of the fish: the head one-fourth the length of the body, the jaws small, the tail forked and symmetrical, the dorsal fin situated over the anal fin. Peering at the tiny fish under magnification, Müller counted the number of rays in the gills (eight to twelve), the number of vertebrae in the spine (forty), and the number of ribs (twenty pairs). Thus scrutinized, the fish appeared to be similar to the genus *Thrissops*, but based on differences in fin and spinal structure, Müller concluded that it was "desirable to establish a new genus for the Siberian fish." Müller created the genus *Lycoptera*, and dubbed the fossil *middendorfii*, in honor of the Russian discoverer.[22]

Later in the nineteenth century, another European observer obtained samples of an almost identical fossil fish from the very same Jehol outcroppings that provided "stone fish" for the Qing emperors. The French networks mobilized in this case, however, decided that the fossil was not *Lycoptera*. The collector was Abbé Armand David (1826–1900), a French Lazarist missionary and naturalist who encountered fish-bearing rocks in 1866 while traveling north of the Great Wall.[23] David's finds were the direct product of global imperial clashes of the mid-nineteenth century. He was not just a wandering priest—he was collecting biological specimens in China for the French National Museum of Natural History, and his journeys were sponsored by the government of the French empire.[24] David's ability to roam around China was facilitated by the recent (Anglo-)French victory over the Qing in the Second Opium War (1857–1860) and the stipulations in the Conventions of Peking (1860) that allowed foreigners to travel in the interior. David's work in the field was linked to the European metropole through continent-spanning networks. The China-based priest sent the sample of his Jehol fossil fish to the French paleontologist Henri Émile Sauvage (1842–1917), then a well-known ichthyologist and director of the natural history museum of Boulonge-sur-Mer in northern France.[25] Sauvage's observations of the fossil, published in 1880 in the *Bulletin de la Société géologique de France*, were even more exhaustive than the descriptions of Müller's report.[26] Like the German scientist before him, Sauvage counted the tiny fishes' ribs, gill rays, and vertebrae, but the Frenchman went even further in describing the size of facial bones, teeth, and the minute structures of the spine. Sauvage ultimately concluded that the fish was not of the genus *Lycoptera* but belonged instead to the genus *Prolebias*, a form of ancient fish commonly found in Europe. He dubbed the new species *Prolebias davidi*, after the Lazarist father who found it, and differentiated the

Chinese species from the European *Prolebias* by virtue of the number of its tiny gill rays. Thus, by 1880, the Jehol stone fish entered into European debates over its identity, debates framed in terms of minute measurements and expanded and mapped onto fissures of empire.

More than fifty years after Middendorff's original "discovery" of *Lycoptera*, the scientific identity of the stone fish was firmly established in the "international" Linnean order in 1901 by Arthur Smith Woodward (1864–1944), curator of the geology department of the British Museum of Natural History and the West's foremost authority on fossil fish at the turn of the century.[27] Woodward's framing of *Lycoptera* was an act not of discovery per se but of synthesis—his identification was the culmination of over fifty years of collection, analysis, and debate involving multiple networks that crisscrossed the Eurasian continent. Smith examined several fossil fish samples in the British museum's collection, along with the related scholarship from Mueller, Middendorf, Sauvage, and others. Triangulating between published descriptions and scrutiny of the samples kept in London, Smith generated the following definitive lineage for the stone fish:

Order II. ACTINOPTERYGII
Suborder IV. ISOSPONDYLI
Family LEPTOLEPIDAE
Genus LYCOPTERA

The genus was stabilized through deployment of standardized, esoteric language describing the fish's anatomy, now in English instead of the botanist's Latin:

Head large; maxilla arched, with a slightly convex deutigerous border; mandible prominent, the dentary gradually deepening from the symphysis backwards without any marked thickening; teeth minute and closely arranged. Opercular and branchiostegal apparatus apparently as in *Leptolepis*. Vertebral centra in form of delicate constricted cylinders. Pectoral fins much larger than the pelvic pair; dorsal fin small and short-based, opposite to the anal fin, which is not longer than deep; caudal fin forked. Scales completely covering the trunk, none enlarged or thickened.[28]

Finally, the identity of the fish was cemented through the addition of a small schematic drawing (fig. 5.3). This banal representation from a catalog of the

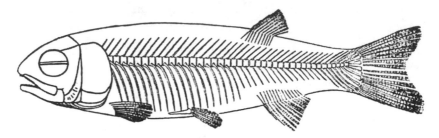

Lycoptera sinensis ; restoration, slightly less than nat. size.

Fig. 5.3 The British Museum's *Lycoptera sinensis*. Arthur Smith Woodward,
Catalogue of the Fossil Fishes in the British Museum (Natural History)
(London: British Museum, 1889), 4.

British Museum contained little of the artistic liveliness of Kangxi's fossil fish
(which "appeared as vivid as if they are still swimming through the billows")
but would allow observers to count ribs and locate fins—the essential means
to establish fish in the "universal order."

According to Woodward, the *Lycoptera* genus included Müller's *middendorfii*, along with another species, *sinensis*, samples of which had been discovered by British subjects working in China. In a parting shot at French rivals,
Woodward dismissed the attempts of Sauvage to create a separate species from
Abbé David's Jehol fossil and further attacked the Frenchman's claim that the
alleged *Prolebias davidi* was the same as that found in Europe. Woodward
concluded that the fossil fish of Jehol was separate from fossil fish found in
Europe, and related in every way to the fossil fish found in sedimentary rock
throughout northeast Asia, from the Amur River to the Shandong Peninsula.

Thus, by the late nineteenth century, the stone fish from Jehol, a corner of
Manchuria known as a place where the sacred dragon veins met, had achieved
a place in the European's universal order as *Lycoptera*. The identification of a
new genus relied on the ability to establish (and defend) the slightest anatomical differences between specimens, differences produced by acute, even obsessive forms of vision and recorded in painstaking detail through use of esoteric
standardized language. The existence of *Lycoptera* was created through new
networks that emerged with the expansion of different European empires into
the Qing empire, networks that now connected Manchuria to St. Petersburg,
Berlin, Paris, and London. At these European centers of calculation, artifacts
were stabilized and standardized and then mobilized to produce new truth
claims about the nature of reality. *Lycoptera* was the product both of meticulous vision and globe-spanning power.

Kangxi's fossil, even with its new scientific designation *Lycoptera*, seemed an innocuous, insignificant artifact; however, the stone fish of Jehol could have a significance that went far beyond its appearance. Within Arthur Smith Woodward's 1901 identification of *Lycoptera* is a telling detail: the British Museum's sample was found in Shandong by one H. M. (Harry MacDonald) Becher, a British mining engineer who worked as an adviser to Qing mining concerns in the late nineteenth century (and who later perished while prospecting for the British empire in Malaysia).[29] It was often the case that where fossils were found, mineral resources were found too. With the dawn of the twentieth century, certain fossils became important because they indicated the presence of the world's most coveted commodity: fossil fuels.

Fossil Fuels, Imperial Wars, and the End of the Dragon Vein (1866–1920)

In the late nineteenth century, the fossils of Jehol became linked to the presence of a vital energy beneath the ground: an energy that was no longer contained in dragon veins but in veins of coal. The ancient environments and geological forces that produced fossilized life-forms also produced the carbon-rich rock known as coal. The most abundant coal deposits were formed during the Carboniferous era, 359 million to 299 million years ago, when tropical climates produced vast wetlands. Buried under sediment, the organic material of these "coal swamps" transformed into peat and eventually into coal.[30] Certain fossils, known as "index fossils," can facilitate the discovery of these deposits. Index fossils are "forms of life which existed during limited periods of geologic time and thus are used as guides to the age of the rocks in which they are preserved."[31] Some index fossils are linked to eras and locations that produced coal, and thus their discovery can indicate the existence of coal seams nearby. With the emergence of coal-driven industry, fossils took on a new significance: no longer just intellectual curiosities, they became associated with the very fuel of modern wealth and power.

The first European to note this connection between the presence of fossils and the existence of fossil fuels in Jehol was the famous German surveyor, Ferdinand Freiherr von Richthofen (1833–1905).[32] Richthofen is best known for his extensive geological studies of the lands of the Qing empire, conducted in the late 1860s and published in his five-volume opus, *China Ergebnisse eigner Reisen und darauf gegründeter Studien*. Although the area was not the main focus of his survey, Richtofen traveled extensively in southern Manchuria on a path that should be familiar to us from the journeys of exiles and emperors:

from the coast of the Liaodong Peninsula to the foothills of the Changbai Mountains, through to the Shenyang area and then back to Beijing via Liaoxi and the Shanhaiguan passage where the Great Wall meets the sea. At the Liaoxi town of Jinzhou, local informants showed him interesting coal-bearing rocks, which, they said, were mined in fossil-rich shale palisades about thirty miles to the north. Given the poor quality of the coal, however, Richtofen decided to skip a trip to see the outcroppings for himself, even though he admitted the fossils that he was shown were of "high geological interest."[33]

If Richthofen had bothered to travel up into the palisades north of Jinzhou in 1866, he might have become the first trained Western scientist to encounter the formations of Yi County, an area that would become the epicenter of Jehol fossil discovery in the twentieth century.[34] Richtofen was, however, focused on the presence of productive, high-quality *Kohlenfelde*: coal fields. His surveying journey across China was fueled by an enthusiasm for the exploitation of a resource which, in his eyes, was grossly underappreciated by the Qing government. Newly freed by Opium War treaties to roam China's territory at will, Richtofen traversed the Qing empire in search of workable coal deposits. In Manchuria, Richthofen did find several promising *Kohlenfelde* at Benxi, Fuxin, and Fushun, places that would become major coal-producing areas by the early twentieth century—although, as pointed out by Shellen Wu in her pioneering study of coal in China, it is clear that Richthofen did not "discover" anything that wasn't already well known to the local population. He dismissively noted these places were primarily mined by small-scale "native" Chinese concerns in order to satisfy local heating and cooking needs.[35]

Richtofen's brush with the Jehol biota is indicative of a profound shift in the meaning and value of subterranean landscapes at the turn of the twentieth century. The German's focus on coal resulted in a missed opportunity to discover a unique treasure of fossils. Richtofen judged the Jehol coal to be of insufficient quality for modern industrial use, and thus simply not worth the effort. But soon, Manchuria would be transformed from the land where dragon veins met to a terrain that produced the world's most coveted commodity: fossil fuel. In the minds of some, possession of this sort of terrain was worth dying for.

Between 1894 and 1905, the quest for fossil fuel would become a central element in two brutal wars fought between empires for control of Manchuria. As we have seen, Russia and China had long engaged in struggles over Manchuria's natural resources, but the emergence of modernity—industrialization, mechanization, and mass armies, all powered by the force of coal—radically shifted the nature of this competition in the late nineteenth century. The Qing homeland of Manchuria (along with the Qing vassal kingdom of Korea) was positioned between two modernizing and expanding imperial powers: Romanov Russia to the north and the newly established and rapidly modernizing empire of Meiji Japan to the east. Japan's rise was particularly destabilizing for the region. Its successful centralization under a Western-looking class of modernizers had launched the island nation into the arena of imperial competition for dominance in East Asia. Japan's burgeoning railroads and industry required food and fuel, and Japan looked to the resource-rich territories of the northeast Asian mainland as a source for agricultural products and mineral resources. As it looked to the mainland, it also saw the troubling presence of looming imperial powers: a Qing empire that seemed on the verge of collapse and an expanding and potentially belligerent Russia. Strategically (and unfortunately) situated between these empires, both the Korean Peninsula and Manchuria became the flash point for wars among these three entities.

The first conflict, the Sino-Japanese War of 1894–1895, was fought between the Qing and Japan over influence in Korea. The war saw the world's first modern naval battles fought between Qing and Japanese state-of-the-art, coal-fueled battleships in the Yellow Sea near the Yalu River. Many of the land battles, however, were located in southern Manchuria: locations such as Anxi, Jinzhou, Yingkou, and Lushun at the tip of the Liaodong Peninsula. The result was a decisive defeat for the Qing, which had to acquiesce to Japanese influence in Korea and saw its sovereignty in Manchuria challenged.[36]

Spurred by the Japanese victory over the Qing, Russia rapidly pushed forward its influence in Manchuria. By 1898, a Russian-built railway cut across the entire expanse of northern Manchuria, Russian-owned mines were opened along the Amur and the Ussuri, and Russian tracks were being laid for a railway down to the southern tip of the Liaodong Peninsula.[37] Russia made further gains in Manchuria during the anti-foreign Boxer Uprising of 1900. In response to Boxer attacks along their rail lines, Russian troops fought against the Qing at places we visited in the previous chapter—Aigun, Harbin, and Qiqihar. Instead of spots for agricultural development and meetings between indigenous peoples and representatives of empires, these locales in northern Manchuria became bloody battlegrounds for yet more Qing losses. By the end of 1900, Russian forces occupied much of Manchuria.

In response to the growing Russian presence in Manchuria (and Korea), Japan launched what was to become the Russo-Japanese War of 1904–1905. Although little understood in the West, the Russo-Japanese War is seen by scholars as a presaging of the horrors of World War I—indeed, some scholars of the conflict have dubbed it "World War Zero."[38] The conflict saw the use of mass modern armies, heavy artillery, modern battleships, and mass casualties: an estimated eighty-eight thousand Japanese and over sixty thousand Russian troops perished in eighteen months of grueling fighting. These tremendous losses haunted both societies long after the cessation of hostilities.[39] Even though the Qing was not a participant in the war, the war's major battles were fought within the Qing's original strongholds in Manchuria—near the old Manchu capitals of Mukden and Liaoyang, as well as on the Liaodong Peninsula and Bohai Gulf cities of Yingkou and Lushun. The conflict between Russia and Japan resulted in the deaths of untold numbers of Chinese civilians. The Japanese victory in the Russo-Japanese War not only opened the door to Japan's colonization of Korea but also gave Japan the power to exploit Manchuria's key resources both above and below the ground. Bought at the cost of two hundred thousand casualties over the course of two wars, by the 1910s, Japan envisioned Manchuria's natural resources as a well-earned "unopened treasure house" that "contained the key to the survival of the Japanese people."[40]

The search for subterranean resources was a constant companion of turn-of-the-century military incursions into Manchuria. This was particularly true for Japan.[41] Japan had been measuring the extent of these natural resources since the advent of hostilities on the mainland. As soon as victory over the Qing became apparent by 1895, the Japanese Imperial Army sent geologists to southern Manchuria's Liaodong Peninsula to conduct a reconnaissance of the region's agricultural, commercial, and mining resources.[42] The Japanese victory in the Russo-Japanese War further opened Manchuria's resources to Japanese scrutiny. Even before the end of the war, Japanese geologists surveyed the lands and resources captured in Manchuria from the retreating Russians, including rail lines and the coal mines at Fushun.[43] Just one year after the signing of the 1905 Portsmouth treaty that ended the Russo-Japanese War, Japanese authorities published a massive survey of Manchuria's resources, the *Manshū sangyō chōsa shiryō*. The survey included a four-hundred-page section on mining products that provided details of the state of gold, silver, iron, and coal deposits in the region, all extensively illustrated with hand-colored maps of the deposits' locations, depths, and qualities.[44] By reorganizing their perspective on nature, imperial powers in East Asia produced what the geographer Bruce Braun has called "vertical territory," a new way of seeing that revisualized the inner structure of a landscape and at the same time devised new ways of calculating the economic and political value inherent in the earth.[45]

The colliery at Fushun encapsulates the complex intersections of geology, natural resources, and contested sovereignty of Manchuria's "vertical territory" at the beginning of the twentieth century. As we saw in chapters 1 and 2, Fushun was originally a strategic fortress that lay between the old Manchu power base at Hetu Ala and the first Qing capital at Mukden. Nurhaci captured the walled city in 1618 in one of his first major victories against the Ming. Its position between the Yongling mausoleum (where Nurhaci had interred the bones of his ancestors) and the Fuling tomb of Nurhaci himself meant that Fushun was at the heart of the dragon vein that nurtured both the Qing ruling house and the entire Qing empire.

As detailed by Victor Seow in his pioneering study of the Fushun mines, the discovery of major coal deposits at Fushun at the turn of the century portended the eventual unraveling of the Qing imperial dragon vein.[46] The area's coal had been excavated on a small scale by local residents for centuries, but fearing harm to the crucial dragon vein and sacred geomancy of the region, the Qing court issued prohibitions against coal mining. The Qing court's prohibitions could not stop foreign incursions into the region in search of mineral wealth. At the beginning of the twentieth century, the modernizing Qing court allowed Han Chinese investors to establish mines at Fushun, a task undertaken with a large infusion of Russian capital. During the Russo-Japanese War, Japan took over operations at Fushun. Under the control of Japan's new South Manchuria Railway Company (Minami Manshū tetsudō kabushiki kaisha), the Fushun colliery was dramatically expanded to become one of the world's largest open-pit mines. This massive gouge in the surface of the earth would eventually grow to over four miles across and over a thousand feet deep and produce hundreds of millions of tons of coal. Within years of the demise of the Qing empire, the dragon vein of the Qing imperial house was, quite literally, ripped asunder by the excavation of Manchuria's soil in the search for fossil fuels.

By the first decade of the twentieth century, the dragon vein of the Qing empire held little significance for Han Chinese elites who were increasingly disillusioned by the Manchu government. What mattered most were the mineral resources that lay beneath the ground and the rail lines on the surface of the ground that were used to transport them. Indeed, the Chinese recovery of rail and mining rights from a foreign-debt-hobbled Qing government helped spark the revolution that overthrew the Manchu imperium in 1911, and a new Han Chinese dominated republican government emerged with the "state control of mining rights as a pillar."[47] By the time the Republic of China was established in 1912, however, it was already too late to recover the largest rail lines that crossed Manchuria: the China Eastern Railway remained under Russian ownership, and the South Manchuria Railway remained a possession of Japan. The largest mines in Manchuria, particularly the Fushun colliery, were to remain under foreign control for decades. Thanks to the search for fossil fuels, the land of Manchuria transitioned from dragon vein to coal vein, from the center of Qing sacred sovereignty to the crossroad of imperial competition. Jehol would become a site at which the multiple meanings of fossils played out on a radically changed map.

The Snails of Jehol: Divisions of Empire, Visions of a Unified World (1920–1933)

In a 1923 article on fossil mollusks of China, the American paleontologist Amadeus Grabau made a very important claim about the fossils of Jehol.[48] He hypothesized that the fossils of Jehol came from a very special "horizon": a unique geological formation whose ancient past could change the way that humans understood the deep history of the earth. This moment, based on examination of humble snail shells, is considered the birth of the Jehol biota. Embedded in Grabau's discovery, however, was another layer of truth: Jehol's political history, combined with the political history of China's early republic, had profoundly shaped the way that scientists saw the fossils of this "unique horizon." Scientists such as Grabau saw the fossils of Jehol as keys to understanding the geological unity of the world. Many of Grabau's Chinese colleagues saw them as resources necessary for national survival in a world bent on violence and domination. Jehol was a microcosm of the linkages and divisions that characterized scientific communities and scientific knowledge in an age of hyperimperialism.

The conditions that brought the paleontologist Grabau into contact with the fossils of Jehol is indicative of the complex terrain of scientific collaboration in early twentieth-century China. Amadeus William Grabau (1870–1946) was a foreign employee of the Chinese Geological Survey, the new republic's premiere institution of surveying and prospecting. As detailed in pioneering studies by Grace Shen and Shellen Wu, geology emerged at the turn of the twentieth century as the science at the center of China's self-strengthening efforts, and it was seen as "an important means of resistance against imperialism."[49] China's first generation of geologists, including Ding Wenjiang (1887–1936), Weng Wenhao (1887–1971), Li Siguang (1889–1971), and Zhang Hongzhao (1877–1951), establish modern geology as a scholarly field and profession in the early years of the republic.[50] This fledgling community aspired to connect China to international scientific networks and inevitably relied on the support of foreign experts. In 1912, the Geological Survey hired a Swedish geologist, Johan Gunnar Andersson, to serve as its chief consultant, and in the first decades of the twentieth century, other foreign advisers from Europe and America were frequently involved in fieldwork and the training of young Chinese geologists.[51] By the beginning of the 1920s, the government's Geological Survey was staffed with over two dozen graduates of domestic Chinese programs, young Chinese men who formed the core of China's geological profession. The emergence of a Chinese geological community was shaped by the political realities of the time: geology was a manifestation of Chinese sovereignty in the face of imperial competition, but at its core it was reliant on foreign personnel and international collaboration.

The American paleontologist Amadeus Grabau seemed an unlikely leading contributor to Chinese geology, but global politics in the wake of World War I eventually brought him into contact with the fossils of Jehol. Born in Wisconsin in 1870, Grabau grew up in a German American neighborhood in Buffalo, New York.[52] His early curiosity about geology developed while exploring the dramatic landscapes of his childhood environs in the area around Niagara Falls. Grabau studied geology at Massachusetts Institute of Technology and then went on to achieve a PhD in paleontology from Harvard in 1900 with a dissertation on fossil mollusks. By the beginning of the twentieth century, Grabau had embarked on a promising academic career as professor of geology and paleontology at Columbia University. The outbreak of World War I changed the course of his career. Grabau maintained a zealous sympathy for Germany, a position that put him at odds with much of American society, including his employer. Grabau's outspoken support for Germany resulted in his dismissal from Columbia in 1919.[53] Always on the lookout for foreign talent, China's leading geologist, Ding Wenjiang, made an offer to Grabau to join the China Geological Survey. The fifty-year-old Grabau, stripped of his Columbia professorship and physically hobbled by painful arthritis, left the United States to become professor of geology at Peking University and chief paleontologist of the Geological Survey of China.[54]

When Grabau arrived in Beijing in the summer of 1921, he found a thriving international community of earth scientists, including Chinese, Swiss, American, French, and British paleontologists and geologists. These men frequently met in Beijing to share their findings at meetings of the Geological Society of China, established by Ding Wenjiang and his colleagues in Beijing in 1922. The society's membership list showed that most of the society's leaders were Chinese, but many foreign scholars were also included. Grace Shen has vividly portrayed the meetings, "tiffins," and "smokers" of this cosmopolitan group, where Chinese and foreign scientists delivered lectures, debated geological theories, dined, and engaged in what appeared to be good-natured "old boy" ribbing. Within the international space of the society's meetings and in the pages of its *Bulletin*, Chinese geologists became "a host of nations," interacting with foreign scientists on what appeared to be an equal footing.[55] In Latourian terms, Beijing had become a new "center of calculation": scientists who had previously been part of networks based at imperial metropoles had now joined a new collaborative community of science located in a non-Western republic. This sentiment of international goodwill, however, could not entirely avoid the political realities of colonial hierarchies and imperial competition.

A speech by the Russian geologist Eduard Anhert given at the inaugural meeting of the Geological Society in 1922 highlights the tensions that potentially lurked beneath scientific collaborations in republican China. In his high-minded speech (delivered in French, and translated into English for the mostly Chinese audience by Amadeus Grabau), Ahnert held up Russia as a shining example of scholarly impartiality that transcended nationality: "In the study of countries bordering their own native land, Russian naturalists have, like all true scholars, freed themselves from the bias of political separatism, and in the same spirit they have always accorded the freest facilities to the scholars of other lands who desired to make a study of the vast territory of Russia. For to the true man of science the earth is a unit, and cooperation is a sacred duty."[56]

While advocating what Shen has termed an "open door" policy of cooperation for science in China, Ahnert also clearly invoked the specter of imperial competition. His speech included a not-so-subtle reminder that certain parts of the vast "natural" territory of China were no longer part of the Chinese nation, and that these territories had been freely explored and exploited by representatives of the Russian empire: "In the past not a few Russian geologists have consecrated the best years of their lives to the study of those regions, which from their nature and geological history form a natural if not political entity with the territory of China . . . Because of their studies in the neighboring territories of Turkestan, Sungaria [sic], Mongolia and Manchuria as well as in China proper, Russian geologists take a deeper interest in work in China than most others."[57]

After this potentially awkward remark noting the Russian dominance over what had once been Qing lands, Ahnert ended his speech with a rousing call to international cooperation. "Explorers and naturalists," Ahnert exhorted his audience, "should form one family, linked together by the love of science. To the true man of science, the earth is a unit, and cooperation is a sacred duty." There is no record of reactions from Chinese geologists at the meeting, but it is possible to imagine the complex emotions that the mixed message of international collaboration and colonial competition must have inspired. Perhaps the earth was a unit in the eyes of the Russian geologist, but political divisions were impossible for the Chinese to ignore.

The impact of the political divisions of empire on science were most evident in Manchuria. When viewed on a map, Manchuria appeared as a sovereign part of the Chinese republic, but control on the ground was divided among multiple power holders. Chinese warlords nominally loyal to the republic in Beijing maintained government functions, fielded standing armies, and exploited natural resources across wide swaths of territory. Central rail lines, mines, and infrastructures on the Liaodong Peninsula and northern Manchuria were managed by foreign powers and policed by foreign militaries. These political realities clearly limited the geographical possibilities for Chinese geologists in Manchuria. While they ranged far and wide within the republic from the eastern coast to the foothills of the Himalayas, a line seemed to exist that dissuaded Chinese scientists from venturing into the northeast. A perusal of the Geological Survey's *Bulletin* makes this clear: in spite of the economic and geological importance of Manchuria, there was a dearth of studies on the region made by Chinese scientists, but an abundance of reports by Russian and Japanese researchers. There are occasional mentions of Chinese investigations conducted around the Japanese-owned mines at Fushun in southern Manchuria, and one Chinese geologist made a rare trip to investigate Russian coal mines along the Amur and Ussuri. Nevertheless, most reports about Manchuria's mineral resources shared with the Geological Society of China are authored exclusively by Russian and Japanese scientists.

Jehol, however, was different. Its mines and minerals were more open to Chinese exploration because of Chinese sovereignty. After 1911, Jehol emerged as a separate "Special Region" of the new Chinese republic, a sort of buffer zone positioned between the contested expanse of Manchuria and China south of the Great Wall. By the first decades of the twentieth century, the home of the emperor's stone fish had become a land wedged between empire and nation.

One of the Chinese geologists' favored destinations in the 1920s was Je-hol's Beipiao mine. Located approximately 150 miles northeast of the old im-perial summer villas at Chengde and just 50 miles north of Jinzhou, this was the region that Ferdinand von Richtoffen had passed by in the 1860s after he judged its coal resources not worth the trip. Like Fushun, the mine had been opened at the turn of the century by Qing officials who wished to expand on the small-scale "native" coal excavation in the area. But unlike Fushun, which had been taken over by Japan after the Russo-Japanese War, Beipiao remained under Chinese ownership, with investors drawn from warlord and local offi-cials. In 1921, the mine's warlord owners invited the renowned geologist Ding Wenjiang to join as managing director.[58] Ding used his position to create a "safe space" for the young Chinese geologists he had trained and encouraged them to come to Beipiao to conduct research at the mine and its geologically intriguing environs. The result was the first sustained series of studies on the geology and paleontology of Jehol by Chinese scientists.

The area around the Beipiao mine was the origin of the snail fossils that inspired Grabau to theorize a "unique horizon" in Jehol, but Grabau had not been a party to their excavation. The American professor's severe arthritis had ended his forays into the field. Instead, his Chinese students and the younger members of the Geological Survey were the ones who ventured into the field on his behalf, collecting specimens and taking extensive notes on the appear-ances of geological strata. Their careful fieldwork made Grabau's insights into the Jehol biota possible—but like the native guides who supported the work of Karl Maximowicz and many other Western scientists, the contributions of these men went for the most part unacknowledged in Grabau's work.

The most frequent Chinese surveyor in Beipiao was Tan Xichou (1892–1952), a graduate of the first class of the government Geological School (1916) and one of the most promising of the Geological Survey's young scientists. Tan (whose name was romanized in English publications as H. C. T'an) was a native of a poor village in Hebei province. The government-sponsored Geo-logical School had provided him with an opportunity for practical education that led to direct employment. Early in his career, Tan demonstrated a talent for conducting and processing field surveys. Among his first assignments was a survey of the Western Hills outside of Beijing: the resulting maps detailing the terrain, stratigraphy, and mineral resources of the area are considered the first geological maps produced solely by Chinese that were entirely in line with international geological standards.[59] Tan's forays into Jehol began in the summer of 1923 and continued throughout the decade. Fieldwork in Jehol in

Plate 1 Dragons rampant above the throne in the Chongzheng Hall,
Qing imperial palace at Shenyang. Photograph by author.

Plate 2 View from the summit of Mount Paektu/Changbai, from the PRC side.
Photograph by author.

Plate 3 Satellite image of Mount Paektu/Changbai, with DPRK/PRC border.
Google Earth.

Plate 4 View of Mount Paektu from approach on the North Korean side.
Photograph by Roger Shepherd.

Plate 5 Mount Paektu tethered to Korea by a line of *ki*. Nineteenth-century map of South
Hamgyŏng Province, Chosŏn chido, Hamgyŏng-namdo, Library of Congress.

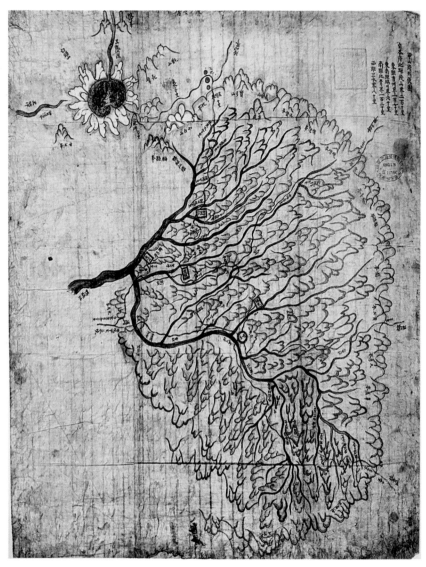

Plate 6 Mount Paektu depicted in map of Kapsan prefecture, nineteenth century. Kapsan chido, Ancient Map 61–51, National Library of Korea.

Plate 7 The "five-flavor berry" plant, *Schisandra chinensis*. Grigorev Mikhail/Shutterstock.com.

Plate 8 Botanical sample of *Maximowiczia amurensis*. Harvard University Herbaria Digital Collections 00039145.

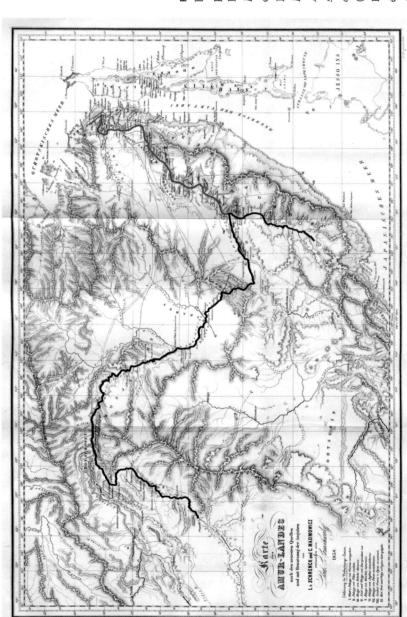

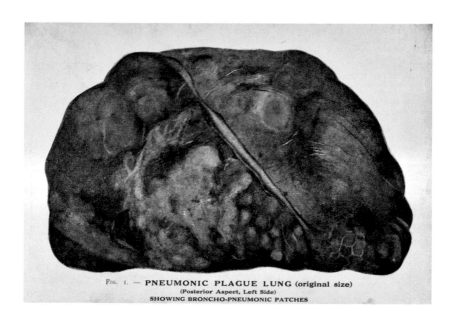

Fig. 1. — PNEUMONIC PLAGUE LUNG (original size)
(Posterior Aspect, Left Side)
SHOWING BRONCHO-PNEUMONIC PATCHES

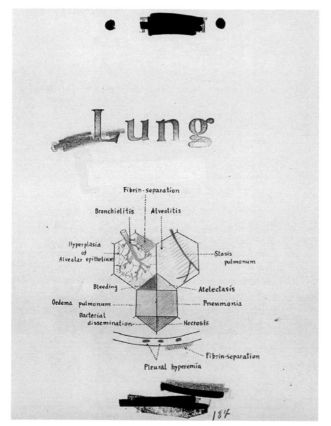

Plate 10 Illustration of an autopsied lung from a pneumonic plague victim. Wu Lien-teh and League of Nations Health Organisation, *A Treatise on the Pneumonic Plague* (Geneva: Nancy, 1926)

Plate 11 Lung autopsy schema from Unit 731's "Report of Q." Document #56-FDTS-197, JWC 253, National Archives.

Plate 12 Soviet advisers help open up the black soil of the Friendship Farm. Nongye dianyingshe, *Guoying Youyi nongchang* (Shanghai: Shanghai renmin meishu chubanshe, 1957).

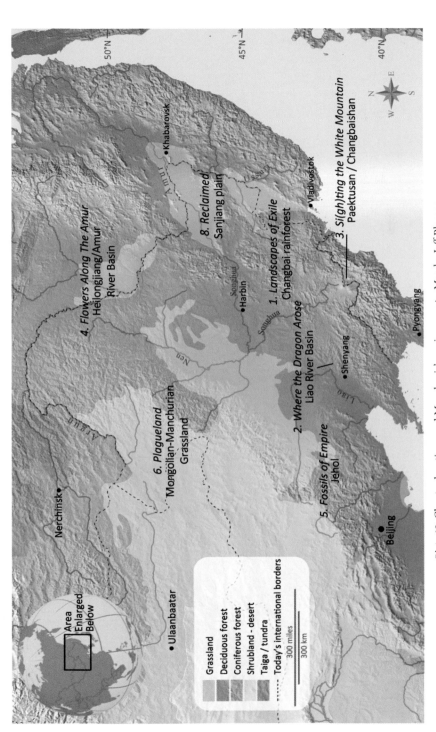

Plate 13 Chapter locations and Manchuria's environments. Map by Jeff Blossom.

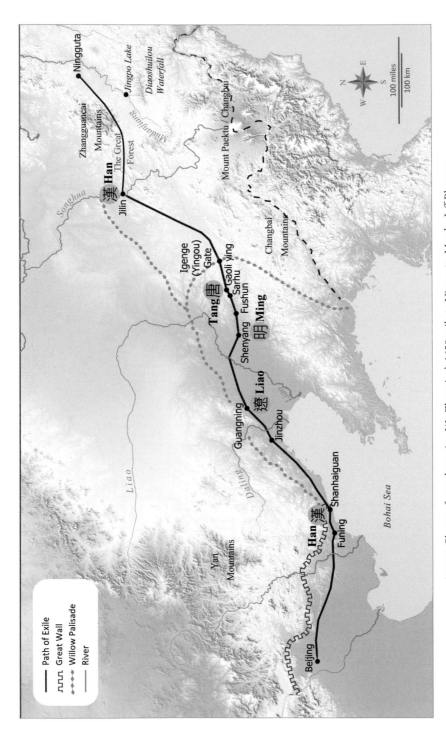

Plate 14 Ghosts of past dynasties: Wu Zhaoqian's 1659 exile to Ningguta. Map by Jeff Blossom.

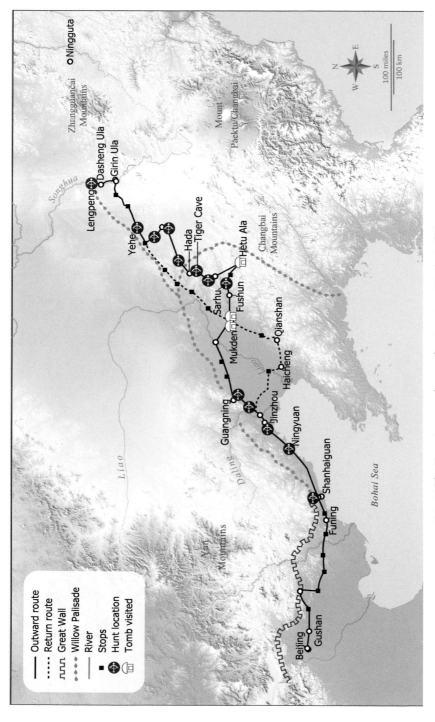

Plate 15 The Kangxi emperor's second eastern tour. Map by Jeff Blossom.

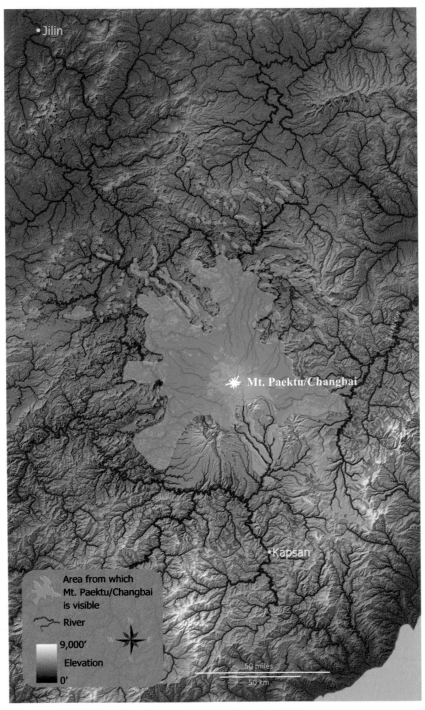

Plate 16 Viewshed map of approaches to Mt. Paektu/Changbai. Map by Jeff Blossom.

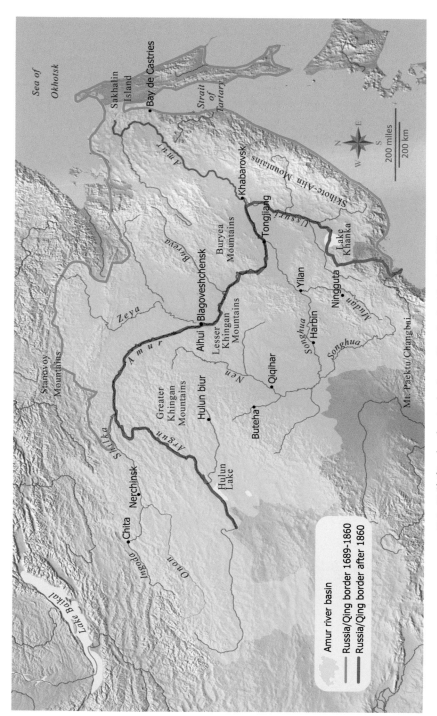

Plate 17 Shifting borders in the Amur River basin. Map by Jeff Blossom.

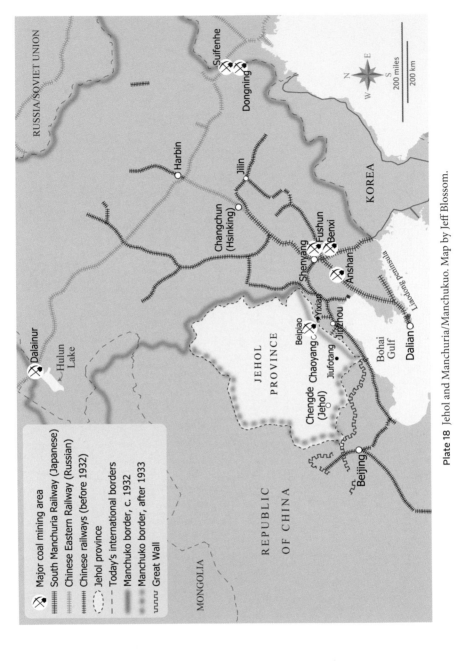

Plate 18 Jehol and Manchuria/Manchukuo. Map by Jeff Blossom.

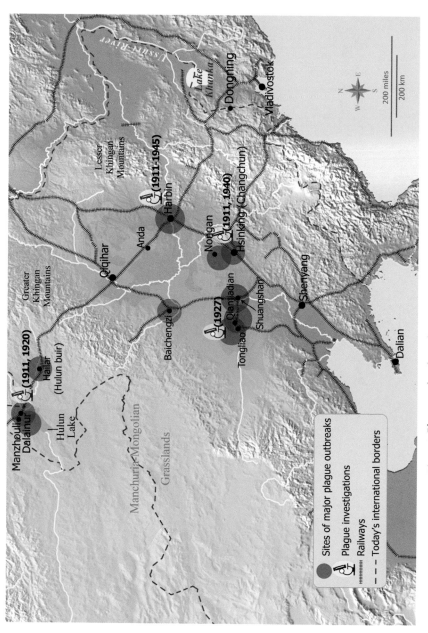

Plate 19 Plague outbreaks and investigations. Map by Jeff Blossom.

Plate 20 Locations of state farms on the Sanjiang Plain. Map by Jeff Blossom.

the 1920s was not an easy task: military struggles between feuding warlords frequently turned the region into a battle zone. It is no coincidence that Ding Wenjiang left the Beipiao mines early in 1925, citing his distaste for dealing with overbearing warlord bosses.[60] But in spite of the unpredictable violence of warlord armies, Tan Xichou continued to work for years on the geology of Jehol, sensing that the stones and fossils there held a significance that went beyond their unassuming appearances.

The snail-shell fossils Tan collected at Beipiao, together with his extensive field notes on the terrain and geological formations of the region, made possible Grabau's 1923 publication on mollusks and inspired his first glimmers into the importance of the Jehol system. Grabau does briefly mention that his conclusions were based on the copious and detailed notes taken by H. C. T'an, but these notes are not quoted or described at length: indeed, were it not for the existence of Tan's own later publications, we may not have had any sense at all of the scope and import of Tan's work from Grabau's own statements.[61]

In his paper on mollusks—the "Ur text" of the Jehol biota—Grabau does make a small gesture toward recognizing the Chinese scientist. In an image from Grabau's publication (fig. 5.4), we see a sketch of five ancient snail shells. Even though several are incomplete (their "whorls wanting"), through the application of intensive visual examination, Grabau was nevertheless able to use the fragments to establish four separate species. Two of the four species are

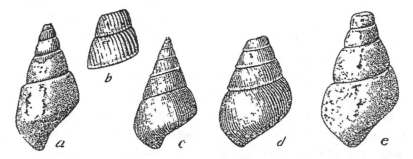

Figs. 2 a, b, *Campeloma clavilithiformis* Grabau, *a*, shell with apical whorls wanting × 2; *b*, uppermost two whorls preserved, enlarged × 6 to show the fine ribbing; *c*, *Campeloma fengtienense* Grabau, entire shell × 2; *d*, *Campeloma yihsiensis* Grabau, shell with apical whorls wanting × 2; *e*, *Campeloma tani* Grabau, internal mold of shell with apical whorls wanting × 2.

Fig. 5.4 Amadeus Grabau's *Campeloma* fossils: The dawn of the Jehol biota. Amadeus Grabau, "Mollusca of North China," *Bulletin of the Geological Survey of China* 6, no. 2 (1923): 195.

named after locations in Manchuria: *fengtienense*, a variation on Fengtian, the name for the environs around Shenyang; and *yihsiensis*, named after Yi Xian (romanized as *hsien*), a county just to the south of the Beipiao mines. On the far right, a somewhat undistinguished fossil snail shell has been given the name *tani*, in recognition of the man who found the fossil in the field. This gesture is a standard one in the production of scientific names, but it is important to note that Tan's work went far beyond simply picking up rocks. Ultimately, the contribution of Chinese scientists to the early discovery of the Jehol biota remains unacknowledged, buried beneath years of unsung fieldwork in the warlord-ravaged coal landscape of southwestern Manchuria.

These two scientists—one American, one Chinese—examined fossils in pursuit of two very different goals. Tan's work was inextricably linked to the pursuit of natural resources for the new republic. In service of this pursuit, fossils were of interest primarily for their ability to date and locate coal seams, and the snails of Beipiao provided Tan with a window into the possible location of China's hidden coal reserves. On the basis of his fieldwork, Tan proposed new theories about the dating and location of coal-bearing strata in northern China regions, including Jehol. Tan detected a novel "Cretaceous coal series" buried beneath a layer of volcanic rock, formed by plant and animal matter that thrived during the last period of the Mesozoic era, from 145 million to 65 million years ago—much later than the Carboniferous period that had given rise to the coal reserves of the United States and Europe. The lush rain-forest environments of these Cretaceous basins now formed workable coal seams of economic value.[62] Looking out over the dry, craggy terrain of 1920s Jehol, Tan could see evidence of fertile, lake-filled basins once teeming with life, life that could become fuel for his struggling nation.

The ultimate goal of Grabau's research differed starkly from that of his Chinese colleagues and students. While Tan sought resources for his own country, Grabau's eyes were constantly fixed on distant locations far beyond China. Over nearly two decades at Columbia, Grabau had become one of the world's top experts on biostratigraphy—the science of categorization and dating sedimentary and volcanic rock layers with the aid of index fossils, fossil species that are characteristic of a specific epoch in geological time.[63] Like Karl Maximowicz, who used his dissections of tiny flowers to generate data for the production of global *Florengrenze*, Grabau drew from minute distinctions in fossil structures to support his emerging understanding of another type of global geography—the location of ancient continents. Every fossil he encountered, every set of observations he received from Chinese geologists in the field, became a small point in an intricate web of data that gave him insight into a vast moving puzzle: pieces of the earth's crust shifting over the course of millions of years. For Grabau, discontinuities in the fossil record from neighboring locales (between north and south China, for example) indicated the presence of ancient seas that had once split a country in two. Continuities in the fossil record across what were now far-flung locations indicated that separate continents were once conjoined landmasses. For Grabau, knowing the fossils of one small corner of Manchuria was simply a data point in knowing the history of the entire world.

Grabau's grand visions led him to tackle the biggest theoretical conundrum of geology in his day: the question of "eustacy," or global changes in sea level that produced discontinuities in the geological record.[64] Grabau dubbed his work "pulsation theory": the idea that the cyclical rise and fall of sea levels was produced by the heating and cooling of the earth's crust beneath the ocean floor. Grabau's pulsation theory has been hailed as "the first comprehensive survey of sea-level fluctuations on something approaching a global scale" and a precursor to theories of continental drift.[65] Through twelve years of work with the Chinese Geological Survey, the fossils and copious field notes brought to him by a generation of Chinese geologists allowed Grabau to achieve his perfect vision of the earth.

Grabau introduced his pulsation theory for large audiences of scholars from around the world during the Sixteenth International Geological Conference held in Washington, DC, in 1933. Many participants from East Asia attended the meeting. A group photograph of conference participants brings to mind the tension between internationalism and empire that characterized the science of the time.[66] Grabau appears in the front row, standing in the far-left corner. Next to him stand two longtime colleagues from China—Ding Wenjiang, the founding father of Chinese geology, and Pierre Teilhard de Chardin, the French Jesuit paleontologist and longtime member of the Geological Society of China who was involved in the excavations of Peking Man in the 1920s.[67] Only a few feet away from the China contingent in the center of the front row, we find several Japanese scientists. First on the left is Kobayashi Teiichi (1901–1996), a paleontologist from Tokyo Imperial University, and like Grabau, an expert on fossil mollusks of China. To the left of Kobayashi stands Yamane Shinji (1885–1962), a professor of geology at Kyushu University and, like Grabau, an expert on the stratigraphy of China. Close by, we find Chitani Yoshinosuke, an expert on East Asian petroleum geology. These were but a few members of a larger Japanese contingent who attended and presented their work at the international congress.[68]

One wonders what interactions were like between the members of the

Chinese Geological Survey and the scientists from Japan at this 1933 conference, given the violent shifting of territory that had just taken place in Manchuria. In September 1931, officers of the Japanese Kantō (or Kwantung) army stationed in southern Manchuria to protect Japanese-owned rail and land concerns staged a coup outside of the old Manchu capital of Mukden. Within a few short weeks, all of Manchuria was brought under Japanese military control. One year after the Mukden (or Manchurian) Incident, Japan established Manchuria as the new "nation" of Manchukuo, in spite of vociferous protests from China and the League of Nations. In the spring of 1933, just a few months before the geological conference, the Japanese military launched a campaign into Jehol, wrested the province at gunpoint from Chinese sovereignty, and incorporated it into Manchukuo, which was soon to become an "empire" under the puppet figurehead "last emperor" Pu-yi.[69] Jehol, the place where the Qing empire's sacred dragon veins once met, had now become a shattered part of the empire of Japan. The next discovery in the Jehol biota was to take place within the empire's shadow.

The "Shining Crocodile of Old Manchukuo" and Other Creatures of the Japanese Empire (1933–1945)

In 1934, Japanese geologists found the complete and detailed skeleton of a four-legged land reptile embedded in outcroppings near a village called Big South Ditch in Jehol province. The fossil was relatively small, approximately fifteen inches long, with short legs, curved, clawed feet, and a long tail that had broken off onto a second fragment of shale. The paleontologist and director of the natural history section of the Manchukuo National Museum, Endō Ryūji, in a 1940 publication of the museum, identified the fossil as belonging to the obscure order of "Thecodontia," a category of crocodile-like protodinosaurs (fig. 5.5). Measurement of the skeleton's vertebrae led Endō to believe that this reptile had raised spines along its back. Rendered in an artist's reconstruction, the creature looks to us today as something like a miniature stegosaurus or a tiny four-footed Godzilla.[70] In honor of Jehol's recent absorption into the newly created nation of Manchukuo, Endō named the specimen *Monjurosuchus splendens*, or "Shining Crocodile from Monju," using the term "Monju" because, as the Japanese scientist explained, it was "an ancient name for Manchukuo."[71]

Today, *Monjurosuchus splendens* is considered one of the major advancements in the establishment of the Jehol biota, and its discovery is included

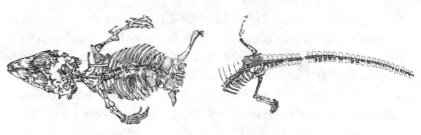

Fig. 5.5 The "Shining Crocodile of Old Manchukuo": Endō Ryūji's *Monjurosuchus splendens*. R. Endo, "A New Genus of Thecodontia from the Lycoptera Beds of Manchoukuo," *Bulletin of the Central National Museum of Manchoukuo* 2 (1940): 4–5.

as "(Endō 1940)" in a seamless chronology along with the earlier discoveries of Woodward (1901) and Grabau (1923). But everything about Endō's discovery—including the Japanese scientist's presence in Manchuria, his exploration of the terrain of Jehol, and even the names he gave to his fossils— was the direct result of the military expansion of the Japanese empire. The "Shining Crocodile" is not only a representative of Jehol's Cretaceous lifeforms; it also is a representative of the tight intertwining of natural knowledge and empire in northeast Asia. Endō's twenty-five-year career as a youth educator, Smithsonian fellow, museum administrator, and postwar employee of a Chinese public university shows a genuine commitment to the promotion of Manchuria's paleontology in an international context, but his work is ultimately inseparable from the conditions of violence and exploitation that characterized the Japanese presence in Manchuria in the first half of the twentieth century.

Endō Ryūji (1892–1968) was a graduate of Tohoku University's geology department, where he studied under Yabe Hisakatsu, a major figure in Japan's imperial-era paleontology. After achieving his PhD in 1924, Endō, like many aspiring young men of the time, found employment in the colonies with the South Manchuria Railway Company (SMRC), Japan's massive institution of colonial resource management and data-gathering on the Liaodong Peninsula. With its neoclassical headquarter buildings in the beautiful seaside city of Dairen and its administrative offices located throughout Japan's leasehold in southern Manchuria, SMRC's many research bureaus were staffed by tens of thousands of young professionals who had taken up residence in towns and cities along the rail line.[72] Endō's career path took a very different (and perhaps less ambitious) direction from that of many of his classmates. While he participated in SMRC geological surveys on occasion, Endō's primary role was as an instructor in the SMRC-run Mukden Teachers' College, where he taught geography and earth sciences to the future middle school teachers of Japan's colonial education system. Unlike other scientists from the home islands who used the colonies as fieldwork sites to further their research and publishing careers back in the metropole, Endō was a permanent resident of Manchuria and became deeply committed to its settler society. He enthusiastically promoted the study of natural history among Manchuria's community of Japanese residents and published popular works about local paleontological finds such as "The Tale of Manchuria's Fossils" and "Excursions to Jehol and Southern Manchuria."[73]

Endō's unassuming "local" career and his participation in international exchanges indicates how Japan's colonial presence in Manchuria had, by the 1920s, achieved a certain acceptance in international scientific circles, even those within China. Japanese academics in colonial Taiwan and Korea as well as geologists of the SMRC were members of the Geological Society of China, and several presented at meetings and published their findings in the society's *Bulletin*. One Japanese geologist famous for his work on the colonial possession of Taiwan (which had been wrested from the Qing in 1895) was even bestowed the honor of lifetime membership in the Chinese society.[74] Japanese paleontologists working in the empire's colonies were also welcomed by the United States. In 1930, Endō was awarded a fellowship to the United States National Museum (the Smithsonian). For two years, Endō lived in Washington, DC, examining the museum's fossil collections and observing the way that the Smithsonian functioned as a public institution. While in the United States, Endō even published a major study of Manchuria's fossils in the Smithsonian's flagship publication, the *Bulletin of the United States National Museum*.[75] Japan's ownership of rail lines in Chinese territory, its exploitation of Manchuria's largest mine, and its stationing of troops to protect these properties formed the backdrop of Japanese science, but these activities all seemed to achieve a certain guise of normalcy in the context of interwar scientific communities.

The upheavals of 1931–1933 constituted a turning point for Japanese geological research in Manchuria and were to have a significant impact on Endō's career. The Manchurian Incident and its aftermath put the Kantō Army on war footing and gave the military the upper hand in decision making in Manchuria. The right-wing officers in the military clearly saw the region as a crucial source of natural resources that had to be exploited to the greatest extent possible as Japan prepared for what they anticipated as an inevitable showdown with the West for control over Asia.[76] As Aaron S. Moore has demonstrated, technology represented a form of fascist ideology that Japan's governing elites used to spur civilian mobilization, and the military put intense pressure on Japan's scientific community to increase research for military and industrial applications in preparation for total war.[77] In spite of some resistance on the part of civilian management in the SMRC, in 1932 the company's research institutes were reorganized and reoriented toward "national defense." On paper, the "nation" that needed "defending" was Manchukuo, the puppet government of Manchuria nominally led by Manchu and Chinese collaborators, but the true driving force behind the search for resources was the Japanese military, and the logic for national defense led back to the home islands.

Perhaps no other discipline felt this new military push for natural resources as strongly as geology. With the establishment of Manchukuo in 1932, the SMRC Geological Survey became involved in several "National Defense Resource Surveys," joint explorations involving military officers, SMRC personnel, and scientists recruited from industry and academic circles in Japan.[78] Geological exploration pushed beyond the Liaodong Peninsula and immediate railway zones to encompass remote outposts in Manchuria in an urgent search for new mineral resources. One example was the series of expeditions to search for oil in multiple sites around Manchukuo launched by the Kantō Army in 1932; the army sent surveyors into the field from Dalainur on the far northwestern border with Mongolia to the banks of the Ussuri River along the border with the Soviet Union, eight hundred miles to the east.[79]

With the Japanese military's invasion and incorporation of Jehol into Manchukuo in 1933, Japanese geologists were dispatched to find new exploitable resources in the region's dry hills. In the summer of 1933, Japanese scientists made a very public trek through Jehol as part of "the First Scientific Expedition to Manchoukuo" (Daiichiji Man-Mō Gakujutsu Chōsa Kenkyūdan).[80] As the historian of science Morris Low has described, the expedition was as much propaganda as science. Decked out in the quintessential "imperial explorer" attire of white shirts, khaki jodhpurs, leather riding boots, and pith helmets, the expedition led by the Tokyo Imperial University paleontologist Tokunaga Shigeyasu (1874–1940) marched across the mountains of Jehol in search of fossils and other items of scientific interest in the company of a platoon of heavily armed Japanese soldiers. An embedded journalist and photographer from the *Asahi Shimbun* newspaper reported the progress of the expedition back for an audience in the home islands. Disregarding decades of scientific work in the area by the Chinese Geological Survey, this staged expedition claimed to "bring science for the first time" to Jehol, a "virgin land remaining untouched with scientific work." In spite of the media fanfare, the expedition uncovered little of scientific value.[81]

Other Japanese expeditions in Jehol in the 1930s, including the Jehol National Defense Resources Survey, the Jehol Coal Survey, and the Jehol Mining Resources Survey, were part of a scramble for resources to support Japan's ongoing military escalation. There was a sense of desperation about these expeditions as they crisscrossed the province, documenting the tiniest of native mining enterprises, exhaustively mapping and photographing terrain, and studying minute details of geological formations.[82] A 1937 volume summarizing the results of geological exploration in Jehol provides a remarkably exhaustive village-by-village catalog of the mineral potential of the province. A list of mineral resources details small-time mining operations in over 240 locales, in places with names like Willow Tree Hollow, Two-Road Village, and Meng Family Hut. Photographs show Chinese miners standing next to their tiny claims under the watchful eyes of Japanese soldiers, a few lumps of coal piled in woven baskets beside them.[83] Such evidence demonstrates that the Japanese military was engaged in an intense search for even the smallest evidence of potential fossil-fuel deposits and could leave no stone unturned. It was these conditions in preparation for total war that led to an important discovery in the Jehol biota.

✳

Endō Ryūji had spent the tumultuous first years of the 1930s ensconced in his study at the Smithsonian, but he could not avoid the pressures of empire for long. Soon after his return to Manchuria in 1933, Endō was conscripted into the Japanese military's National Defense Resources Survey. Fighting had barely subsided in the province of Jehol, but Endō and several other SMRC geologists were tasked with investigating rumors that oil had been found bubbling out of the earth near a Buddhist temple in the vicinity of the mines at Beipiao.[84] Endō was skeptical about these rumors. He was familiar with Tan Xichou's studies from the 1920s and had scrutinized earlier surveys of the area by past Japanese military and SMRC scouting, none of which had mentioned oil. Endō was also aware that the US Standard Oil Company had prospected in Jehol in the 1910s but had ultimately abandoned the enterprise when no substantial deposits were found.[85] In spite of the pessimistic outlook based on previous explorations, the Japanese military demanded resources, and Endō and his team marched into the war-torn region in pursuit of local rumors about gurgling oil.

Their prospecting focused on large outcroppings near a village called Nine Buddha Temple (Jiufotang), located in between the old Qing palaces at Rehe and the mines of Beipiao. The area had been well known for its *Lycoptera* fossils—indeed, the area is the same region known to Kangxi as the "land of the Kaerqin Mongols" famous for its "stone fish." While he collected beautiful examples of *Lycoptera middendorfi*, the paleontologist was searching for something beyond the fossils: Endō knew the "stone fish" could be considered an index fossil for the possible presence of fossil-fuel beds. Finally, after interviewing locals and being led to the site of the alleged gurgling oil, Endō's skepticism was confirmed: while some modest deposits of bituminous oil shale (tar sands) existed at the site, and the ground on occasion even spontaneously burst into flame, there was no evidence of any "gushers." The team had spent several difficult months in the field, traveling almost entirely on foot and in horse-drawn wagons, dodging Chinese resistance fighters, and sleeping in abandoned farmhouses, yet in the eyes of the Japanese military the expedition had been an abject failure.[86]

Endō, however, considered the expedition a paleontological success: the failed search for oil had allowed him to collect hundreds of fossils from what he now called the Nine Buddha Temple Formation.[87] Endō and his colleagues frequently returned to the area around Nine Buddha Temple for further paleontological excavations. During one large-scale excavation in the winter of 1934, an assistant uncovered the remains of a lizardlike animal, its lovely, delicate skeleton seemingly poised midstride across the rocks of Jehol. This would become, with Endō's analysis, *Monjurosuchus splendens*, the "Shining Crocodile of Old Manchukuo."[88] This fossil, considered an important stage in defining the Jehol biota, was entirely the result of imperial Japan's desperate wartime search for fossil fuels, noticed and nurtured through the efforts of a mild-mannered teacher who participated, willingly or not, in the demands of empire.

The Shining Crocodile would get an appropriately "Manchukuoan" display venue when Endō's long-held dreams of establishing a "Manchurian Smithsonian" finally came true. Three years after his return from being a visiting scholar in DC, Endō helped establish the National Museum of Manchoukuo, first in Shenyang in 1935, and then in the new Manchukuo capital of Changchun (Xinjing) in 1939. The initial geological collection for the museum came from the mineral and fossil samples that Endō had used in his geology classes at the SMRC Teachers' College. These joined other displays in the "national museum," including exhibits on natural history, ethnology, and the fine arts.

Due to the new Manchukuo government's consistent budget problems, plans for a grand National Museum building were never realized; instead, exhibits were mounted in a large department store–like edifice at a major intersection in Changchun.[89]

Museums reflect the aspirations of the societies that construct them: as the historian of science Lukas Rieppel has vividly demonstrated, even an entertaining *T. rex* exhibit could stand as an emblem of Gilded Age capitalism and American exceptionalism.[90] The National Museum of Manchukuo was designed to manifest the idea of Manchukuo as a unique and unified society that spanned the human and nonhuman worlds: a vision of a multiethnic Manchukuo "Concordia" where primitive peoples coexisted in harmony with the nonhuman environment.[91] Artifacts of "native" everyday life, such as the cradles, hunting gear, and costumes of indigenous peoples, including artifacts of the Nanai people, stood next to zoological exhibits featuring stuffed Manchurian tigers, falcons, and kaluga fish from the Songhua River. The past and the present were also united in "at a glance" exhibits that collapsed time to represent a timeless nation. Archaeological finds projected a contemporary "Manchukuo" identity into the hoary past. Paleontological exhibits did similar work. A mammoth tusk from the Pliestocene discovered in northern Manchuria was prominently exhibited along with the entire skeleton of a modern elk from the same region.[92] Museumgoers could also find delicate, intricate fossils of fish, plants, and reptiles encased in slabs of gray shale: unique fossils from the Nine Buddha Temple formation and other areas in Jehol. Among the fossils was the prototype of *Monjurosuchus splendens*, the Shining Crocodile of Old Manchukuo, its skeleton displayed in two fractured pieces, its head and torso on one slab, its impossibly long tail on another. The frantic military-driven search for war resources that had produced the fossil was not mentioned here. Instead, *Monjurosorus* stood as mute evidence of the natural wonders of Manchukuo and the scientific ability of Japanese civilization to bring such wonders to the light of day.

Endō Ryūji used the museum to broadcast his local Jehol discoveries to an international scientific community. He served as the director of the paleontological section, directing explorations and identifying fossils, and he also took leadership of the museum's official scholarly publication, the *Bulletin of the Central National Museum of Manchoukuo*. In rapid succession in the early 1940s, Endō published papers identifying fossil species that were native to his adopted home and now proudly bore the name of the new nation it had become. While Endō reached into Manchukuo's past to name his fossils, he also reached across the Pacific to the United States to demonstrate his linkages to international networks of paleontology, formally thanking his mentor at the Smithsonian, Dr. Charles Gilmore, for his "helpful assistance and criticism" in studying Monjusaurus—a gesture still possible in 1940, a year before the outbreak of hostilities between Japan and the United States. These publications would result in another step toward the establishment of the Jehol biota: R. Endō, "A New Genus of Thecodontia from the Lycoptera Beds of Manchoukuo," in the *Bulletin of the Central National Museum of Manchoukuo* 2 (1940): 1–14.[93]

Endō's dreams of a Manchukuo Smithsonian ended in 1945 with the defeat of Japan. In the chaos of the massive Soviet invasion of Manchuria in the last days of World War II, the collection of the National Museum of Manchukuo was scattered, and the Shining Crocodile of Old Manchukuo was lost.[94] In spite of this trauma, Endō continued his dedication to his adopted homeland even after the end of the war. Like hundreds of other Japanese scientists and technicians, Endō chose to stay on in a Chinese-controlled Manchuria after 1945 and joined the faculty of Northeastern University in Shenyang, where he taught classes on paleontology and geology to young Chinese nationals. Endō continued to work in Shenyang for five years, until the ascendance of the Chinese Communist Party put unbearable pressure on foreign expatriates. Endō finally returned to Japan in 1950, after spending twenty-five years of his life creating knowledge about the nature of Manchuria. He finished his career as an instructor at Tokyo's Home Economics College for Women—a quiet denouement to a life spent promoting science in the heart of empire.[95] The fossils he discovered live on as important stepping-stones in the creation of the Jehol biota, their "Manchukuoan" names the only indicators of the desperation and violence of total war that marked their discovery.

Conclusion

If, as Suman Seth has observed, the history of all modern science might be understood as science in a colonial context, then to gaze on the early fossils of the Jehol biota is to witness a concrete manifestation of this history.[96] The fossils themselves are unassuming fish, snails, and lizards, but their significance goes far beyond their benign appearance. This chapter's examination of the sequence of citations in the history of the Jehol biota—Arthur Smith Woodward's 1901 identification of *Lycoptera*, Amadeus Grabau's 1923 identification of *Campeloma*, and Endō Ryūji's 1940 announcement of *Monjurosuchus splendens*—demonstrates that Manchuria's scientific networks were insepara-ble from Manchuria's history as a violently contested space. While some of the paleontologists mentioned here may have seen themselves as participants in a harmonious circle of international scholars or even as members of a local sci-entific community, their presence in China and the functioning of their net-works were deeply imbricated in hierarchies of empire. Excavating the stories of the fossils of Jehol shines an unsparing light on these hidden connections.

Deeper excavation behind the citations reveals the emergence of China's first generation of geologists and paleontologists, men who began their ca-reers in a national terrain that was intersected, both geographically and in-tellectually, by complex webs of empire. Chinese scientists in the republic of China in the first decades of the twentieth century maintained a relatively independent space of activity.[97] Men like Ding Wenjiang were deeply involved in global networks of earth science. They took degrees in Europe and the United States, established their own national geological survey, and hosted, collaborated with, and even employed Western scientists. Yet their scientific practice in Manchuria was circumscribed by the location of mineral rights claimed by imperial powers and the dangers of warlord conflict. Chinese sov-ereignty in Jehol made the region a promising buffer zone of collaboration for Chinese and Western scientists. But as we see with Tan Xichou's unsung contribution to Amadeus Grabau's theories, empire and semicolonial status constrained Chinese science.

The fossils of Jehol further illustrate how scientific scrutiny of life was mapped onto divergent visions of the earth. The Qing emperors were care-ful observers of the intricate, delicate fish fossils found in the vicinity of their Rehe hunting grounds. But in the nineteenth century, ways of seeing fossils underwent an abrupt shift as European scientists sought to extend their "uni-versal order" to traces of ancient life found around the world. The Kangxi emperor would have seen little utility in counting the ribs of a stone fish to compare it with a stone fish from Europe or the Americas. Other eighteenth-century observers were similarly disinclined: a figure no less important to sci-ence than Carl Linnaeus complained that fossil enthusiasts of his time "made a lot of trouble for themselves in setting up a genus for each petrification" and called the close examination of fossils "agreeable but of little use."[98] By the late 1800s, however, this way of scrutinizing minute differences in fossil structures was found to be tremendously useful in ways that neither Kangxi nor Linnaeus could have imagined. Minute differences in fossil structures indicated important facts about the earth. This knowledge could be put toward two radically divergent uses. For biostratigraphers like Amadeus Grabau, index fossils suggested configurations of ancient continents. The swirls of a snail shell could inspire visions of a world with no China, no Japan, no Europe, no Asia: only supercontinents such as Gondwana or Pangaea. Read for different purposes, index fossils could indicate the presence of fossil fuels. By the early twentieth century, Kangxi's "stone fish of Jehol" had become the British Museum's *Lycoptera*, a genus that could signal the presence of Cretaceous coal seams.

This transition in the meaning of fossil fish signaled a radical transition in the meaning of Manchuria. Jehol as the "place where dragon veins meet" harkened to the centuries-old tradition of sensing dragons in the earth, energies in the terrain that gave rise to emperors. We have seen how the Qing court—its Manchu rulers, Han Chinese courtiers, and European scientific advisers—all contributed to the location and creation of the dynasty's dragon vein in Manchuria, a geological formation that emerged from the sacred Long White Mountain, rolled along mountain ridges past the tombs of the dynasty's founders, and launched from the Soaring Capital of Mukden to encompass all parts of the empire. As Manchuria joined the "century of coal," the land of the Qing dragon vein was probed and plundered by men in search of carbon-based energy to build modern nations and fuel modern armies.

A shift from *qi* to fossil fuel is perhaps not what is typically meant by the term "energy transition."[99] Setting aside the idea of the evolution from human labor to wind and water and then to steam, coal, and oil, we might see the demise of the dragon vein and the rise of coal as a certain type of energy transition, a cultural reconceptualization of the invisible energetic value held within the earth. Daniel Zizzamia has noted that the cultural power of such energy transitions frequently results in societies "formulating new visions for their world."[100] According to Zizzamia, the nineteenth-century energy transition in the United States was accompanied by visions of a lush, tropical prehistoric Western environment that gave optimism to investors and developers, "paleo-restorative dreamers" who built their vision "from the otherworldly and presumptive power of fossils."[101] Similar fantasies emerged for Manchuria as a land of boundless resources, but the product of these fantasies could be measured not only in tons of coal but also in the deaths of hundreds of thousands of humans and the literal hollowing out of the land. As Victor Seow has observed, coal production in Manchuria "served as a microcosm of modern industrial society: of hubristic attempts to tame and transform nature through technology; of the relentless privileging and pursuit of production; of the disciplining and degradation of labor; and, above all, of the intensified consumption of animate and inanimate energy."[102] The fantasies behind a scientific way of seeing underwrote some of the darkest history of bloody wars and ruthless extractions in Asia's twentieth century. The Jehol biota is a microcosm of those fantasies and the conflicts they drove.

PLAGUELAND

Pursuing *Yersinia pestis* on the Manchurian-Mongolian Grassland

The happiest, and perhaps after all, the most truly philosophic way of
studying the entozoa is to regard them as a peculiar fauna, destined to
occupy an equally peculiar territory. That territory includes the widespread
domain of the interior of the bodies of man and animals.
—THOMAS SPENCER COBBOLD, *Parasites: A Treatise
on the Entozoa of Man and Animals* (1879)

In the late spring of 1939, Yamagata Miyuki stood in "Eastern Peace" county
near the Manchukuo border with the Soviet Union, listening for the call of the
koma hohojiro (the Rufus-backed bunting, *Emberiza jankowskii*). Yamagata
was a mining engineer stationed in Dongning as part of a coalfield development survey. This area had initially been excavated by Russians, but since
1932 had been a part of the Japanese protectorate of Manchukuo. Now that
all-out war had broken out with China, the Japanese military was intensify-
ing coal extraction throughout the territory, including in this peaceful valley
seventy miles northwest of Vladivostok.

As he scoured the hills near the Suifen River for coal, Yamagata also in-
dulged in his hobby of birdwatching. For the Japanese engineer, this region
was remote and pristine, a haven of untouched nature, and yet even this idyllic
spot was firmly embedded in the modern Manchurian extraction economy.
The east-west Harbin-Vladivostok line of the Chinese Eastern Railroad ran
through the area, and mechanized coal mines had been part of the landscape
since the 1910s. In the quiet of the evening, after the drills and explosions of
the mines had ended, Yamagata strained to hear the call of the *koma hohojiro*:

"bi, bi, bi-ji, bi, bi, bi-ji-jyo," a call that, he reminded his readers, was quite distinctive from the call of the more common Siberian *hohojiro*, whose song was "bi-chi, bi-chi, chi-ro-ri-ri, bi-chi, bi-chi, chi-ro-ri-ri." Yamagata wanted to catch the rare *koma hohojiro* and raise it as a pet songbird, as he had with many other birds in the Manchurian countryside that he had called home since the beginning of the China war in 1937. Once caged, the birds added beauty and joy to his domestic life—at home, his wife lovingly took care of his pet birds in his absence.[1]

To find the elusive *koma hohojiro* required an intimate knowledge of its favored habitat, and Yamagata was a keen observer of the bird's interactions with its environment. Over the course of months of observation, he noted that they preferred plateaus and grasslands near plowed fields and were seldom found in the forests. They stayed close to the ground in the branches of bushes at a height of six feet or lower and came to the river valley only to drink water in the evenings. Repeated observations of multiple elements in the field had provided Yamagata with a masterful knowledge of the complex ecology of this one small corner of Manchuria.

Finally, one spring afternoon, at a spot about 650 feet from the river, at an elevation of about one thousand feet, in a field full of lily of the valley, peonies, and bellflowers, Yamagata found a nest of a *koma hohojiro* in the branches of a flowering apricot tree. Inside the nest were three chicks that had just hatched, their eyes not yet open. Yamagata was able to capture the female of the nesting pair first and caught the male the next day but regretted having to leave the hatched chicks to an uncertain future.[2] Yamagata ended his memoir of birding in Manchukuo with several poems in which he recounted bittersweet vistas of the Manchurian landscape and exquisite emotional encounters with dying birds.

Among these tasteful works is a poem about a very different subject, one inspired by Yamagata's professional calling, entitled "The Coal Miners and the Rat." The poem gives insight into Yamagata's approach to understanding the entanglements of the human and nonhuman environments of Manchuria. Yamagata describes the Chinese miners hard at work in the dirty, noisy mines of Dongning, their faces black as coal, their eyes and teeth gleaming white—to the Japanese engineer, the Chinese actually look as if they have "become coal" themselves. Yamagata notices a rat outside the entrance of the mine and is about to kill it when the Chinese "coolies" come running out of the mine and beg him not to kill the rat. The rat, Yamagata learns, has become their "god." Within the mine, the rats and the miners together face the same dangers: "roof collapses, methane explosions, spontaneous ignition of coal seams, floods." These are dangers that humans cannot foresee, but the rat is "wise and knows these things in advance" and so the miners feel secure as long as the rats are there. Yamagata refrains from killing the rat, and smiles to signal that he understands the "coolies."[3] By mentioning this knowing smile, Yamagata conveys that his grasp of the habitat of Manchuria's humans is as thorough as his grasp of the habitat of Manchuria's bird life. He understands the nature of Manchuria—both above and below ground—as a complex ecology that inextricably intertwined land, animals, and man.

On the other side of the Manchurian rail network in the prairies of western Manchukuo, this intertwining of land, animals, and man frequently played out in a radically different way: here rodents were not life-saving "gods" but deliverers of death. One typical example comes from the South Manchuria Railway Company Health Department annual report of 1935. In May of that year, police reports noted suspicious deaths in settlements of Shuangshan (Twin Mountains) County, an area along the newly constructed rail line linking the Manchukuo capital of Hsinking to the Mongolian grassland town of Baichengzi (White Town). In tiny villages with only a few dozen families each, villages with names like Xie Family Grotto and Apricot-Viewing Hamlet, victims had succumbed within a few days of the onset of plaguelike symptoms: high fever, headache, delirium, and swellings in the groin.

On word of these suspicious deaths, infectious disease control teams from the Manchukuo government and the Japanese military poured into the villages. Doctors performed on-the-spot open-air autopsies of the dead, plunging knives into infected buboes to retrieve tissues from which bacterial samples could be cultivated, checking to see if hearts, livers, and lungs had been infected with *Yersinia pestis*. Police blockaded entire villages, halting all traffic in and out. Suspected cases were removed from families, while exposed family members were imprisoned and subjected to daily inspection for signs of plague. To prevent the spread of plague, public health personnel fanned out to over one thousand towns and villages in the surrounding area and injected over 170,000 "Manchurian people" with plague vaccine. Villagers who attempted to escape into the surrounding fields were tracked down and captured by police. To find the origin of the plague outbreak, inspection teams also tracked down the rodents of Twin Mountains county, capturing 14,197 animals representing nine different species, including thousands of rats. Inspectors sacrificed the animals and scoured their bodies for signs of plague. Of the 14,197 animals captured, 54 were found to harbor the plague germ. In spite of these control efforts, the disease continued to sprout up in the grasslands, and the same entourage would return to this land of plague again and again.[4]

For some Japanese colonial settlers like Yamagata Miyuki, life in modern Manchuria offered a chance to contemplate the beautiful interconnectedness of nature.[5] An hour's stroll away from the eastern rail lines that ran through the Changbai rain forest, one could find fragile blossoms and lovely songbirds worthy of study and appreciation, a complex but benign ecology that not only included *Emberiza jankowskii* and *Rattus rattus* but *Homo sapiens* as well. Along the western rail lines that ran through the Manchurian-Mongolian grasslands, however, the "interconnectedness of nature" inspired fear and instigated systems of hypercontrol, all because of the presence of one species of bacteria: *Yersinia pestis*. The disease this bacterium causes—the plague— presented a serious challenge to Japanese colonization efforts in East Asia from their very inception. Ultimately, the very same rail lines that facilitated Japan's exploitation of Manchuria's natural resources also exposed humans to *Yersinia pestis*. If Manchuria was the jewel in the crown of Japan's empire, it was a jewel placed in a very precarious setting: the threat that plague could travel down the rails and devastate Manchuria's population centers seemed ever present.[6]

While Yamagata may not have had a consciousness that he was birdwatching in the homeland of a deadly pathogen, the hundreds of microbiologists, entomologists, zoologists, epidemiologists, and physicians stationed in Manchuria in the first half of the twentieth century saw Manchuria as a veritable plagueland. The knowledge these men created about the nature of Manchuria had one urgent goal: to reveal the complex invisible pathways that linked the environment, *Yersinia pestis*, animals, and man. Knowing the plague was central to knowing the nature of Manchuria. To know the plague required subjecting both the landscape and the interior of bodies alike to a penetrating visual scrutiny—a form of vision that had major implications for the identity of Manchuria itself.

＊

This chapter explores plague research conducted by scientists in Manchuria from the 1910s to the end of World War II. As we have seen from the previous chapter, these years were ones of chaos and militarism, encompassing warlord battles, the rise of Chinese nationalism, Japan's occupation of Manchuria, and the violent rise and fall of the puppet state of Manchukuo. During this time,

a distinctive form of modernity took hold in Manchuria: one that was characterized not only by aggressive resource extraction and colonial domination
but also by the perpetual threat of plague. The scientists charged with controlling plague, regardless of their political allegiances, all saw *Yersinia pestis* as
deeply embedded in a complex Manchurian ecology. This chapter traces how
modernity and its attendant ways of knowing nature transformed Manchuria
from "the land where the dragon arose" into a "plagueland," a transformation
that had devastating consequences for the region and its inhabitants.

Recent scholarship has emphasized the intersection of medicine and environment in early twentieth century approaches to the study of disease. This
approach emphasized seeing all organisms from man to microbe existing in
a complex "web of life," a web that, once adequately probed and understood
through science, could be delicately balanced and maintained through policy.
Born out of European expansion into territories from California to Uganda
to the Philippines, this ecological approach was most vigorously applied in
colonial settings.[7] Helen Tilley has described how Western researchers in
the interwar years hoped to turn colonies into "living laboratories," places
where field sciences focused on problems as they existed in all their complexity within a specific terrain.[8] This place-centered emphasis fostered an interdisciplinary approach, bringing together multiple sciences—meteorology,
zoology, bacteriology, anthropology, medicine—to grapple with disease as a
phenomenon of local ecology. This ecological approach emphasized seeing all
organisms from man to microbe as existing in a web of life. Scientists working on complex diseases such as sleeping sickness in Africa strove to make
this complex web visible by mapping multiple relationships across terrestrial
space: "fly to climate, fly to fauna, fly to flora, fly to parasite, fly to human."[9]
To visualize these relationships required combining laboratory and field sciences, understanding the interrelatedness of biological systems, and thinking
of humans as an integral part of the world of biological phenomena.

Manchuria adds a unique spatialized element to the history of human-
environment-disease interactions in the first half of the twentieth century.
Plague in Manchuria was most certainly deeply imbedded in the environment, carried by local animal vectors. However, in the first half of the twentieth century, plague in Manchuria demonstrated a frightening ability to
shift between the more familiar bubonic form, transmitted from rodents to
humans via the bite of the flea, and the rarer pneumonic form, transmitted
directly from human to human through the air. The location of the microbe
within the space of the human body determined, in essence, what the disease

was and how it was transmitted. Before the advent of antibiotics, all forms of plague had a near 100 percent mortality rate. But if the bacillus that caused plague, *Yersinia pestis*, lodged in one particular location within the human body—the lungs—it was particularly terrifying. Transmitted through the activity of breathing itself, death tolls from pneumonic plague could quickly reach into the tens of thousands.

Desperate to understand the conditions that gave rise to different forms of the plague, scientists in Manchuria attempted to visualize the spatial terrains occupied by *Yersinia pestis* by exploring animals in subterranean communities, tracing the transmission of the bacillus across the surface of the earth, and above all, determining the location of *Yersinia pestis* deep within the bodies of its mammalian hosts—rodent and human—after death. In trying to map the ecological entanglements of bacteria, vectors, and hosts, scientists in Manchuria moved seamlessly across the divide between field and laboratory, between microbes and more complex organisms, and across that most fraught divide, from the bodies of animals and into the bodies of man.

In this pursuit, scientists in Manchuria, no matter what their nationality or ethnicity, were applying standard principles of infectious disease study that had its origins in European colonial tropical medicine. As the British parasitologist Thomas Spencer Cobbold put it in his classic 1879 text (quoted in the epigraph to this chapter), pathogenic organisms were to be regarded by the scientist as "a peculiar fauna, destined to occupy a peculiar territory." This territory not only was terrestrial but also included "the widespread domain of the interior of the bodies of man and animals."[10] By thinking in spatial terms without regard for distinctions among spaces, all abodes of a pathogen could be considered "terrains" or "territories." In what Cobbold called the "happiest and most philosophic" way of thinking about disease, the surface of the land and the interior of the human body achieved a certain uneasy equivalency. This chapter follows plague scientists in Manchuria as they sought to visualize *Yersinia pestis* in a variety of such terrains: from the Manchurian-Mongolian plains to the "visceral territory" of its human hosts.

Plague science in Manchuria is often told in two distinct narratives. One narrative in a heroic mode tells the story of Dr. Wu Lien-teh (Wu Liande, 1879–1960), the much-lauded Chinese "plague fighter" of the North Manchuria Plague Prevention Service (1911–1931). Wu's Cambridge-trained expertise in modern biomedicine, coupled with his loyal dedication to the Chinese state, have made him the center of a triumphant nationalist narrative of Chinese-administered modern medicine used to save Chinese lives.[11] Another, much darker narrative characterizes examinations of Japanese colonial plague science in Manchuria. Here the focus has been on the medical atrocities perpetrated by Unit 731, the organization of the Japanese Imperial Army that developed and deployed bacteriological weapons in Manchuria and elsewhere in the Pacific region from 1932 to 1945. From their headquarters near the north Manchurian city of Harbin, Unit 731's doctors and scientists conducted laboratory research on plague using captured human subjects and unleashed biological-weapon field experiments on Manchuria's civilians.[12] The atrocities perpetrated by Unit 731 have been compared to Nazi scientific experiments and are remembered today as a central element of China's suffering at the hands of Japan: part of what some have called China's "forgotten Holocaust."[13]

By following the specific research work conducted by both Chinese and Japanese scientists, this chapter suggests that the science of both Wu and Unit 731 shared something in common: a single-minded pursuit of the location of *Yersinia pestis* across all the environments that the bacillus inhabited, from the grasslands of the Manchurian-Mongolian plains to the interior of the region's human bodies. Because of the complex ecology of the plague bacillus and the way it intersects with its Manchurian environment and with multiple hosts, plague research took on a complex interdisciplinary approach that sought to understand plague "in its place." This goal required visualizing the progress of the bacillus by tracking movements of mammalian vectors in space and on the surface of the land. As they followed the travels of *Yersinia pestis* across a given terrain, researchers produced maps that demonstrated the movement of plague from one town to the next, "cartographies of disease" that visualized the relationship between death and space.[14] As Marta Hanson has noted for disease maps in China, these disease maps served both as "analytical tools" intended to visualize the relationship between space and disease and also as "technologies of power" that legitimated various forms of colonial control.[15]

Because of the perpetual terrifying specter of pneumonic plague, researchers were compelled to follow *Yersinia pestis* into the "visceral territory" of the human body. Autopsy—the cutting open of human bodies to expose them to the view of researchers—has emerged as a primary symbol of the atrocities inflicted on China by Japan during World War II, detailed in books, films, and museum exhibitions as the very essence of Unit 731's evil. This chapter views postmortem studies as an integral part of knowing the ecology of plague, a project that required the spatial tracking of *Yersinia pestis* across all terrains. This visualization was an essential part of knowing the nature of

Manchuria, regardless of the nationality of the scientist. Like their European and American counterparts, Chinese and Japanese researchers were trained to see microbes as a "peculiar fauna, destined to occupy an equally peculiar territory," a territory that included "the interior of the bodies of man and animals." This mode of visualizing territory was part and parcel of the violent modes of modernity that engulfed Manchuria in the first half of the twentieth century.

Under conditions of intense imperial competition and total war in northeast Asia, what had been called the "happiest and most truly philosophic way" of studying disease ultimately resulted in medical atrocity, perpetrated by men who were supremely confident of their ability to know the nature of Manchuria. In trying to unravel the entanglements of bacteria, vectors, and hosts, scientists in Manchuria moved seamlessly across spatial boundaries. While there is undoubtedly an ethical line that Japanese researchers crossed—intentionally infecting human subjects with pathogens—the bacillus itself traversed multiple spaces in ways that drew researchers in pursuit. This way of seeing the interconnectedness of nature cemented Manchuria's identity as a tragic plagueland.

Visualizing the Ecology of Plague

The transmission of plague is often summarized in neat schematic images, with arrows conveying straightforward movement of bacteria from rodents to flea and then to humans. Such simple graphics not only eliminate many of the variables involved in the transmission of plague but also belie the immense and still-ongoing research that brought these invisible variables to light. The transmission of plague, when presented from a scientific perspective, appears as a set of unalterable realities. This chapter instead presents plague knowledge as the result of a process that unfolded in a specific place and time: a process that required exposing invisible connections between multiple elements within the Manchurian environment.

To understand these connections, we need to start by envisioning Manchuria on multiple scales: an aerial view of the landscape that takes in thousands of square miles at a glance, a ground-level view that looks at patterns of human movement across the face of the earth, a below-the-ground view that perceives the subterranean society of animals, a microscopic view of bacteria, and a deep visual dive into the tissues of the human body. We begin with the widest lens, an aerial view of the environment that is home to the plague, the

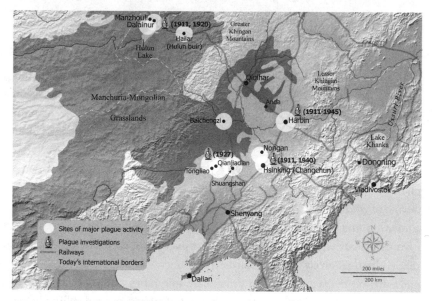

Fig. 6.1 Plague outbreaks and investigations in Manchuria, 1911–1940.
Map by Jeff Blossom. See plate 19 for color version.

Manchurian-Mongolian grassland. Fig. 6.1 helps us begin to place the plague within this significant environment.

A satellite view from hundreds of miles above the earth shows that Manchuria seems to comprise two different environments. To the east and north, near the mountainous borders with Korea and Russia, the region appears as a dense green forest. To the west, near the border with Mongolia, dull browns and sandy beige dominate the satellite image, indicating plains, grasslands, and desert. If we remove national and internal political boundaries from our view of the earth, we see that these "brown" areas within Manchuria are an eastern extension of the Mongolian plains. Sometimes known as "Eastern Mongolia," the western parts of Heilongjiang, Jilin, and Liaoning (along with parts of today's Inner Mongolian Autonomous Region) correspond to what is defined by today's environmental science as the Mongolian-Manchurian grassland, a massive ecosystem that has a unique relationship with *Yersinia pestis*.

The Mongolian-Manchurian grassland covers more than six hundred thousand square miles in northeast Asia, bridging national territories within China, Mongolia, and Russia. The ecosystem also bridges fairly significant changes in terrain. The gentle two-thousand- to three-thousand-foot slopes of the Lesser Khingan Mountains between China and Russia define the northern reaches of the region. The more rugged Greater Khingan Range runs southwest to northeast, bisecting the ecosystem and forming a sort of "rift valley" that abruptly separates the high Mongolian Plateau on the west from the lower Manchurian Plain. In spite of being a transmontane system with significant differences in altitude and accompanying river systems, the environments share enough commonalities to form a distinct ecosystem. Grasses thrive on this flat, dry land, giving ecological continuity to both sides of the transmontane divide, and the landscape presents itself as a vast plain, punctuated by undulating rivers, occasional lakes, and low hills on the distant horizon.[16]

If we view the landscape from a less distant vantage, as if from a low-flying plane, we would see that on the surface these prairies may seem relatively devoid of wildlife.[17] Once home to herds of antelope, deer, and bighorn sheep, some wild ungulates such as gazelle still exist today in isolated pockets. Carnivores such as wolves and foxes range the plains, but their nocturnal habits render them relatively invisible. Birds, including water birds and birds of prey, pause by the rivers and lakes, but many are migratory birds on their way to elsewhere. The most important fauna are out of sight—some dwelling below the surface of the earth, some too small to see.

To understand this fauna, our vision needs to penetrate the surface of the earth. Beneath the flat landscape of the plains are networks of hidden tunnels created by colonies of the many rodents of the Mongolian-Manchurian grassland. Diverse species of gerbils, voles, lemmings, hamsters, jerboas, ground squirrels, and marmots populate this subterranean ecosystem.[18] Many of these species are intensely social, engaging in cooperative strategies to seek out food sources, warn of dangers, and group together to maintain warmth during long winter hibernations. It is in part this intense sociability—and the dense populations it supports—that allows these rodent communities to act as a sustained host to the microbe that causes the plague: *Yersinia pestis*.

Worldwide, *Yersinia pestis* has been found to exist naturally within over two hundred different types of rodents and over a dozen species of rabbit. But among this staggering variety of animals, only a few species in very specific locations are considered significant long-term hosts, or reservoirs. While enzootic plague is found in specific geographical territories (or foci) across Eurasia, Africa, and the Americas, the PRC has the dubious distinction of having the largest number of different foci (sixteen) within its borders.[19] Indeed, genetic studies and computer modeling have allowed scientists to trace the origin of all plague epidemics from ancient times to the present to territories that are now encompassed by the People's Republic.[20] Most of these foci are in border regions: *Yersinia pestis* can be found in golden marmots in the highlands of Tibet, among jumping jerboas in the desserts of Xinjiang, and in red-backed voles from the rain forests of Yunnan.[21] A glance at the borders along the northeastern corner of China reveals that four closely proximate and unusually large foci converge within the Mongolian-Manchurian grassland ecosystem, making the region one of the most congenial locales for *Yersinia pestis* on the planet. We should, of course, not equate the nation of China itself with sickness. The invisible microbe exists among underground colonies of burrowing rodents that are unconcerned with political borders, and these foci extend beyond the territory of today's China. Nevertheless, contemporary plague-focus maps reveal that the borders of what was once Manchuria was a homeland of sorts for *Yersinia pestis*.[22]

Continuing with our sliding visual scale, we need a microscope to reveal the existence of *Yersinia pestis*, perhaps the most intriguing, complex, and important organism in the Manchurian environment. The pathogen that causes the plague—a word synonymous with devastating, terrifying infectious disease—*Yersinia pestis* is a member of a much more banal group of bacteria, the Enterobacteriaceae, a family that includes *E. coli* and *Salmonella*, bacteria more often associated with frequently encountered (though rarely fatal) bouts of gastrointestinal distress in humans.[23] Under the microscope, *Yersinia pestis* looks similar to *E. coli* but without the dramatic flagella: it appears as a pinkish, rod-shaped bacillus. Unlike other enterobacteria, *Yersinia pestis* is not found in soil or water but thrives in the warm environment of mammalian bodies.[24]

Yersinia pestis is uniquely capable of traveling deep into the inner visceral terrain of its host. Molecular biology studies have shown that the bacillus produces an anticlotting agent that allows it to slip along easily within the bloodstream. Other proteins allow the germ to resist the immune system's attempts at dissolving or engulfing it.[25] As it gains the upper hand against the blood's defenses, the bacteria multiply at an astounding rate. It can move into the lymph nodes, where the bacteria produce the swollen buboes typical of the bubonic plague. In pneumonic plague, *Yersinia pestis* lodges within the lungs, cramming the alveoli, causing painful, bloody cough and ultimately leading to the failure of lung function. While continuing its journeys within the blood, the pathogen can lodge and reproduce within multiple organs of the host: liver, spleen, heart, kidneys, intestine, and even uterus and testes. Once *Yersinia pestis* finds its way into the body of a mammal, whether it lodges first in the lungs (pneumonic plague), the lymph glands (bubonic plague), or continues on to contaminate the bloodstream (septicemic plague), death can be only a few days away. To survive, then, *Yersinia pestis* must be transmitted from host to host, passing across terrestrial terrain and into the internal terrain of mammalian bodies over and over.

Yersinia pestis can travel between the bodies of its mammalian hosts in a variety of ways. This diversity in "modes of transportation" makes the plague one of the most complex of all the infectious diseases. Most frequently the bacteria journeys from host to host via a bite from the tiny, agile, and ubiquitous companions to mammals, the flea. Over eighty species of fleas have been found to be able to transfer *Yersinia pestis*, but some species are more effective than others. Other types of ectoparasites, including lice, are also capable of transmitting the disease. The ability of an insect to function as a plague vector depends on a wide variety of environmental and situational contingencies, including temperature, time of infection, and the anatomy of the insect's mouth, throat, and gut.[26] A focus on insects alone misses other ways that the bacillus can move from body to body. Direct contact with the blood of an infected host can result in infection. This might happen during a violent encounter with the body of an infected host, such as a fight or during the butchering or skinning of an animal. Most importantly for our understanding of the plague in Manchuria, if *Yersinia pestis* lodges in the lungs, then the bloody sputum expelled in coughs might travel through the air and can be inhaled directly into a new host's lungs, resulting in the transmission of pneumonic plague.[27] It is through this airborne mode of transmission that plague takes its most contagious and deadly form, a form that was to haunt Manchuria for decades.

In sum, the location of *Yersinia pestis* within a body's geography creates radically different implications for disease transmission. Transmission of the bubonic plague requires a specific environment and a specific cast of characters: proximal interaction of rodents, fleas, and humans. Once *Yersinia pestis* resides in the lungs, however, all that is needed for transmission is the proximity of one human being to another. Halting the spread of pneumonic plague requires direct interventions in the most basic of human activities: moving, gathering, and even breathing. The fear of pneumonic plague drove researchers to pursue and know the travels of *Yersinia pestis* from the broad plains of the Manchurian-Mongolian grasslands to find its specific location deep within the inner recesses of the human body.

In the absence of antibiotics, the plague has an almost 100 percent mortality rate for humans, but *Yersinia pestis* in and of itself has no specific designs on human life. We may think of plague occurring when *Yersinia pestis* invades the environment of man, but in the case of Manchuria, it is more accurate to say that plague occurred when man invaded the environment of *Yersinia pestis*.

Plague and Manchurian Modernity

The same processes of modernity that eviscerated Manchuria's dragon vein also created the conditions for the spread of deadly disease. Intensive pursuit of Manchuria's natural resources at the turn of the century pushed humans into the territory of *Yersinia pestis*. While humans were always a part of the ecology of the region, by the beginning of the twentieth century, people had intruded into the territory far more often than before, and engaged in behaviors—digging into the ground, congregating in large clusters, building shelters, and killing great numbers of fellow mammals—that made them become hosts for the plague. Plague might be thought of as an ancient disease, but it was modernity that made Manchuria a plagueland.

The pursuit of coal and the penetration of railway networks into the area at the turn of the twentieth century triggered a radical change in the environment of the Mongolian-Manchurian grassland. We have seen how the Russian empire, with Qing backing, constructed the Chinese Eastern Railway in the late 1890s. Like the Amur itself, this nine-hundred-mile-long rail line connected the Pacific Ocean to Inner Asia, but in a much more efficient way. It created a direct line across northern Manchuria from the Pacific port of Vladivostock to Manzhouli, a town on the northern edge of the Manchurian-Mongolian grasslands. West of Manzhouli, the railroad continued northwest into Russian territory and linked up with the Trans-Siberian Railway, thus linking the Manchurian plains to Moscow and on to Europe. Rail lines constructed by the Japanese empire extended this network from the western grasslands into the Korean Peninsula. The Japanese-owned South Manchuria Railway then linked the grasslands to maritime East Asia, channeling the movement of goods and people from the Manchurian plains via rail and then steam to Tokyo, Hong Kong, Singapore, and points beyond. Manchuria's rail lines also connected directly with Chinese-owned rail lines to the south, forming a link to the metropolises of Tianjin and Beijing and the densely populated North China Plain (To visualize this network, see the rail lines detailed in figs. 5.1 and 6.1).[28]

As we saw in the previous chapter, the pursuit of land, coal, and railway rights in Manchuria sparked the conflicts that would lead to the Russo-Japanese War, the subsequent victory of Japan, and the devastating rise of the Japanese empire on mainland Asia. But imperial competition, resource exploitation, and the sudden growth of railways also facilitated an invisible epidemiological disaster. Rail lines penetrated into plague-endemic areas, while the excavation of coal led to the burgeoning of towns and increased human population density, thus furthering the opportunity for *Yersinia pestis* to find a home within *Homo sapiens*. These phenomena could not remain local. Within the short space of a few years at the turn of the twentieth century, the once-remote northern stretches of the Manchurian-Mongolian grassland ecosystem became connected to a global transportation network. The entire world was now linked, through coal and steel, to the home territory of *Yersinia pestis*.

Coal was not the only natural resource that opened the gates to plague. As William Summers and others have detailed, the expansion of rail lines brought together two powerful forces that converged on the grasslands in the early twentieth century: the global demand for fur and the labor of hundreds of thousands of sojourners who sought to escape poverty by supplying this demand.[29] With the drastic decline of wild fur-bearing animals as frontiers closed around the world, more and more humble species were sought to satisfy a global demand for fur. This included the groundhog-like Siberian marmot, or tarbagan, a burrowing mammal similar to a groundhog that was native to the Mongolian grasslands.[30] Once the railroad penetrated their territory, the resulting "harvest" of tarbagan was astonishing: an estimated one million skins were shipped out of Manzhouli in 1910 alone.[31] But this indiscriminate slaughter of the animals—the mass skinning, gutting, and butchering of the host body, the splattering of its blood—released the *Yersinia pestis* bacterium from the confines of its primary host to infect the bodies of a new host: man.

The devastating Manchurian pneumonic plague epidemic of 1910–1911 and the world's response to it has been detailed by numerous scholars.[32] The first cases of plague became apparent among tarbagan hunters around the town of Manzhouli on the Russia-Mongolia-China border in November 1910. The Chinese New Year of 1911 brought fear and panic as a terrifying illness swept through northern Manchurian cities. One day individuals would appear healthy, then they would suddenly come down with high fever and headache. Victims would develop a violent, bloody cough and die within a few days. Fleeing death, Chinese trappers traveled the rails for points east and south, bringing the illness with them to railway towns in central Manchuria. The city of Harbin, a newly established rail hub divided into Russian-controlled and Chinese-controlled sections, was particularly hard hit.[33]

Manchuria's position as a contested borderland made management of plague particularly fraught. Russian authorities fought plague in territories along Russian-controlled rail lines, and Japanese authorities tried to control plague in the territories adjacent to their rail lines. Korea had just been occupied as a formal colony of Japan, bringing the full force of a Japanese government to Manchuria's neighbor. British, French, and German interests were rapidly expanding in treaty ports throughout northern China, and even the United States had been expanding its economic interests in the region. The Qing still governed the vast majority of the empire's territory, but foreign powers seemed poised to expand their political control in the name of plague control.[34]

Facing threats to its autonomy, the Qing government rushed to establish its own organization to contain and study the plague in areas under its jurisdiction. A Cambridge-trained MD from the British-controlled Straits Settlements of Malaya, Gnoh Leen-tuck, known by his Mandarinized name Wu Lien-teh (Wu Liande), was appointed by the Qing court to be the head of the newly created North Manchuria Plague Prevention Service (NMPPS). During his many years in Europe, Wu had studied pathology with Sims Woodhead, tropical medicine under Ronald Ross, and bacteriology at the Institut Pasteur.[35] Wu was a British colonial subject, fluent in the King's English and with little facility in spoken Mandarin Chinese; nevertheless, the Qing government eagerly hired him to oversee its new military medical college and called upon him to become the empire's first "plague fighter." Wu became a central figure in plague research and control, ultimately hosting an international conference of researchers in Mukden to share information gathered from the plague experience.

Wu's major contribution to plague research was his discovery of the pneumonic nature of the 1910–1911 epidemic. At the onset of the epidemic, Japanese and Russian researchers had assumed that the plague was transmitted through the typical rat-to-flea vector. Wu's careful work with autopsy and pathology led him to believe that the lung was the seat of disease for this plague, and thus it was an airborne threat.[36] As described by the historian of medicine Sean Hsiang-lin Lei, Wu wielded the microscope to make visible the location of the infection in plague victims, a discovery with important implications for plague-control strategies: humans instead of rats would become the focus of public health interventions. Wu's discovery also had important implications for medicine in China. Chinese medicine, with its theories of *qi* and focus on individual treatment, had proved ineffective for managing the plague. Only Western medicine, with its microscopes, laboratories, and sense of discrete, visualizable pathogens, could isolate and identify the microscopic cause of disease, and thus direct government action.[37] Through visualizing the plague,

Wu had helped both to establish Western medicine in China and to establish China's fledgling modern medical community on an international stage.

By the spring of 1911 the plague had subsided, but not before it had caused the death of an estimated sixty thousand people. In a few short months after the end of the epidemic, the Qing empire would be overthrown by Han Chinese revolutionaries: men who embraced science and industry as the sole foundation for modernity. Just as these revolutionaries saw the development of mining industry as a basis for a modern republic, so was the development of the medical sciences seen as a necessary cornerstone for national sovereignty—a way to overthrow China's reputation of being the "Sick Man of the Far East."[38] As the Qing dynasty gave way to the Chinese Republic, *Yersinia pestis* remained. While it never again struck with the same deadly force as it did in 1911, plague occurred in the Manchurian-Mongolian grasslands with worrying frequency during the republican era. As coal-fueled changes transformed the region, the threat of plague intensified: manifesting with varying degrees of virulence, emerging and then inexplicably disappearing at different times and in different locales, causing different symptoms in its victims, and happening in the presence (or notable absence) of different potential animal vectors. The remarkable complexity of plague ecology provoked numerous questions about its origins and modes of transmission.

Like the plague, imperial competition for Manchuria continued after the demise of the Qing. The complex ecology of the plague played out on a complex political stage. For Chinese researchers, investigating plague was a crucial exercise in national sovereignty. For Japanese researchers who came to Manchuria in ever-larger numbers, first with the South Manchuria Railway and then with the Japanese establishment of Manchukuo, investigating plague was central to the entire project of empire. But in spite of their different political motivations, both Chinese and Japanese researchers pursued the same goal: to reveal the invisible webs that bound man and *Yersinia pestis* together on the Manchurian Plain.

Wu Lien-teh and the Body-Microbe-Environment Web

Wu Lien-teh, the British subject who had emerged as a heroic "plague fighter" at the behest of the Qing court, continued to fight and study plague for China under the new republic. After the end of the 1910 epidemic and the demise of the Qing, the North Manchuria Plague Prevention Service constituted the regional public health authority in Chinese-controlled areas of Manchuria:

running hospitals, doing public health education and outreach, working with local warlord authorities, and collecting vital statistics. During times of plague outbreak, they enacted procedures to halt the plague: conducting house to house searches, isolating cases and contacts, inspecting passengers at rail stations, overseeing the disposal of corpses and the disinfection of houses. Plague outbreaks also provided crucial moments and materials for another key aspect of their work: conducting field studies and laboratory experiments to reveal the invisible webs that bound plague to the nature of Manchuria.

Wu approached plague ecology as a multilevel, multiscale *terra incognita*. The grasslands abounded in multiple species of rodents: Which ones were most susceptible to plague? What were their symptoms? How could you tell if an animal had plague? Plague bacilli could be transmitted in multiple ways: through the air, via contact with blood, via the bites of ectoparasites. Which vector or mode of transmission was responsible for which form of plague? How readily could the bacillus be transmitted in these various ways? Were insects besides fleas responsible? Scientists also probed *Yersinia pestis* within the environment of the body. What were the seats of the disease, the nature of the lesions? What made some infections pneumonic, others bubonic? Did their course through the body depend on the mode of infection? These numerous questions generated a complex matrix of observations and experiments, ranging from field to clinic to lab, all designed to visualize the invisible webs linking environment, bacteria, animals, and man.

Wu's research began soon after the great plague ended, when he and his colleagues returned to the site of the original outbreak a few months later to reinspect the environment. In what Wu proudly called "the first scientific experiment sent out by the Chinese government," researchers from the NMPPS traveled the China Eastern Railroad five hundred miles from Harbin to Manzhouli in the summer of 1911 to investigate claims that tarbagan were again dying from plague in the vast grasslands on the border with Russia (Manzhouli can be found in the upper left corner of the map in fig. 6.1, and Wu's investigations are indicated with a telescope icon).[39] Wu traveled with railroad cars that were especially outfitted with mobile labs, hoping to bridge the gap between the laboratory and the field.[40] They also hoped to be able to probe the connection between humans, animals, and the environment at the moment when an outbreak was occurring.

NMPPS researchers first attempted to map out the plague-specific links between mammals and their environment. At their first stop in the frontier town of Manzhouli, Wu's team surveyed the local inns in which human-to-human plague transmission had taken place during the previous winter. The inns were ramshackle structures made of mud and wooden planks, or crude cellarlike dugouts with living quarters underground. The research team measured and photographed the crowded sleeping quarters, mapped the space, calculated the density of human bodies, and supplemented their calculations with estimations of the smell—"indescribable, being made up of a mixture of foul breath, the vapours of the old dirty fur garments and decomposing pelts which were lying alongside the men."[41]

From the habitat of humans, the scientists then turned their attention to the habitats of the tarbagan. They studied the complex environment of tarbagan burrows, extensive tunnels that ranged like city streets beneath the surface of the Mongolian grasslands. The scientists exposed this hidden world to visual scrutiny. They detailed the size and shape of the openings of the marmot's burrows, noted the presence of sand, piled rocks, and feces outside each "door." The team then stripped the topsoil to expose the extent and depth of the networks, the structure of the tarbagan dens, and the number, sex, and approximate age of the animals they found within their nests. They also noted the complex connections among tunnels and the existence of "backdoor" escape routes. They expressed their findings in photographs and line drawings that mapped the extent of these subterranean cities. By exposing the hidden world of the tarbagan, the NMPPS scientists hoped to create a clearer picture of how these rodent "societies" functioned as reservoirs for the plague.[42]

Wu's team then shifted from exposing the animal's external environment to probing the animals' internal spaces. Along with this shift in terrain came an increase in violence. With the help of skilled Russian trappers, the team captured forty wild tarbagans and transferred them from the field to the railcar-laboratory cages. One of the main goals of Wu's research was to establish the normal body temperature of marmots. In humans, the first sign of the plague was a dramatically elevated temperature. But since a baseline normal temperature for marmots had not been established, it was difficult to know what constituted an elevated temperature for a tarbagan. The team devised a technique for taking tarbagan temperatures: one team member would hold the hapless rodent immobile with a combination of wires and tongs while another inserted a thermometer into the rectum of the marmot and took a reading. The scientists took multiple readings at different times of day in an attempt to establish normal baselines. Photographs of this work show a placid, well-equipped field laboratory poised in the middle of a vast Mongolian prairie. In between taking the marmots' rectal temperatures, Wu and his colleagues even found time for afternoon tea.[43]

However, probing the marmot in the field required the application of force and violence, and the animals resisted it. Scientists dutifully recorded the details of the marmots' struggles as they were captured, restrained, splayed, and probed. These observations make for gruesome reading. Researchers noted the animals' cries, whether or not they "resented provocation," described their convulsions and paralysis of limbs, and observed as marmots "attempt to escape using forelegs only, hind legs being dragged."[44] Given the stress-inducing nature of the experiments, it is difficult to say if the researchers actually established a "normal" body temperature. Since most animals expired after the course of a few days of temperature taking, the experiments primarily established a narrative of marmot death in the absence of plague.

After death, the researchers continued to probe the animals, conducting postmortem examinations to detect lesions and establish a cause of death:

> Tarbagan No. XVIII: "attempted to escape while being brought to camp in sack," convulsions while lying on ventral surface, appeared to be biting at cage in attempt to escape. Temperature 107.6, had convulsions while temperature was being taken, numerous fleas. 4:30pm gave a few gasps and died. Post-mortem, nothing abnormal found.
>
> Tarbagan No. XIX: Young animal. August 23rd killed and cooked.[45]

The NMPPS research of the 1910s, conducted in the absence of any active *Yersinia pestis* infection, began the process of visually pursuing *Yersinia pestis* across the multiple borders between the environment and the interior of the mammalian body. To achieve this, the researchers engaged in what the historian of science Robert Kohler has called a "field physiology," crossing the borders between field and lab in order to understand the ecology of plague.[46] For the tarbagans, this process spanned the borders between life and death: scientists scrutinized the terrain of their underground habitats and probed the interior terrain of the animals' bodies after they were sacrificed. The scientists of the NMPPS would need to wait another ten years before they could cross over another border into the bodies of men.

✻

In 1920–1921, the pneumonic plague returned to Manchuria. In the winter of 1920, reports out of Dalainur near Manzhouli indicated that plague had revisited the Sino-Russo (by then the Sino-Soviet) border area where it had begun back in 1910. Predictably, the disease spread along Manchuria's rail lines to Harbin, Changchun, and Shenyang, and it eventually made

Fig. 6.2 Plague and modernity: Grappling corpses near coal mines at Dalainur, 1921. Hong Kong University Library, Special Collections, "Organisation of Second Pneumonic Plague Epidemic, 1920–21." Unpublished album U 614.42518 M26 o. Reproduced by permission of the University of Hong Kong Libraries.

its way east all the way to Vladivostok. Photographs from Dalainur taken by Wu and his colleagues during the plague outbreak show how railroads and coal had transformed even this remote location on the fringes of the Manchurian-Mongolian grasslands. One image shows plague workers collecting corpses from a stark, frigid field (fig. 6.2). They wrangle the bodies with primitive tools: hooks attached to long poles. The workers' outfits—dark cloaks with peaked hoods—seem like something from a medieval scene. But the horizon behind the workers represents a Manchurian modern: full of tall smokestacks, warehouse complexes, and mine shafts. At Dalainur, the radical change brought to Manchuria by modern resource exploitation and political violence in the early twentieth century was also reflected in the lists of the dead: in addition to one thousand Chinese workers in the coal mines there, plague victims included hundreds of White Russian soldiers who had escaped into China after being defeated by the Bolsheviks.

The presence of numerous human victims during the active outbreak in 1920–1921 allowed scientists to add clinical observations to their field and lab work, giving them new opportunities to pursue *Yersinia pestis* across multiple borders.[47] This work was facilitated by the addition of a new formal research space: the plague isolation unit at the NMPPS Harbin Hospital. Harbin was a logical place to establish a research headquarters for the Plague Prevention Service. As a central hub where Manchuria's north and south rail lines met, Harbin was a crossroads not only for northeast Asia's cosmopolitan populations (Russian, Chinese, and Japanese)—as the great epidemic of 1910–1911 had demonstrated—but also for *Yersinia pestis*. Sketches of the layout of the NMPPS Harbin Hospital show that its interior environment made it an ideal setting for plague research, with spaces designed to facilitate connections between medicine and the laboratory, between clinical observation and the observation of the interior of bodies. The immediate proximity of patient wards to assay equipment allowed for the bacteriological studies of effluvia from living patients, while the proximity of beds and autopsy rooms provided a specialized space for pathological studies immediately upon the death of the patient. Even the physical progress of the disease in sick patients became an object of scrutiny—the quarantine wards were equipped with high windows that allowed physicians to safely observe the entire course of the illness as their patients lay dying.[48]

With these facilities in place, the plague bacillus itself became an object of observation, and plague victims functioned as a window onto the bacillus. Researchers sought to visualize the invisible paths that the microbe traveled through the air and across objects. How long could the bacillus last outside of the body of the host? How far could it travel? How might it journey from one host to another? To answer these questions, NMPPS doctors obtained bacilli directly from living plague victims by having the patients expectorate into petri dishes. The sputum was then placed in different artificial environments that mimicked conditions in natural environments: on cotton, wool, or soybean cake; exposed to sunlight, cold, dryness, and various disinfectants, including Lysol, alcohol, carbolic acid, and lime. Each medium was then scrutinized under the microscope for the presence of *Yersinia pestis*. To understand how plague might travel through the air, petri dishes were placed at different distances in the room with a coughing patient to see how far their sputum could travel. While pneumonic plague was understood to be spread through coughing, scientists sought other possible means of travel away from the human body and attempted to culture plague bacillus from the clothing, money, and even the urine of plague victims: urine specimens were collected from a total of thirty patients, twenty-three from the living, seven from the dead.[49]

Researchers also sought to visualize invisible pathways of infection between hosts. NMPPS researchers assumed that animals infected humans, and experience with outbreaks indicated that humans could infect one another. But was it possible for humans to infect animals with pneumonic plague? To test this proposition, the doctors placed healthy guinea pigs in open buckets within the isolation wards together with plague victims and shifted the buckets' positions from the floor, to an adjacent bed, and then to the same bed as the patient. These experiments created macabre scenarios of dying patients lying close to rummaging rodents. Sometimes the guinea pigs escaped their confines as buckets were overturned by patients in fever-induced delirium. While these experiments directly tested the question of human-to-animal transmission, they clearly were designed to understand a larger, more ethically fraught question: the distance required for airborne transmission of plague from one infected human to another healthy one. Of a total of fifty-five guinea pigs thus placed, five died of plague.[50]

NMPPS researchers mapped the travels of *Yersinia pestis* from host to object and from host to host, but they also sought to map the entire journey of the bacillus within the body of the host. This meant infecting experimental animals with *Yersinia pestis*, observing the manifestations of disease symptoms from the outside, and then opening the interior surfaces of the body through autopsy after death. To accomplish this process, Wu and his colleagues first attempted to infect marmots through direct inhalation. This required the construction of a makeshift but brutal apparatus to confine the animal and expose it to a *Yersinia pestis*-laden suspension. As described by Wu: "The animal was firmly strapped upon the prepared stage with the nose held inside an iron muzzle. It was then covered with an oblong metal box without a bottom and having a small circular aperture at the head end for the introduction of the nozzle of a spray. An emulsion of the culture was sprayed from a graduated cylinder fitted with a fine atomiser."[51]

The goal of the experiment was to follow the *Yersinia pestis* into the interior of the marmot body by conducting postmortems of aerosol-infected animals, but the struggles of animals against the iron muzzle frequently resulted in bloody hemorrhages in the very organs of greatest interest to the researchers—the nasal passages, trachea, and lungs. Instead of the anticipated massive devastation of *pestis*-inundated organs, the infected animals displayed minimal growth of the bacillus in the respiratory tract. Wu concluded that marmots, when infected through inhalation, developed a sort of chronic and possibly asymptomatic plague that bypassed the organs of respiration.[52]

This conclusion of "silent carriers" was backed up by experiments that Wu conducted with the apparently healthy contacts of human plague victims. These were relatives and friends of those who had died of plague, now forcibly quarantined in an isolation ward to see if they too would develop symptoms of the disease. While they waited, tested twice a day for fever, these men and women were made to expectorate into petri dishes. Researchers then made cultures from the expectorate and injected it into healthy guinea pigs. To Wu's surprise, the guinea pigs succumbed soon after the injections, even as the human subjects remained healthy. Postmortem examination of the guinea pigs revealed the cause of death to be the plague.[53] In reporting these experiments, Wu did not seem to consider the possibility of contamination. Instead, he concluded that both humans and rodents could be asymptomatic carriers of plague. For those concerned with plague control, the idea that the Manchurian environment contained invisible carriers of the disease was an absolutely terrifying proposition.

The scientists of the NMPPS had tried to make visible the invisible travels through space of the plague germ: from the hosts into the environment, in the space between hosts, and then deep into the bodies of victims. Wu's experiments were like an attempt to unravel Ariadne's thread, clues to the complex unknown pathways linking *Yersinia pestis* to man through the maze of Manchuria's environment. In following these threads, the scientists crossed the borders between animals and humans multiple times and from multiple directions, coming very close to (but never crossing) what appear to us now as ethical borders that should not be breached. In many ways, these border-crossing experiments, designed to expose the webs of disease, only deepened the mysteries of the origins, transmission, and nature of the plague in Manchuria.

*

The last major plague outbreak that Wu and his team encountered occurred in 1927–1928 around Tongliao, a city on the edge of the Mongolian grasslands approximately 140 miles northwest of the old Manchu capital of Mukden/Shenyang. This outbreak was, in essence, a different disease: while most of Wu's career had been spent researching pneumonic plague, the bodies of the victims in this outbreak all manifested swollen lymph glands, or buboes, marking this epidemic as bubonic plague. Bubonic plague indicated that transmission was different—not from human to human via the air as was the case with pneumonic plague, but rather from animal host to human via an insect vector. The time and place of this outbreak were departures as well. While previous epidemics had originated along the Sino-Russo-Mongolian border in the dead of winter, this plague originated further south, during the summer, in an area that was closer to Manchuria's vital population centers. Confronted by vector-carried bubonic plague in a new environment, the NMPPS generated a complex series of studies combining epidemiological mapping, rodent natural histories, clinical observations, insect-vector experiments, and postmortem pathology observations. The result was a tour de force of visualizing the spatial webs that entangled nature and man on the Manchurian-Mongolian grasslands.[54]

Upon hearing of the outbreak, Wu and his colleagues traveled along rail lines to this barren flat landscape, a poor region that was home to Mongol herders and Han Chinese agriculturalists who were attempting to eke out a living on the dry plains. As they approached Tongliao, the scientists of NMPPS began by surveying the geography of death at multiple scales.[55] They first charted the progress of the disease through cities and towns linked by the railway: Baichengzi, Fuyu, Changchun, Nongan. NMPPS then focused its work on Qianjiadian (Qian Family Store), a small town between Changchun and Tongliao that had suffered the highest proportion of deaths in the region. The scientists arrived when the town was in the throes of the plague and bodies of victims were still lying where they had died. Wu and colleagues mapped the locations of the inns where travelers had died and the locations of the bodies within the interiors of those inns. They made detailed descriptions of the buildings' architecture, noting rat holes in the walls and thatched roof construction. Unlike pneumonic plague outbreaks, where mapping the movement of infectious humans was of utmost concern, here the main concern was visualizing the terrain that brought rodent and human into proximity.[56]

The 1927–1928 bubonic plague outbreak also gave NMPPS the opportunity to focus on the question of insect vectors, an endeavor that linked field observations with studies with the laboratory. Upon arriving in Qianjiadian, the researchers' attention was drawn to the presence of fleas. In one inn where they discovered eight bodies, the physicians "immediately started to search for human parasites in the clothes and blankets of the deceased, an ordeal that lasted over one hour." In another inn that held thirteen bodies, they found "fleas jumping about merrily, eager to attack any fresh-comer." The researchers examined four fleas under the microscope and were surprised to find that they were human fleas, *Purex irritans*, and not the rat flea, *Xenopsylla cheopsis*. Indeed, even the rats the researchers examined were infected not with rat fleas but "human fleas."[57]

As they had during the pneumonic plague outbreak in 1921, in 1927 NMPPS placed guinea pigs in the same rooms with dying plague victims to see if the animals would catch plague—this time, the transmission would not be from expelled sputum but from the mobile fleas (and perhaps, Wu mused, even lice) present on the bodies of the human victims. This experiment ended when the guinea pigs "were killed by a delirious patient and eaten by dogs," a horrifying scenario that suggests the perils of attempting to cross the border between animal and human in a live plague situation. In spite of this brief setback, Wu concluded that the human flea was an important vector of plague, as they had caught them "red handed" in the act of transmitting infection from human to rodent, a scenario that admitted the possibility of flea-vector transmission between humans as well.[58]

While Wu and his colleagues had solved to their satisfaction the question of the responsible vector, they were unsure of the original host in the Tongliao outbreak. Wu described Tongliao as a place of unknown threats and potential carriers: he even seemed suspicious of the regions' camels.[59] NMPPS scientists conducted extensive field research on nine different species of burrowing rodents found in the region. First, they constructed a natural history of different varieties of gerbils, mice, hamsters, bats, susliks, jerboas, and voles, observing the habits of each in the wild. Scientists recorded the location, size, and construction of their burrows, noting, for example, that jerboas had the cleanest sleeping chambers, hamsters created the most complex underground tunnel networks, and susliks had the habit of digging into grave mounds and stealing scraps of cloth off of human corpses.[60]

NMPPS researchers then captured healthy specimens of these species and transported them to the laboratory, introducing animals of the field to the world of artificially induced plague. A series of experiments on hamsters were designed to see the operation of various modes of infection: skin contact, subcutaneous injection, contact with the conjunctiva (eye injection), ingestion, and inhalation. Echoing the previous "guinea pig in a bucket" human-to-animal transmission experiments, NMPPS researchers placed infected hamsters in cages together with healthy hamsters to see if transmission would take place. Researchers stumbled upon the possibility of infection through ingestion of body parts when healthy Hamster 17 "partly ate up the corpse of infected Hamster 11 by gnawing off all the ribs and clavicles" and subsequently died. Following up on this intriguing lead, scientists forced Hamsters 18 and 19 to "devour the spleen of the plague-infected Hamster 16." Only one of the animals died as a result.[61] Through a bewildering and often macabre variety of methods, researchers introduced the plague bacillus into the bodies of hosts, replicating the journeys within and between species they imagined could happen in the natural environment of Manchuria.

Once the pathways of animal-to-animal transmission were understood, researchers mapped the locations of the bacillus in the interior of the animals' bodies. The numbers assigned to each hamster in life were maintained in the autopsy data, ensuring direct linkage between the specific experimental procedures for each animal and the resulting state of the viscera. Given the importance of pneumonic plague, researchers paid particular attention to the involvement of lungs in infected animals, describing the overall appearance of the diseased organ, detailing which lobes were affected, noting the abundance or paucity of bacilli in the aveoli and blood vessels, and measuring the amount of "pink colored exudate" that issued from the dissected lung.[62]

NMPPS also scrutinized and mapped the bodies of human victims at Tongliao, from the surface to the interior. Charts of "bubo incidence" represented the frequency of skin-level swellings vis-à-vis the geography of human anatomy: inguinal, femoral, axillary, cervical, popliteal, cubital.[63] To accompany his tabulation of buboes locations, Wu included full-body photographs of plague patients. These men, women and children stand facing the camera, their clothing pulled entirely away from the body to reveal the location of buboes, whether in the armpit, neck, or groin. Some point to their buboes, as if prompted to do so. Most stare directly at the camera: some look off into the distance. The inclusion of detailed photographs of the naked dying, while jarring to our sensibilities, was standard procedure for journals of the time, bolstering NMPPS claims to objectivity and scientific modernity.[64]

Wu's team also plunged their vision into the inner recesses of human victims' bodies. Most autopsies in past outbreaks were conducted in the field and were thus of necessity hurried and cursory: a swift plunging of a knife into the spleen of victims to retrieve enough tissue to confirm the presence of *pestis* germs under the microscope. Wu himself performed several dissections during the 1920–1921 outbreak of pneumonic plague at Dalainur. As described by Wu, not only did the "very cold weather and strong wind" rush the researchers; far more perilous was the threat of violent resistance from onlookers who were horrified by the sight of the procedures.[65] At Qianjiadian in 1927, Wu and his colleagues devised a way to do more leisurely open-air autopsies of human bodies out of sight, at a spot near the cremation pits where the corpses would ultimately be disposed. Dressed in double layers of rubber overalls, with hoods and goggles covering their faces and eyes, and two pairs of long postmortem gloves covering their hands, Wu and his colleagues carefully dissected fifteen humans outdoors. Photographs taken by NMPPS personnel capture the macabre scene of bodies lying prone on the ground while researchers sliced open ribs and flesh (fig. 6.3).[66]

Stained and preserved on slides, samples taken from these "primitive" field autopsies were taken back to the more "civilized" environs of Cambridge,

Fig. 6.3 Field autopsy performed by North Manchurian Plague Prevention Service personnel, Tongliao, 1928. Hong Kong University Library, Special Collections, "Plague Research in Southern Manchuria, Tongliao, 1928." Unpublished album U614.42518 W95p. Reproduced by permission of the University of Hong Kong Libraries.

England, where Wu and his mentor G. Sims Woodhead conducted extensive observations on the pathology of the plague. Some aspects of the identity of each human victim at Tongliao, including age and place of death, were known to the researchers, and as a result, each autopsy had a specific number assigned to it. NMPPS autopsy reports provided detailed observations of multiple organs, beginning with observations of the interior of buboes, then the trachea, heart, lungs, liver, and spleen, noting swellings, color, presence of blood, and under the microscope, presence or absence of *pestis*. The greatest attention was lavished on the lungs, since researchers were determined to see if bubonic plague, spread by the bite of fleas, could somehow become pneumonic plague and spread through the air. Using the standard pathology terms he had mastered at Cambridge, Wu created extraordinarily lengthy descriptions of the organs, precise and vivid narrations of the changes that *Yersinia pestis* wrought while making its home in the interior of the human body.[67] This was a lexical cartography: Wu's research generated dozens of pages of postmortem prose that was only rarely accompanied by images.

Of all the results of Wu's research, one image in particular, however, does stand out—a "true to life size" color illustration of an autopsied lung that Wu saw fit to feature in two of his major publications: his 1926 *Treatise on Pneumonic Plague* and the 1934 *Memorial Volume* commemorating the history of the NMPPS (plate 10). This large, graphic illustration of the diseased lung, a deep necrotic purple with blotchy red patches, was proudly featured as the frontispiece of both works, standing as a symbolic culmination of all of Wu's investigations and a visual embodiment of plague itself.

Taken as a whole, the scientific research of the Chinese NMPPS was a multilayered, multiscale project to visualize the complex ecology of plague. This was at its core a project of mapping the invisible journeys of *Yersinia pestis* across the Manchurian terrain: from its home within the bodies of rodents living in tunnels beneath the Mongolian grasslands to contact zones between different mammals (rodents and humans, or human and human); categorizing and tracing the various routes taken by *pestis* aboard different vector vehicles, whether tiny insects or bodily effluvia; and ultimately mapping the multiple terminuses of death, from the locations of expired bodies where they lay on the surface of the earth to the colonies formed by *pestis* within the interior of bodies. This was not a unidirectional journey but a three-dimensional transportation network of multiple pathways linking multiple hubs. To reveal this network required the transcendence of borders between the field, the lab, and the clinic. It also required thinking of humans and animals in a similar mode: both were "territories" that the plague bacillus could occupy. The task of knowing the nature of *Yersinia pestis* in Manchuria required the unveiling of a complex ecology of death of which humans were an inextricable part.

Making this ecology visible required a certain blindness to the brutality of modern scientific research. Wu Lien-teh often expressed pejorative views of the humans of Manchuria's plague zones.[68] In Wu's accounts, Manchuria's frontiers were inhabited by "coolies," soldiers, gamblers, morphine addicts, and other undesirables, men whose primitive and ignorant lifestyles seemed to predispose them to the ravages of plague. They also seemed remarkably ungrateful for the heroic efforts made by science to save them. The stories of body snatching, poisoning, and "germ warfare" that circulated among the populace were, in Wu's eyes, the irrational ravings of the uneducated and uncivilized. While he critiqued the Manchurian underclass for its overactive gothic imagination, Wu seemed immune to the horror lurking within his own actions: family members separated from their homes at gunpoint and taken off to isolation wards, never to return; white-garbed physicians who crept up to corpses on the street, plunged in a dagger, and cut out pieces of spleen;

autopsies that left bodies flayed and disemboweled; macabre experiments involving humans and animals; and, when all was done, mass cremation fires that "burnt fiercely in the open" due to the "fatty constituents of the cadavers" and left behind "only white crumbled bones as residue."[69]

The Japanese Empire and the Creation of Manchuria as Plagueland

At the same time as the North Manchurian Plague Prevention Service sought to understand and control plague in areas under Chinese control, an extensive Japanese-dominated epidemic control network also struggled against plague in areas under its jurisdiction. Epidemic prevention was central to Japanese efforts to establish and protect its massive economic and political investment in Manchuria. Indeed, from the beginning of Japan's expansion in East Asia at the turn of the century, it was immediately clear that the success of its colonial enterprise relied on control of the unseen microbes that threatened resource extraction and settlement.[70] Controlling the plague required knowledge of the routes through which *Yersinia pestis* intersected with man. Japan pursued this knowledge in Manchuria through extensive research networks, a tremendous investment of capital and personnel, and through the application of violence. Even as they tried to control the plague, their presence made Manchuria into a land of plague.

The management of epidemic disease formed a central element in Japan's emergence both as a modern nation and as a modern empire. Public health infrastructure and the training of medical personnel was a priority of the Meiji government from its inception. A Japanese medical and bacteriology profession emerged as early as the 1870s with the return of young men trained in European (and especially German) universities and laboratories. The career of Japan's own "microbe hunter," Kitasato Shibasaburō (1853–1931), who competed with the Swiss scientist Alexandre Yersin to discover the plague germ in Hong Kong in 1894, was a product of this early development. Kitasato, a protégé of the famous German bacteriologist Robert Koch, had isolated bacilli from the blood of Hong Kong plague victims, but a misidentification of the bacillus meant that the laurels went to Yersin, after whom the plague germ was eventually named.[71]

The story of Kitasato's endeavors in Hong Kong reminds us that plague was present at the very birth of the Japanese empire. The globe's third plague pandemic emerged from its mainland China origins in 1894, the same year

Japan battled the Qing empire for supremacy in East Asia in the First Sino-Japanese War. The plague arrived on Taiwan just as Japan gained the island as its first formal colony after the Qing defeat. As Liu Shiyung has demonstrated, when faced with the challenge of plague and other epidemic catastrophes, Japan's colonial administration on Taiwan developed a highly comprehensive and highly invasive policy that measured, inspected, and injected the bodies of residents with the goal of making the tropical island into a "hygienic zone" safe for the habitation of Japanese citizens.[72]

Plague also shadowed Japan's colonial presence in Manchuria from the very beginning. As we have seen, in 1905, Japan's victory in the Russo-Japanese War led to Japan's acquisition of the South Manchuria Railway and its "Kantō" (Kwantung, or "east of the Great Wall") territories on the tip of the Liaodong Peninsula. Only five years later, the great pneumonic plague epidemic of 1910–1911 broke out along rail lines in north Manchuria. As Robert Perrins has wryly observed, the railroad that formed the very foundation of Japan's empire in Manchuria had, ironically, become the one thing that most threatened the health of its colonial project.[73] Through strict quarantine and draconian inspection conducted by the South Manchuria Railway Company and the Kantō police, Japanese-administered cities such as Dairen (in Chinese, Dalian, the far southern terminus of the South Manchuria Railway, located at the tip of the Liaodong Peninsula) managed to avoid the worst ravages of the Great Manchurian Plague.[74] In spite of being spared, the lesson of the outbreak was deeply felt: Japan's existence in Manchuria required constant vigilance against epidemic disease, particularly Manchuria's most devastating disease, plague.

From the beginning of Japan's presence in Manchuria, the establishment of medical institutions was considered paramount—a strategy put into place by Gotō Shinpei (1857–1929), a physician trained in Germany who was the first civilian governor of Taiwan and subsequently served as the first director of the South Manchuria Railway Company (SMRC).[75] Under Gotō's guidance, the SMRC began to construct a biomedical and life sciences research infrastructure in southern Manchuria that would eventually grow to include two agricultural experimental stations, a central laboratory, the Mainland Institute of Science, two institutes for the prevention of cattle disease, a network of thirty hospitals, and two hygiene institutes modeled after the Institut Pasteur. The largest of these was the SMRC Hygiene Institute in Dairen, established in 1926 to be the crowning glory of Japanese colonial hygiene in Manchuria.[76] This complex covered over nine acres and included laboratories, vaccine and serum production rooms, dissection rooms, rooms for experimental animal cages, a stable for larger animals, pastureland, and housing facilities for employees. The institute's work was divided into six different sections, including bacteriology, pathology, chemistry, and vaccine research. A separate wing of the Dairen Hygiene Institute was dedicated entirely to research on plague. While other epidemic diseases were far more common for Japanese in Manchuria, plague was a threat that seemed to be embedded in the land itself, one requiring constant scientific vigilance. Japan's colonial science institutions produced a large corpus of medical works that took up the question of health and disease in Manchuria.

Nowhere was the perception of Manchuria as "plagueland"—a singularly diseased environment—more apparent than in these writings on colonial hygiene. Taken collectively, these publications express an almost Hippocratic obsession with the potential effect of Manchurian air and land on Japanese settler health. Even though these writings are clear in their embrace of the germ theory, local environment nevertheless appears as a crucial determinate of disease. Japanese settler hygiene texts begin by enumerating the factors of Manchuria's climate, including temperature, humidity, barometric pressure, sunlight, rainfall, and snowfall. This data was mapped onto the territory; numbers pegged to discrete points on the land would help settlers navigate the specifics of Manchuria's multiple environments. Charts provided readers with average high and low temperatures for different cities and towns. Detailed isobars winding their way across maps of northeast Asia showed variability within the Manchurian climate between regions of plentiful rainfall and arid deserts. Some texts even offered comparative insights based on the invisible lines of latitude: while Manchuria's position on the globe might seem relatively benign (Manzhouli, for example, is revealed to be on the same parallel as Paris), authors were unanimous in warning settlers that there was nothing at all benign about Manchuria's climate.[77]

These writings also suggested that there was something in the air and land itself that threatened health. While epidemics were clearly attributed to bacteria, authors noted that diseases fluctuated with the change of seasons. Manchurian summers gave rise to gastrointestinal diseases, such as dysentery, amoebic dysentery, typhoid, and cholera. The frigid and sunless winters of Manchuria resulted in numerous respiratory illnesses, such as bronchitis, pneumonia, and tuberculosis. Diphtheria, measles, scarlet fever, smallpox, and typhus were other common scourges that made seasonal visitations, seemingly shifting with the change in the air. Where Qing observers once saw a place infused with the *qi* of kings (*wangqi*), Japanese observers saw Manchuria as a "land of many evil epidemic miasmas."[78]

Although these texts divided disease into infectious epidemic diseases (or *densenbyō*, literally "diseases spread from person to person") and diseases endemic to the land (or *chihōbyō*, literally "diseases of the local land"), the borders between diseases viewed as particular to a local environment, diseases spread by insect vectors, and diseases passed between individuals was quite fuzzy.[79] There was also concern about *fūdobyō*, a category Sakura Honda has discussed as encompassing both local environment and local culture (literally "diseases of custom and soil"). For Japanese researchers, these diseases in Manchuria included "Manchurian typhus," Keshan disease, Kashin-Beck disease, "Mongolian" goiter, Kala-azar, and hemorrhagic fever.[80] Japanese researchers probed the environment in an attempt to understand the origins and modes of transmission for these new diseases. While authors over and over again noted that the overwhelming majority of cases of these "diseases of the local land" were "Manchurians" who were "native" to the land itself, there was still a palpable underlying anxiety about the webs of connection cast by the Manchurian environment. In the absence of a clear sense of etiology, the rational, scientifically guided planning for mass settlement of Manchuria was without a clear means of defense, and Japanese settlers might still fall victim to these mysterious entities.

While settler hygiene texts were universal in seeing dangers lurk within Manchuria's climate, texts varied in their appraisal of the dangers presented by "Manchurians" themselves. Some authorities placed the people of Manchuria and the people of Japan on the same trajectory of civilizational progress, with Manchurians lagging somewhat behind in hygienic advancement. In these texts, the "natives" of Manchuria may have been less healthy, but they were not singularly diseased or beyond hope. Through technology, science, and education, colonial *eisei* would result in uplift for both "Manchurians" and Japanese alike, with a natural evolution toward a unified level of hygienic advancement shared by both.[81] By contrast, other authors focused on the "Manchurian" people as an unmitigated threat. Over and over they warned Japanese settlers that chains of infectious diseases transmission invariably led back to natives: if not to local denizens of Manchuria themselves, then to carriers who were among the huge waves of migrants who regularly arrived in Manchuria from points south in China proper, like Shandong and Shanghai. The primary way to avoid illness, then, was to limit contact with Chinese people. In these treatises, "Manchurian" humans were natural reservoirs of disease, inseparable from the diseased environment of Manchuria itself.[82]

Of all the myriad diseases that threatened Japanese settlers in Manchuria, the most terrifying of all was *pesuto*, or plague. Plague combined all the elements that made Manchuria a dangerous, insalubrious place. The disease quite literally originated in the land itself, among animals that silently burrowed beneath the earth, scurried across fields, and crept into the rafters of houses. While the disease seemed to originate in specific areas, it was not contained to them and could potentially travel to penetrate Japanese residential enclaves. And while it was usually understood to be a vector-borne disease, transmitted from animals to humans by the bites of fleas, plague in Manchuria had taken a far deadlier and more terrifying pneumonic form, spread in the air from human to human. Given its varying etiology, plague was simultaneously a *chihōbyō* (local or endemic disease) and a *densenbyō* (infectious epidemic disease)—a danger that emanated from the invisible connections between the land and its people. Japanese researchers strove to clarify these connections under conditions of increasing suspicion, anxiety, and violence.

Japanese research on the 1927–1928 bubonic plague in Tongliao offers interesting insights into Japanese attitudes about the connections among humans, the environment, and the plague in Manchuria. In the summer of 1928, researchers of the newly established SMRC Hygienic Institute in Dairen were dispatched to investigate reports of suspicious deaths in the area of Tongliao. This was the site of Wu Lien-teh's extensive clinical and autopsy studies detailed in the previous section. Since some of the towns and villages hit by plague were located within the Japanese-administered rail-line easement, Japanese researchers were able to conduct what were, in essence, competing studies on the same plague epidemic. Personnel who joined in the investigation included the Dairen Hygienic Institute bacteriological section chief Andō Kōji, and Kurauchi Kikuo, the institute's expert on serums and vaccines.[83]

Unlike the NMPPS researchers at Tongliao who photographed the bodies of victims, focused on detailed observations of buboes, and made extensive autopsies, the Japanese studies of the 1927–1928 Tongliao outbreak are relatively devoid of clinical observation. Instead, researchers focused on what Christos Lynteris has called the "ethnographic plague": cultural and behavioral factors that were seen as contributing to plague incidence or plague prevention.[84] Andō and his colleagues mused that the region had probably been plague-free when it was primarily a Mongol region, since, he argued, the local Mongols had no understanding of a disease like bubonic plague and the Mongolian language had no word for the disease. The researchers described how influxes of Han Chinese traders and agriculturalists had altered the nature of the Mongolian plains, which in turn may have prompted the emergence of plague.

Although they noted that the Han farmers were "industrious," in their opinion, the Han, along with the local Mongols, led "a rather primitive existence" in small, miserable villages. The larger towns fared no better in the opinion of the Japanese scientists: while they provided a few more amenities, their inhabitants were similar to village dwellers "in their stupidity and primitiveness." For Andō, this "primitiveness" was manifest in the native's response to the disease itself: superstitious and frightful, they left their sick unattended and their dead unburied, behaviors that rendered them subject to the gaze (and the knife) of the more rational (and clearly more advanced) Japanese scientist.

Japanese researchers probed bodies at Tongliao, but they were the bodies of thousands of rodents that were suspected of being carriers of plague. Here the Japanese scientists placed themselves in direct competition with Wu Lien-teh's research. Wu had concluded that the common rat was the host for the deadly Tongliao outbreak, but Japanese researchers were suspicious of this conclusion, as they "could not find a single suspect among 5,000 rats that were examined." To pinpoint the responsible mammal (and, it seems, determined to outperform Chinese plague researchers), Japanese researchers purchased a mind-boggling total of 57,216 small mammals from locals who were incentivized with monetary rewards to scour the landscape for possible suspects. The Japanese investigators performed autopsies on over five thousand mice and rats, hundreds of voles and shrews, and over eight thousand susliks, searching the tiny bodies for swollen lymph nodes at the internal intersection of the limbs and abdomen: telltale signs of the plague. Of the thousands of mice and rats examined, none was infected with plague. Instead, the plague appeared in a total of thirty-six (of eight thousand) susliks. After cutting open tens of thousands of animals, the culprit, it seems, had finally been identified.[85]

In spite of this monumental dive into the bodies of rodents, Japanese researchers remained suspicious of "Manchurians" themselves. Even though they confirmed that the suslik was a reservoir of plague in the Tongliao area, they suggested that humans were the primary vector and theorized that humans infected the susliks, and not the other way around. Attacking the position of Wu Lien-teh's Manchurian Plague Prevention Service, Andō asserted that "the condition in which the uncivilized natives live, with innumerable blood-sucking insects such as flea, louse, and fly, etc., make superfluous the assumption of Dr. Wu that the rat epizootic preceded human outbreak." Ultimately, Andō concluded, "most of the infections seem to have been carried by the ignorant natives themselves."[86] In the debate over which was the more dangerous element—the land of Manchuria or the people who lived there—

Andō positioned himself firmly in the camp that feared humans most, and saw them, like animals, as carriers of the plague.

*

The Mukden Incident of September 18, 1931, marked the beginning of fifteen long years of brutal war between China and Japan—a war that would eventually lead Japan into its fateful global involvement in World War II. As we saw with the case of the earth sciences from the previous chapter, Japanese medical and public health sciences in Manchuria became militarized and intensified during the war, their activities characterized by increased manpower, increased coordination, and increased invasiveness. The Mukden Incident also significantly changed the administrative space of plague control and research. With the Japanese takeover, Wu Lien-teh was forced to leave Manchuria and folded the North Manchurian Plague Prevention Service into the Chinese Health Ministry in the Kuomintang-controlled south. Japanese authorities, once responsible for the health of only the southern tip of the Liaodong Peninsula and a narrow band of land on either side of the SMRC tracks, now presided over the entire environment of Manchuria. The establishment of the client state of Manchukuo in 1932 added another organizational layer to Manchuria's biomedical infrastructure. In addition to the SMRC network, the Kantō leased-territory government, and the Kantō army, the "imperial" government of Manchukuo moved to take over private hospitals and medical schools as well as Chinese municipal health bureaus, forming a state apparatus that included some "Manchurians" (Chinese) together with Japanese physicians. Although ostensibly under the leadership of the last Manchu emperor Puyi and a "Manchurian" administration, all levels of the Manchukuo government were overseen by representatives of the Japanese Imperial Army.

The medical infrastructure of Manchuria burgeoned under occupation, but the health of its people plummeted as Japan extracted harvests to feed its military, dispossessed residents of their land to make way for Japanese settlers, brutalized labor as it plundered mineral resources, forced thousands of women into prostitution, and fostered opioid addiction on a mass scale: the combined sum of which the historian Marc Driscoll has labeled "necropolitics," an "unrepentantly exploitative and profoundly inhumane modernity" in which life became (quoting Fanon) "a permanent struggle against an omnipresent death."[87]

Plague was central to the "profoundly inhumane modernity" of Manchuria during the war years. Plague outbreaks occurred in multiple locations almost every year of the Japanese occupation of Manchuria from 1932 to 1945. In each outbreak, deaths ranged from as few as one hundred up to several thousand. Death by plague was indeed an "omnipresent threat." Rather than associate this increase in plague with the nature of Manchuria, it is important to see the role of Japanese imperialism. The violence and chaos of Manchuria during the occupation years, combined with the intensive military and administrative surveillance over Manchuria's space and people, directly created this measurable increase in the plague. Japan's colonial authorities may have conceptualized Manchuria as a "land of many evil epidemic miasmas," but it was Japan's exploitation of the land through violent colonial occupation that made Manchuria into a plagueland.

Under the occupation, plague appeared in its old haunts with alarming frequency, especially in the far northwest around Manzhouli and in and around Tongliao and bordering counties. Most disconcerting to the new authorities, however, was the intensification of plague outbreaks in Nong'an, a poor county in the liminal environment between the Manchurian-Mongolian grasslands and the Songhua River valley whose name, ironically, means "Rural Peace." Plague had been a constant low-level presence in Nong'an, with small outbreaks occurring over a dozen times, but the occurrence of a few plague cases in Nong'an was never seen as exceptional news.[88]

After 1932, however, Nong'an was in the vicinity of a very special place: the city of Changchun. Changchun was a major transportation center, originally the "last stop" on the South Manchuria Railroad before trains had to switch to the larger-gauge Russian tracks of the Chinese Eastern Railroad to travel on to Harbin (this juncture can be readily seen in the map in fig. 5.1). In 1932, Japanese authorities designated Changchun as the capital of the new nation of Manchukuo. As we have seen, Manchuria had been the site of numerous capitals, including Nurhaci's original "capital" at Hetu Ala (later known as Xingjing, or "Rising Capital") and the Qing "Soaring" capital at Mukden (called Shengjing in Chinese). Compared to the cosmologically significant places along the "dragon vein" of the Qing empire, Changchun would be a different sort of capital. Christened "Hsinking" (Shinkyō in Japanese, or Xinjing in contemporary Chinese romanization), this "new capital" was to stand as a symbol of Japanese-led Pan-Asian modernity: a planned city with broad boulevards, marble-halled public buildings in monumental Sino-Japanese architectural style, efficient rail facilities, well-manicured parks, and world-class museums. Hsinking was even slated to become "the first city in

Asia in which all [new] residential, commercial, and industrial buildings were equipped with water closets."[89] But these utopian colonial projects, planned by a Japanese intelligentsia to incorporate the best things that modernity had to offer, faced a particular challenge in Manchuria. Among them was the fact that Hsinking was only a one hour train ride away from Nong'an county, itself a "capital" of sorts, but a capital of the land of *Yersinia pestis*. Manchukuo's dreams of modernity were haunted by the specter of the plague.[90]

The contrast between the glittering vision of a modern, ethnically harmonious Hsinking and the plagueland-realities of Nong'an were starkly embodied in the comments of Hiroki Masaji, an SMRC Medical University physician who participated in grueling efforts against the plague during an outbreak in Nong'an in the summer of 1933, a year after the establishment of Manchukuo. In the conclusion to his extensive epidemiological report, the weary Dr. Hiroki recommended a variety of suggestions that he hoped might improve response to the inevitable future outbreaks of plague in Nong'an: the addition of more automobiles to facilitate the travel of medical teams between desperately poor, remote villages; the importation of more wood for coffins so that corpses of plague victims would not be eaten by wild dogs; and the employment of more interpreters, or better yet, Japanese physicians who could speak "the Manchurian language" (Chinese) to better communicate with terrified, suspicious villagers who evaded and resisted the Japanese doctors' every attempt to help them. Without increased attention to these matters, Hiroki warned, it would be only a matter of time before the plague in Nong'an moved on to threaten the capital of Manchukuo, only thirty miles away.[91]

Plague in the Capital

In 1940, the inevitable happened: plague struck in the Manchukuo capital, killing dozens of people and turning parts of the city into zones of terror. By 1940, Japan had already been prosecuting an all-out war in China proper for three years. While fighting remained confined to south of the Great Wall, deprivation and unease hung over the Japanese-occupied northeast. The management of the 1940 Hsinking outbreak provides a comprehensive overview of the complex process of knowing the plague in Manchuria under wartime conditions. As it had before the war, tracking *Yersinia pestis* meant scrutinizing the environment at multiple levels—from the geographical layout of neighborhoods to the microscopic terrain of internal organs. It also meant pursuing the bacillus across multiple boundaries with a single-minded ferocity. In 1940s Manchukuo, knowing the plague was ultimately inseparable from the brutal techniques of total war.

The first victims of the 1940 Hsinking plague were two children who lived above a pet hospital. The Tajima Dog and Cat Clinic was located in Sanjiaodi (Triangle District), a bustling commercial neighborhood in Manchukuo's capital city, just southeast of the train station. Areas close to Sanjiaodi featured large modern department stores, broad boulevards, and the picturesque Nihonbashi (Bridge of Japan) Road, named after the famous bridge in Tokyo from which all distances in the empire were measured. Sanjiaodi itself was a bit of a mixed bag: a Japanese neighborhood in the "old city" that also included thousands Chinese and Koreans, its narrow streets lined with small local shops and crowded apartment buildings.[92] The presence of a "dog and cat hospital" in the area showed that the neighborhood, even if not as prosperous as others, was a place where well-established residents raised their children, shopped, and enjoyed the simple pleasures of family pets. Sanjiaodi was also the spot where the 1940 Hsinking bubonic plague epidemic began.[93]

The first victims of the plague were very young: two-year-old Tajima Tadako, the youngest daughter of the Japanese vet clinic owner, and Wang He, a thirteen-year-old Chinese girl who was employed as the toddler's babysitter. Both the children had become ill on September 23, 1940. The Tajimas sent Wang He home, and then took their daughter Tadako to a local hospital, where she was admitted and diagnosed with the flu. The child's symptoms— high fever, headache, restlessness—grew worse, and she died at the hospital on September 29. The family's calamity was compounded when Tadako's eight-year-old sister, Tensenko (literally "child of Tianjin"), became ill and died just four days after the death of her sister. Wang He, in the meantime, had died in her nearby home, and two other men (surnamed Han and Song) who lived in the same apartment building as Wang He also became severely ill.

The disease finally came to the attention of the authorities when Ōta Ōji, an employee of the Japanese army who lived next door to the pet clinic, checked himself into the nearby Hsinking SMRC hospital with a high fever. Doctors immediately transferred Ōta to the Kantō Army hospital, where his condition continued to deteriorate. On the morning of September 29, after enduring six days of high fever, Ōta complained of a painful tightness in his chest and began coughing up bloody, foamy sputum. He died at one in the afternoon, less than twenty-four hours after being admitted to the army hospital. Ōta's sputum, together with smears made from his organs during the army hospital's autopsy, were transferred to the Hsinking Municipal Hygiene Laboratory. There, the laboratory chief, a Dr. Miyagi, inspected the samples under a microscope and observed bacilli that resembled the plague germ. Suspicious that there might be other cases, Miyagi visited the buildings near the pet clinic, and found Han and Song in advanced stages of a disease with plaguelike symptoms. He examined the already-encoffined corpse of little Tadako and discovered troubling signs of buboes. Back at the Hsinking lab, a culture of the bacilli from Ōta's sputum confirmed Miyagi's suspicions: plague had come to the capital.

Soon the full force of Manchukuo's plague prevention system descended on the Sanjiaodi neighborhood. Military personnel cut off all traffic in and out of the area. The streets were surrounded by steel-plate fences, anchored several feet deep into the ground to prevent rodents from escaping underneath them and topped with barbed wire to prevent humans from escaping over them. An all-out war was waged on rats and mice: citizens were called upon to catch pests in the streets, and poisonous gases were pumped into the sewers beneath the city. Police and special hygiene teams doused infected apartments with sanitizing chemicals. Teams of police and medical inspectors went door-to-door looking for the sick. Over four hundred contacts of confirmed cases were forced into quarantine for observation at the military-run Chihaya Hospital on the west side of the city.[94]

It was at Chihaya Hospital that Li Dejin, a ten-year old Chinese boy who lived next door to the pet clinic, died after four days in quarantine. The chief physician of Chihaya Hospital observed in detail Li's condition and decline:

10/6 PM: Developed a 103 degree fever
10/7 AM: Developed a small swelling in his right groin painful upon palpitation, fever 100.3, pulse 100
10/8 AM: became confused and delirious, as if in a terror—patient's father, (quarantined together with boy), cried out "My son has gone insane!"—began cough that produced bloody, foamy sputum (which tested positive for *pestis*).
10/8 PM Patient expired later this evening.[95]

The interventions exercised in Hsinking were fairly standard anti-plague measures, experienced in many of Manchuria's cities multiple times since the 1910–1911 great plague. What made this episode fairly unique, however, was the personnel involved. The efforts against plague at Hsinking in 1940 included representatives from many different organizations, including the Hsinking municipal government, the Manchukuo state, the Kantō civil administration, the South Manchuria Rail Company, and the Kantō Army, but the entire project was overseen by the army's Epidemic Prevention and Water Supply Detachment, also known as Unit 731: the Japanese Imperial Army's germ warfare organization.[96]

Today, Unit 731 is synonymous with medical atrocities perpetrated by Japan against China during World War II.[97] Under the leadership of Dr. Ishii Shirō (1892–1959), Unit 731 established a massive biological and chemical weapons development facility in Manchuria, in Pingfang, a suburb of Harbin. In addition to state-of-the-art bacteriological laboratories, the complex included comfortable living facilities for hundreds of researchers and their families, complete with dining facilities, a movie theater, and other recreational opportunities. Two structural elements symbolized the evil nature of the research that went on in the complex: a large, square-shaped prison at the middle of the facility that housed human captives, and three tall smokestacks that marked the presence of a large incinerator. It is estimated that between 1936 and 1945, a minimum of three thousand people lost their lives in the laboratories of Unit 731, victims of human experimentation at the hands of Ishii and his colleagues. Euphemistically referred to as "logs" (*murata*), the prisoners (mostly Chinese, but including some Russians and Koreans) were intentionally exposed to pathogens as part of the unit's quest to develop effective germ warfare techniques. After infection, many of the bodies were opened and exposed to the gaze of Unit 731 researchers. Some were used in surgical practice; some were vivisected—that is, dissected while still alive—to visualize the course of the disease; some were autopsied after death in order to obtain pathology information. All were murdered, their bodies disposed of like laboratory waste in the complex's incinerators.

Unit 731's work was not limited to the Harbin facility alone. As we have seen in the previous chapter on the earth sciences, the creation of Manchukuo and the emergence of planning for "total war" brought a wide range of civilian scientific institutions and personnel under the control of the military. This is clearly illustrated in the case of Unit 731. The commander, Ishii Shirō, recruited a large network of physicians and researchers from mainland Japan's most prestigious universities and laboratories. The Army Medical College in Tokyo, Ishii's alma mater, conducted germ warfare research. Multiple branches of 731, units with an assortment of numerical designations (e.g., Unit 100, Unit 1644), were distributed across territories occupied by the Japanese empire in the 1930s and 1940s from Shanghai to Manila. In Manchuria, modern research facilities such as the Dalian SMRC Hygiene Institute and the SMRC Medical School in Shenyang were reorganized as branches of Unit 731, woven into the web of investigation and production that would ultimately be used in the development of biological weapons, even as they continued to treat disease and produce vaccines for legitimate medical purposes.[98]

This interpenetration of both legitimate and morally reprehensible medical activities has made understanding the nature of Japanese wartime research a difficult and contentious undertaking. The 1940 plague in Hsinking is one case in point. Present-day historians in China contend that this plague outbreak was the direct result of the intentional use of biological weapons by Ishii Shirō and Unit 731—after all, Ishii and his Epidemic Prevention Detachment oversaw the epidemic control work in the city for two weeks. Even more damning is the fact that one of the primary reports chronicling the outbreak was penned by Takahashi Masahiko, the head of Unit 731's plague research division.[99] On the other hand, some Japanese historians, including Tsuneishi Keiichi, one of the scholars first responsible for exposing the atrocities of Unit 731, have concluded that the 1940 episode was a natural outbreak, albeit one that served as an inspiration for the unit's subsequent research and actual germ warfare attacks.[100]

My goal here is not to probe the origins of the Hsinking outbreak but to place Unit 731's studies at Hsinking and elsewhere in the larger context of creating knowledge about the nature of the plague in Manchuria. Many of the questions asked and the protocols employed by Unit 731 researchers were part of a long legacy of plague research going back to even before the 1910 Great Manchurian Plague, and would not have seemed out of place in the laboratories of the Chinese plague fighter, Wu Lien-teh. Scientific approaches to both plague control and plague propagation used by Unit 731 were aimed at revealing the complex webs that linked *Yersinia pestis* to its rodent hosts, insect vectors, the physical environment, and man. As we have seen, the interventions employed to uncover these links could be revolting, violent, and savage. Plague research undertaken at times of active outbreaks always included human victims placed under conditions of confinement and compulsion, subjected to experiments that seemed to be constructed as much to satisfy scientific curiosity as to seek a cure. But all research adhered to the classic approach, cultivated since the early days of infectious disease science, that pathogens were a "peculiar fauna" that was "destined" to occupy a territory that included "the interior of the bodies of man and animals." The political control of territory required the ability to cross all boundaries and see into this most intimate of terrains.

The obvious difference between Chinese and Japanese research is that

Unit 731 crossed these boundaries not only to control disease but also to inflict disease on a mass scale. This work took place under conditions where researchers dehumanized their victims and considered them as simply another form of mammalian host for the plague bacillus. Such dehumanization saw humans in Manchuria as nothing more than an isolatable variable element in Manchuria's vast environmental web, simply another part of plagueland that was always already susceptible to a fatal infestation by *pestis*.

✳

The investigation of the 1940 Hsinking outbreak (led by Unit 731's plague expert, Colonel Takahashi Masahiko) reveals a great deal about the nature of plague research in Manchuria, but it reveals even more about the nature of imperialism there: its secrets, hatreds, suspicions, and violence. The first challenge was to understand the origins of plague in the Manchukuo capital. In the past, isolated plague cases in Hsinking had been brought in from the outside, from locations on the outskirts of the city that were within the endemic Manchurian-Mongolian grassland plague focus. Was this outbreak of plague imported, as it had been in the past? If so, what was the vector and route for its introduction? Or had the plague bacillus been lurking within the environment of the capital city, an invisible part of its native ecology unleashed by an unknown trigger?

Takahashi's investigations clearly reflect the extreme political and military anxieties of the day. One of the first theories that Takahashi considered was conspiracy: in other words, given what they called "the current state of affairs," authorities investigated the possibility that plague germs had been disseminated by enemies of the Japanese empire. Hinting at the extensive secret police network that existed in the capital at the time, Takahashi's report tersely states that "foreigners," "highly suspicious people living in Hsinking," and various "suspects" were investigated for any clues, but none was found. The personnel and conditions at Hsinking's medical labs were also investigated—an interesting aside that demonstrates these laboratories were likely conducting research with *pestis* bacillus. No security breaches were found. At the same time, Takahashi's team completed a spectacularly detailed spatial and chronological map of the progress of the plague in Hsinking through extensive interviews with family members, friends, and employers of the first victims. None had been in contact with suspicious materials or people, and none had been to Nong'an or had contact with anyone from Nong'an.[101]

Takahashi's suspicions then focused on the Tajima Veterinary Clinic itself.

His team captured, killed, and examined dozens of rats from the streets and sewers around the clinic. Some were found to be infected with plague—but how these rodents came to be infected with plague was unclear. At the clinic itself, investigators counted the number of fleas on the dogs and cats, probed the pet cages for flea samples, and even scraped dirt up off the floor looking for tiny flea parts. After examining the fleas under the microscope and scrutinizing their anatomy, they found a total of only six fleas belonging to *Xenopsylla cheopis*, the rat flea species that is considered the most efficient vector of bubonic plague. Researchers pored over Dr. Tajima's account books, searching for clues about diseased animals—especially animals that had recently been in Nong'an. Takahashi's final hypothesis was that infected fleas from Nong'an—or perhaps even just one infected flea—had taken a ride on a cart pulled by a sick horse or ox that had arrived in Hsinking pulling goods from Nong'an. Takahashi imagined that when the owner took the sick animal to the Tajima Veterinary Clinic, the flea had jumped off its mode of conveyance, and the seeds of the epidemic were planted.[102] By focusing his mind's eye to the level of one flea, Takahashi envisioned an almost unimaginably small event that prompted a citywide disaster.

In Takahashi's estimation, however, the "Manchurian" people were as much to blame for the epidemic as the infected flea. Harkening to discourses on settler hygiene in Manchuria, this cutting-edge Unit 731 bacteriologist conducted an "environmental hygiene" inspection of the neighborhood, seeking out stagnant air and a lack of windows as culprits in fostering disease. Upon inspection, investigators found the apartments to be poorly ventilated and dark, places where health-giving ultraviolet light did not penetrate. Even in the Tajima household, the apartment interior showed food was improperly stored and beds were piled with thick dirty blankets. Ultimately, however, people and not the environment were responsible for the plague in the capital. As Takahashi concluded: "It is not difficult to imagine that the unclean animal enclosures at the veterinary clinic . . . could provide a congenially warm place for rodents and insects; moreover, we must consider that living within the same household [with the first Japanese victims] were unhygienic Manchurians (Chinese), which provided excellent conditions for the spread of the plague."[103]

In addition to scrutinizing the layout of the terrestrial environment and its potential vectors, Takahashi's team also probed the interior environment of human bodies for clues about the nature of the plague. By conducting extremely detailed postmortem studies of dozens of victims, investigators learned that the vast majority of cases were primarily bubonic plague, with lung involvement visible only in those who had survived a bit longer, giving *Yersinia pestis* an opportunity to infect the lungs through the bloodstream. There had been no person-to-person transmission of plague through the air. In the end, the introduction of the (alleged) one infected flea had resulted in a total death toll of at least twenty-eight people.[104]

Takahashi's investigations were multidimensional, multidisciplinary observations that combined environmental, zoological, clinical, pathological, epidemiological, and bacteriological methods. Equipped now with this masterful overview of plague ecology, Unit 731 gained precious insights into the nature of plague in Manchuria—and the potential for its weaponization. An effective large-scale outbreak of plague among humans required two things: the introduction of a very large number of infected *Xenopsylla cheopis* into a human living environment and/or a technique that could infect humans with *Yersinia pestis* directly through the lungs, thus creating the highly contagious pneumonic plague. After the Hsinking outbreak, Unit 731 endeavored to create plague-based biological weapons that could improve upon the inefficiencies of nature.

Conclusion: Unit 731, Plagueland, and "Visceral Territory"

In their laboratories at Pingfang, the scientists of Unit 731 investigated the weapons potential of many diseases, including typhus, anthrax, and hemorrhagic fever, but the one disease that absorbed the majority of their time and resources was the plague.[105] Three separate divisions perfected the cultivation of large amounts of *Yersnina pestis*,[106] bred massive amounts of fleas,[107] and devised delivery systems for both fleas (exploding bombs) and germs (airborne aerosol dispensers).[108] Unit 731 used these plague weapons in documented attacks on several locations in the war with China, including Ningbo, a large city near Shanghai on the eastern coast of China, attacked in 1940, and during the brutal battle of Changde in central China in 1943.[109]

The one activity of Unit 731 that forms the core of the unit's legacy of atrocity, however, is its use of human beings in medical experiments designed to develop and test the effectiveness of biological weapons. Perpetrated on thousands of captives, including many political prisoners brought to Pingfang by the secret police, the experiments have been seen as the epitome of evil conducted in the name of science, the Asian equivalent of the atrocities perpetrated by Nazi doctors during World War II. The suffering of the victims, who were injected, probed, eviscerated, and incinerated like medical waste, forms the central narrative of Unit 731 atrocities in film, museums, and on the internet. Unit 731's experiments are typically presented as bizarre and irrational, designed more to cause suffering and satisfy the sadistic desires of craven men than to generate findings of scientific value. And yet while the experiments of Unit 731 were by any definition craven and inhumane, they were not unimaginable.

By placing the plague research of Unit 731 in the context of plague research in early twentieth-century Manchuria, certain ready similarities become apparent between Unit 731's research protocols and the experimental designs of other researchers, including those of Wu Lien-teh. Standard laboratory experiments were designed to visually manifest the transmission of plague, and Unit 731 utilized similar procedural logic. Spatial configurations in Unit 731's laboratory experiment designs showed uncanny similarities with the experimental designs of previous researchers. Their crossing of the boundaries between field, clinic, and laboratory as well as the crossing of boundaries between the living and the dead was characteristic of decades of research in Manchuria's plague zones. This comparison is not designed to remove Unit 731's research from the realm of atrocity. Rather, this comparison helps to throw into the clearest light possible the ways that dehumanizing ideologies combined with the desire to understand human-microbe interactions in Manchuria's unique plague environment, leading the researchers of Unit 731 to treat humans as animals as they sought to manipulate these interactions for purposes of total war.

Much of what we know about Unit 731's plague experiments comes from testimony given by high-ranking Unit 731 personnel to Soviet war-crimes tribunals or to US occupation forces in Japan after the war. Written in brief statements of fact or military-style "executive summaries," the reports communicate "results" and lack the narrative richness of Wu Lien-teh's copious publications. The reports do make it possible, however, to envision the laboratory settings and procedures for Unit 731's plague experiments, and it is to these we now turn.

In highly intimate and highly repulsive experiments, Unit 731 researchers directly injected human captives with plague germs in order to test vaccines. The first goal of these experiments was to determine the "minimum infectious dose" at which 50 percent of test subjects would develop plague (designated "MID50"). Unit 731 researchers developed three different vaccines that produced 50 percent protection against a "challenge subcutaneously with 1000MID": in other words, vaccines were tested on human subjects who had been injected with plague doses twenty times higher than the "minimum" dose that would produce fatal plague in most normal subjects.[110] Ever since their late nineteenth-century experiences with plague in the first colony of Taiwan, Japanese interventions against the plague had relied on the use of plague vaccine, but the stability, effectiveness, and safety of the vaccine were always in question. From the perspective of Unit 731 researchers, using human subjects as substitutes for laboratory animals provided better data for the development of vaccines: data used not to ameliorate civilian suffering, but as a military strategy to provide for the defense of Japanese troops in a total-war world where biological weapons were readily conceivable.

Unit 731 placed a great deal of effort on finding delivery mechanisms that could infect large numbers of people at once. Several types of plague experiments were carried out by placing human subjects in a large room thirty feet by thirty feet and exposing them to *Yersinia pestis* through different types of vectors and delivery systems. While the specific features of the room are not clear from descriptions, the chamber was either partially made of glass or was equipped with large windows. While smaller than the research wards of the North Manchurian Plague Prevention Service hospital in Harbin, this chamber shared similar features, particularly the presence of a glass window that allowed scientists to observe experiments involving humans "in progress" while simultaneously protecting them from the possibility of infection. In experiments that harkened back to Wu Lien-teh's technique of letting healthy guinea pigs roam together in the wards with quarantine human plague victims, Japanese researchers sought to uncover whether plague transmission could be effected as "mammals" ranged about in an enclosed space with infected fleas—although in this case, the direction of infection was not from man to experimental animal but from experimental animal to man. The results were tersely reported as follows: "It was found that if subjects moved freely around a room containing a concentration of 20 fleas per square meter, 6 of 10 subjects became infected and of these 4 died. It was also found that one flea bite per person usually caused infection."[111]

These observation chambers could also be retrofitted to conduct tests on the transmission of pneumonic plague in ways that are reminiscent of NMPPS pneumonic plague inhalation experiments on marmots. In one Unit 731 experiment (described to US military personnel as representing a "typical" procedure), four human subjects were placed in a room about 350 cubic feet in size. An aerosol suspension of *Yersinia pestis* was introduced into the room using an ordinary disinfectant sprayer. Results indicated that 300 centiliters of a 1 milligram per centiliter suspension was "highly effective," resulting in a 100 percent infection rate and a 60 percent mortality rate in victims over the course of repeated trials. Records do not indicate how many such trials were conducted.

In experiments designed solely to rate the effectiveness of weaponized vectors, Unit 731 scientists subjected humans to prolonged exposure to fleas that had been "exploded" within the chamber: "Bomb trials were carried out using the 'UJI' porcelain bomb with primacord explosive. The fleas were mixed with sand before being filled into the bomb. About 50 per cent of the fleas survived the explosion which was carried out in a 10 meter square chamber with 10 subjects. 8 of the 10 subjects received flea bites and became infected and 6 of the 8 died."[112]

Following a logic that adhered to the ecology of plague, Japanese researchers scaled up these indoor laboratory results and expanded them in outdoor field trials. The Unit 731 "proving grounds" were located near the town of Anda, located in the prairies of Heilongjiang between Qiqihar and Harbin: a land once familiar to the nineteenth-century Qing administrator Xiqing as a place where local people gathered reeds from the nearby marshes. The way Japanese researchers at Anda created knowledge of the nature of Manchuria would have been beyond anything that Xiqing could have ever imagined. Here Japanese researchers sought to understand how Manchuria's environment, including temperature, light, wind speed, and humidity, might affect the ability of fleas and *pestis* to infect human subjects.[113] In one field test conducted at Anda, researchers staked fifteen prisoners to the ground and then guided airplanes to drop two dozen plague-flea bombs over the site. The victims remained staked outdoors for "a long interval," after which they were taken back to Pingfang for observation to see if they would develop plague. This particular experiment was apparently not very successful, resulting in more refinements to the flea-bomb design and repeated "trial runs" in the field. Similar airplane runs involved victims staked on the ground and strafed by plane-dispensed *pestis* aerosol.[114]

The mysteries of how invisible fluctuations in climate, temperature, or some other unknown and unperceived environmental factors influenced the emergence, spread, and maintenance of *Yersinia pestis* in reservoir populations had long been a central puzzle in the ecology of the plague, prompting researchers to scour the countryside for rodents, measure the depth of jerboa tunnels, take humidity readings in hibernation dens, and even insert thermometers into the rectum of marmots. Unit 731 experiments at the Anda

Proving Grounds sought to parse and control these environmental variables, using human subjects and artificially introduced vectors to create plague in the field where it had not naturally occurred. Ironically, while Japanese medical experts had long blamed the "primitive" and "unhygienic" behaviors of Chinese residents in Manchuria for their susceptibility to plague, in these horrific experiments, Japanese medical experts rendered Chinese bodies absolutely susceptible, incapable of taking any action at all to defend themselves or avoid infection, stripping them of their humanity and forcing them to become like inanimate parts of the environment.

One aspect of plague study in Manchuria that remained remarkably consistent across the course of the twentieth century, up to and including Unit 731, was the role of extensive autopsies and postmortem pathology studies. Pathology studies were a logical conclusion to experiments in the laboratory and observations in the field, allowing the researcher to follow *Yersinia pestis* from the external environment into the internal terrain of the body of its host. Infectious disease studies had, since the late nineteenth century, entailed thinking of pathogens as "a peculiar fauna" that were "destined" to occupy territories that included the interior of the body of man. As we have seen, ever since the 1910 Great Manchurian Plague, detailed postmortem observations of the progress of the disease through the organs of the body, particularly the respiratory organs, had been central to explorations of the mystery of pneumonic plague in Manchuria. The position of the bacillus within the body's terrain determined the disposition of the plague in the outer world.

Autopsy and its way of mapping the body produced a document that has been seen as the ultimate manifestation of Unit 731's barbarity: the "Report of Q." Over seven hundred pages long, the "Report of Q" is the longest research report provided by Unit 731 scientists to US military investigators after World War II.[115] The document is a meticulous presentation of data generated through postmortem examinations performed on fifty-seven plague victims. The report is a tour de force of organization and detail. It begins with an exhaustive chart noting the initials, age, sex, and location of each case. The report identifies each of the dozens of plague victims as an individual. An overview of the autopsy for each individual follows, detailing the general appearance, in order, of the heart, aorta, tonsil, pharynx, epiglottis, lung, liver, stomach, intestines, kidney, spleen, pancreas, thyroid, pituitary, testis or ovaries, lymph nodes, and skin. The data is then reshuffled and recategorized, and what follows is a treatise on each organ, with fifty-seven descriptions of fifty-seven hearts, fifty-seven spleens, and fifty-seven pairs of lungs. Each organ is described in great detail, with lungs receiving particular attention, lobe by lobe, in the exacting language of modern science which echoes Maximow-icz's "autopsy" of flower ovaries performed almost one hundred years earlier: "Right, inferior: Lobar pneumonia in the gray hepatisation, with numerous leucocytes, some serous exsudates [*sic*], at some places slight hemorrhages or at some places some bacterial accumulation, all over the pulmonal tissues. Intense fibrinous swelling of alveolar walls. Some bacterial accumulation at perivascular portions." These detailed histological observations are accompa-nied by meticulous, highly stylized schematic illustrations, rendered in col-ored pencil, representing the microscopic changes observed in various organ tissues, particularly of the lung (plate 11). Lungs were represented as multiple hexagonal shapes (representing each lobe), spleen as concentric circles, liver as squares, kidneys as lightbulb shaped. In these illustrations, prose descrip-tions are conveyed in a concise color-coded visual shorthand: "hemorrhage" in red, "degeneration" in pink, "necrosis" in purple, and so on. The beginning of each organ section contains the color and shape key for subsequent draw-ings of each organ, like a legend of complex symbols for this map of the body. Each organ section ends with a comprehensive chart summarizing findings in an "at a glance" format, with types and locations of pathological changes listed on the left (defining each row) and the number for each case at the top (forming the columns). The number and intensity of lesions for each case are keyed in to each square using a shorthand tally method.

Overall, the "Report of Q" is a remarkable example of modern scientific observation. With its delicate hand-drawn artwork, the report is aesthetically pleasing, even eerily beautiful. It has also been seen as the ultimate represen-tation of the horrors of Unit 731, a savage depiction of singular atrocities car-ried out on victims of human experimentation. "Discovered" as a "secret doc-ument" over and over by researchers focused on the history of Unit 731, the "Report of Q" has been taken as clear proof of Japan's war crimes in China.[116] Indeed, the hundreds of pages of the report, each dramatically suspended in the air behind glass, form a central exhibit in China's new state-of-the-art Unit 731 museum in Harbin.[117] While the document is certainly a testament to the brutal efficiency of Ishii Shirō and his colleagues, the report can also be seen as the culmination of decades of visualizing and mapping the terrain of Man-churia as a plagueland.

The "Report of Q" stems from an event that we have already encountered: it is the pathology report of the investigation of the 1940 plague outbreak in Nong'an and Hsinking that was overseen by Takahashi Masahiko, Unit 731's plague expert.[118] This pathology report, when coupled with Takahashi's reports on epidemiology, zoological surveys, and bacteriological examinations, mapped the journey of *Yersinia pestis* across the territories of Manchuria, both human and terrestrial. The "Report of Q" is of the same genre as Wu Lien-teh's many reports on the pathology and histology of plague, including those drawn from the many anonymous victims that the North Manchurian Plague Prevention Service found dead in the mines at Dalainur in 1921, or those prepared from bubonic plague patients who expired after being cataloged and photographed while in NMPPS quarantine in Tongliao in 1927. The difference in these 1940 reports is one of thoroughness and detail, provided by the massive Japanese police and military presence, combined with the wartime "public health" concerns that made collection of information so comprehensive in Japanese-occupied Manchukuo.

Indeed, through the Unit 731 reports, it is possible to follow the entire course of *Yersinia pestis* through several cases who had been admitted to hospitals or who had been placed in isolation as contacts of plague victims. Such is the case of little Tajima Tensenko, the second daughter of the Japanese vet clinic owner who died of plague in a local hospital. She is listed in Takahashi's epidemiology report as case number S-2, the second death to come to the attention of authorities in Hsinking (romanized as Shin-kyo, hence "S-2"), designated by the initials "T.T." The ten-year old Li Dejin, the Chinese boy who died in a state of delirium while in quarantine, is listed in Takahashi's report as case S-8 (Shin-kyo 8), with the initials T.L. (romanized as Te-chin Li). The same numbers and initials link cases in Takahashi's epidemiological investigation with the pathology report. Every heart, lung, and spleen described in the "Report of Q" can be traced back to a living human being who had a name, a family, an address within a city, and, in several cases, a history of suffering recorded by Japanese physicians who watched as healthy individuals succumbed to *Yersinia pestis* over the course of a few days.

If the Nong'an-Hsinking outbreak was caused by Unit 731, then the research that followed certainly belongs in the annals of the world's worst medical atrocities. If the Nong'an-Hsinking outbreak was a natural occurrence, the reports that followed were a standard scientific response to plague, a way of creating knowledge about the nature of Manchuria honed over the course of decades. Extensive dissection of human victims was nothing new. Such postmortems formed a centerpiece of Wu Lien-teh's North Manchurian Plague Prevention Service research results. Like the victims at Tongliao, the victims at Hsinking were given coded numbers, but their identities were known, and the results individuated. The prose style employed in both NMPPS and Japanese reports is exactly the same, their terms of morbid anatomy part of the universal language of science. Each used visual aids to convey data. The frontispieces of the NMPPS annual reports were adorned with large lurid drawings of the infected lungs of plague victims, their swellings and necroses detailed in vivid reds and purples. The Japanese scientists devised a different way to convey their findings, creating a highly schematic set of symbols that could present a large amount of dissection data at a glance. But the reports were simply one element of a larger interdisciplinary research program that emerged after the 1910–1911 Great Manchurian Plague, an interdisciplinary program that influenced both Wu Lian-teh and the scientists of Unit 731. Each was engaged in making visible the complex web of environment, bacteria, and man in Manchuria; each was dedicated to mapping the journey of *Yersinia pestis* from the terrestrial terrain of the Manchurian-Mongolian grassland to the visceral terrain of man. In their pursuit of total war, the scientists of Unit 731 used these maps, constructed in such painstaking detail, as a blueprint to re-create that journey at will.

Unit 731's atrocities were the result of a violent colonial occupation that was buttressed by an ideology of civilizational superiority and carried out in the morality-destroying context of total war. They were also the result of a decades-long attempt to know the nature of Manchuria by visualizing and mapping the complex ecology of the plague. In pursuit of this invisible ecology, researchers crossed multiple boundaries: between the field, the clinic, and the laboratory; between the exterior and the interior of bodies; and between the human and nonhuman worlds. Many scholars have demonstrated that medical studies have rationalized brutal actions in the name of discovery, and a seamless movement between animal and human studies was typical of twentieth-century science from Tuskegee to Tanzania.[119] But in Manchuria, the Japanese bacteriologist Ishii Shirō and his colleagues took this propensity one step further by eliminating the boundary between human and nonhuman altogether. In doing so, they transformed natural knowledge of Manchuria into a crime against humanity.

~ 7 ~

SCIENTIFIC REDEMPTION

The Flying Voles of Gannan Revisited

When plague-infected voles fell out of the sky over Gannan county in the spring of 1952, only seven years had passed since the end of the Japanese occupation of Manchuria. While plague had been a frequent presence in Manchuria during the occupation, there was something very different about this event.

The first difference was that Gannan had not been part of the occupation era's plague outbreaks. Perhaps the county's remote location had spared it in the past. Even though it was located just forty miles north of Qiqihar in the heart of the Manchurian-Mongolian grasslands, Manchuria's rail lines had bypassed Gannan, leaving this part of the prairie relatively safe in its geographical obscurity. In 1952, then, the plague was new to Gannan.

The voles that fell from the sky over Gannan in 1952 also fell on a "New China." Using the Manchurian countryside as its base, the Chinese Communist Party had established the People's Republic in 1949, and the region that included Gannan was now administered under the communist-led Northeast People's Government.[1] The names of the villages where the voles were found in 1952—People's Masses Village (Minzhongcun), Workers and Peasants Village (Gongnongcun), New People's Village (Xinmincun) and Justice Village (Gongyicun)—reflected this political shift. These "villages" were actually settlements in a network of state-run farms (*guoying nongchang*) newly established by the PRC government and inhabited by personnel—mostly soldiers from the People's Liberation Army—who had been relocated to the area to produce grain for the new state.[2]

And when the voles fell, New China was in the middle of a new war with a new enemy: the United States of America. In other parts of the world, this conflict was called the "Korean War": a United Nations "police action" designed to protect the integrity of South Korea against aggression from North Korea. But in Manchuria, this was the "War to Resist America and Aid Korea," an effort to defend the Chinese motherland (and its neighboring state, North Korea) against yet another invasion by yet another imperialist enemy. In support of the war effort, New China's masses had been mobilized to participate in the "Resist America, Aid Korea" campaign, and exhortations to participate in everything from state-mandated factory production to domestic chores were pitched with an anti-American message. Among the more dramatic mobilizations was the anti-American "patriotic hygiene" campaign that instructed New China's citizens how to respond in case of germ warfare attacks from US forces.[3] This fear of germ warfare was not unfounded. After all, Japan had used biological weapons against China during the war, and now the United States was using Japan as a base for its military forays onto the Korean peninsula. Moreover, after World War II, the US military had cut a deal with the Japanese scientists of Unit 731, offering them immunity from prosecution in exchange for data from their biological weapons experiments.[4]

It is easy to imagine, then, that when 717 strange rodents were found scattered across Gannan county on the morning of April 5, 1952, germ warfare was a logical conclusion. Even though Gannan was located five hundred miles away from the Korean front, its residents had already been mobilized to search out and destroy alien pests that might threaten the health of the homeland. When the plague voles came, the people were ready.

The reader may remember the Gannan incident from the beginning of this book, where it was used to introduce the challenge of knowing Manchuria's environments. The strange tale of the flying voles of Gannan can be found in the *Report of the International Scientific Commission for the Investigation of the Facts Concerning Bacterial Warfare in Korea and China*, a compilation of investigations into alleged incidents of US bioweapons use in Manchuria during the Korean War. The reports of disease-infested rodents being thrown out of enemy aircraft may seem bizarre, but regardless of their veracity, they represent a pivotal phase in the long history of the making of natural knowledge about Manchuria. Manchuria may once have been a land of marvels, but in the first half of the twentieth century, it become a land of plague. The 1952 germ warfare investigations can be seen as an attempt to claim Manchuria for the Chinese nation after decades of imperial competition for the region. Through these investigations, Chinese scientists both demonstrated their mastery over the nature of Manchuria and rehabilitated Manchuria from its stigma as a singularly diseased and tragic place. Knowing the nature of Manchuria through science under the new Chinese state could transform the northeast from a subjugated "plagueland" to a place of modern hope and salvation for the Chinese people.

The *Report of the International Scientific Commission* (hereafter *Report*) was a collection of research papers authored by Chinese scientists who had been mobilized by the PRC state to investigate evidence of germ warfare. What made the report "international" was the fact that the papers were compiled and published in English under the auspices of the World Peace Council, a Soviet-backed international organization of leftist intellectuals that positioned itself as an alternative to US-influenced international organizations such as the United Nations and the Red Cross.[5] Both North Korea and the PRC had begun making public allegations that the United States was using germ warfare in the Korean War as early as the spring of 1951. The United States categorically denied the allegations but was compelled to respond as the accusations became part of the ongoing and tortuous armistice negotiations between the warring parties that were being held at P'anmunjŏm.[6] The United Nations initially suggested sending a Red Cross team to the area to investigate, but both North Korea and the PRC rejected this plan. Instead, the PRC lobbied the World Peace Council to form its own committee of "independent scientists" to bring the news of the Korean War germ warfare atrocity to the world.

The six-man committee, formed in May 1952, included scientists from the Soviet Union, Italy, Sweden, Brazil, the United Kingdom, and France.[7] Its most famous member was Joseph Needham, the Cambridge biochemist and scholar of the history of science in China who had spent time in China during World War II and maintained close connections with China's scientific community after the communist victory in 1949. During the committee's six-week visit to China, members did not conduct investigations in the field but instead received reports authored by PRC scientists. Through Needham's tireless efforts, the Chinese scientists' investigations were compiled into the English-language *Report* and made available throughout the world. After it was issued, however, the *Report* was mostly rejected in the West and languished in obscurity. Since then, when they have been considered at all, the germ warfare allegations (and their investigations) have for the most part been dismissed as a hoax perpetrated by China and North Korea to gain leverage in stalemated negotiations to end the Korean War.[8]

Many of China's top scientists participated in producing the report. Several of these scientists had advanced degrees from American and European universities, and they used this Western training to expose the West's science-based attacks against the people of North Korea and China. Dai Fanglang, the director of the Institute of Plant Pathology at the Beijing College of Agriculture and a graduate of Cornell University, identified spots on plant stalks disseminated by US planes north of Pyongyang as *Cercospora sojina* Hara, a fungus that produces an infection that can wipe out soybean crops, the mainstay of agriculture in the northeast.[9] Hu Xiansu, a Harvard PhD and one of the founders of botany in China, confirmed that leaves disseminated by American planes in North Korea and northeast China were an alien species, variants of *Lindara glauca* (a type of laurel) and *Quercus aliena* (a type of oak) that were "strictly confined to south of the 38th parallel of latitude" in Korea.[10] The chief entomologist of the Chinese Academy of Sciences, Chen Shixiang (whose romanized name "Sicien" Chen reflected his Paris training), identified multiple species of flies and spiders collected from "American bombs" that fell in multiple locations from Kuandian in the Changbai Mountains to Siping on the edge of the Manchuria-Mongolia grasslands.[11]

Each report included detailed expositions of laboratory methods demonstrating Chinese scientists' mastery of the techniques of modern science. Fungus spots and bacteria strains were examined under the microscope, their minute structures (stromata, conidiophores, flagella) measured in microns and translated into the standard descriptive language of anatomy and taxonomy. Cultures of pathogens (derived from infected leaves, spores, or spider slurry) were painstakingly isolated, purified, prepared, and injected into soils, or mice, or dabbed onto the vulnerable skins of fruits. The resulting diseases were noted in detail through prose and photographs: shriveled brown cotyledons emerging from infected soil; rings of rot expanding on otherwise perfect pears; diseased lungs and spleens cut out of laboratory animals that had died of plague and anthrax.

The goal of so much exacting observation, translated into precise language, numbers, and images, was not only to make an airtight case proving American use of biological weapons; it also served to make an airtight case that science in New China had a commanding knowledge of the nature of Manchuria. Reports were designed to establish that PRC scientists knew exactly which species were native to the region and which species were anomalous "alien invaders" introduced by the enemy. The region was no longer a place of natural marvels and invisible mysteries, with so many forms of life that they could not all be identified or named. Unlike the work of the Russian botanist Karl Maximowicz whose *Florengrenze* swirled across northeast Asia with no regard for political boundaries, or the fossils that Amadeus Grabau hoped would spark a vision of a world with no countries, PRC scientists used their knowledge of "universal standards" to fit local life into rigidly defined national territories. Certain trees adhered to political borders and only appeared below the thirty-eighth parallel. Statistically significant differences in the length of a hind foot made a rodent a foreign species. Even though most of the scientists mobilized in the investigation had not worked in the northeast before, they used their expertise in scientific literature and in scientific methodologies, obtained through Western training and from years of experience working in China south of the Great Wall, to make definitive pronouncements about the nature of Manchuria as an incontrovertible possession of New China.

Not only did these scientists know the nature of Manchuria; they also redeemed this nature from its legacy of disease and death. Many experiments were designed expressly to refute charges that the diseases allegedly disseminated by US planes—anthrax, encephalitis, dysentery, cholera, plague—were part of the "native" environment of the region. In one such set of experiments, five different species of flies were collected from various locations throughout Shenyang (Mukden). The flies were dipped in sterile solution and the solution cultured to identify any bacteria from the surface of the fly bodies. The flies were then further washed and purified in mercuric chloride before being ground into a slurry to allow investigation of the bacteria on the interior of their bodies. Cultures of over seven hundred flies produced through meticulous bacteriological methods all turned out negative: according to the *Report*, Shenyang's flies did not inherently carry the germs of epidemic disease.[12] Contrary to accusations of disbelieving imperialists, New China's Manchuria was no longer a plagueland.

*

With the *Report*, China's best-known scientists had claimed definitive knowledge of the nature of Manchuria, but a close reading of individual papers reveals that many of the scientists' investigations did not take place in Manchuria. Instead, these knowledge claims emerged from laboratories in the metropole. Plant and fungus samples were sent to directors of botanical institutes, who then studied and identified the samples in their Beijing offices. National Academy of Science zoologists scrutinized rodent skulls which had already been cleaned, dried, and preserved as scientific specimens. Central government bacteriologists confirmed that the fleas and chicken feathers presented to their labs did indeed hold plague and anthrax germs, but they received the vector objects in test tubes ostensibly sent from the Manchurian field.[13]

Only a small number of the reports detail investigations carried out in the Manchurian field itself, and of these, the investigation of the voles of Gannan (appendix M) is the most vivid and dramatic of all. It is also the longest, and gives the most detailed insights into how direct knowledge of the nature of Manchuria may have been constructed in the early years of the PRC. The investigation's techniques—bringing laboratories into the field, mapping the location of vectors, scouring the environment for representatives of rodent species, culturing bacteria, and performing autopsies on mammalian bodies to pinpoint the existence of plague—were all familiar. Although the events detailed in the Gannan report take place in New China, a close reading reveals uncanny similarities to plague investigations from the "old China" before Liberation.

Immediately after villagers discovered the voles the morning of April 5, 1952, the People's Government of Northeast China dispatched a mobile laboratory to investigate. The mobile laboratory was manned by experts from the Northeast People's Government's Plague Prevention Institute, a name that harkened back to Wu Lien-teh's North Manchurian Plague Prevention Service of the late Qing and early republic. While the plague prevention experts conducted investigations, the villages were subjected to swift and strict plague containment measures reminiscent of past methods. Armed militia of the People's Government surrounded the villages and enacted cordon sanitaire, preventing by force any travel in or out of the area. The residents of the villages were examined for signs of plague, while their homes and clothing were inspected and disinfected with Lysol and DDT. No human cases of plague were found, but as a precaution, the interiors of their simple homes were burned by igniting straw on top of sleeping platforms and floors to kill germs and fleas. As a further precaution, the Plague Prevention Institute personnel, accompanied by the local militia, injected 7,148 residents ("84.4% of the population") with plague vaccine. As in past episodes elsewhere in Manchuria (recall the 57,216 rodents collected for Japanese researchers at Tongliao in 1927), government authorities encouraged local residents to scour the environment and collect any rodents they could find so that the scientists could identify the extent of plague infection in local species. In spite of the intensive search, the residents retrieved only a handful of mice: except for the 717 flying voles (almost all of which had been killed and burned before the arrival of government representatives), Gannan was apparently remarkably devoid of rodents.[14]

In procedures familiar from investigations conducted by Wu Lien-teh in Manzhouli, Dalainur, and Tongliao decades before, the plague prevention team meticulously mapped the locations of the suspected plague vectors found in Gannan's human habitats. The majority of voles had somehow made their way into the interior of homes after their drop: 74.4 percent of the voles were found on sleeping platforms, 17.2 percent found in yards, and only 8.4 percent on rooftops. Vole locations were specified in detail: "*kang* (heated sleeping platform) on the north side of the house"; "in front of a haystack," "in a vegetable cellar." Although the report did not include a map of these locations, the prose report contained a vivid spatial portrait of Gannan as a potential epidemic site.[15]

The scientists' efforts also included zoological identifications and bacteriological assays. Four voles that had managed to evade extermination were turned over to Plague Prevention Institute experts. The bacteriologist and zoologist Ji Shuli compiled a comprehensive and painstaking examination of the voles' morphology, including observations of the fur color, fur density, number of foot pads, and measurements of the skull, tail, hind feet, and ears. This precise identifying data was then compared with the scientific literature on the region's indigenous rodents. Ironically, all the literature used to identify Manchuria's indigenous rodents had been written by foreign scientists. The bibliographies of the Gannan studies echoed the past imperial competition for the land and resources of Manchuria. The most comprehensive rodent scholarship was the 1941 *Revised Monograph of Japanese and Manchou-Korean Muridae*, compiled by the Japanese zoologist Tokuda Mitoshi at the height of the Japanese occupation of Manchuria. Information on Manchuria's rodents was also derived from the publications of British sportsmen (Arthur Sowerby's 1923 *The Naturalist in Manchuri*), Russian zoologists (Loukashkin's 1937 "Mammals of North Manchuria"), and even from *Report of the First Scientific Expedition to Manchukuo*, the Japanese media-event expedition to Jehol led by the paleontologist Tokunaga Shigeyasu in 1934. Following the standards for native species originally generated by these colonial scientists, the mysterious voles of Gannan were determined to be a unique nonnative species, as judged by the unusual length of their tails and hindquarters and the number of pads on their tiny feet.[16] Close visual scrutiny of anatomy had provided empirical proof of imperialist perfidy.

The report concluded with a detailed description of the bacteriological studies performed on the Gannan voles, or rather, vole, since after their morphological confirmation only one vole from the attack remained in viable condition for further study. Laboratory assays were performed by the Plague Prevention Institute scientists Ji Shuli, Xin Jun, and Zhang Jiefan and overseen by the institute's director, Cui Jizheng. Step by meticulous step, the report describes the perfect "Koch's postulates" sequence performed by the researchers, a procedure that oscillates between autopsy and microscopy. First the vole was cut open and its internal organs examined. Material was obtained from the spleen and liver and made into a saline slurry. This slurry was then injected into laboratory animals, which were then autopsied after their death to obtain pure cultures. A series of tests then were performed on these cultures, including injecting the purified strains into laboratory animals to see if the plague could be induced. In exhaustive assays that took place from April 10 to May 2 in the Plague Prevention Institute's laboratories in Harbin, the scientists concluded that this one vole preserved from the hundreds of Gannan's flying voles had been carrying *Pasturella* (*Yersinia*) *pestis*, the germ that causes the plague.[17]

Images included in the report provide visual confirmation of the scientists' work. In one particularly impactful collage (fig. 7.1), multiple photographs encompass morphological, anatomical, and bacteriological studies. In the upper left corner, a dorsal view of a dead Gannan vole is compared with a specimen of an indigenous vole (*Microtus gregalis* Pallas)—the Gannan vole's hind feet seem stunted and small compared to those of the *gregalis*. A sequence of dried skulls of the Gannan vole pictured to the right provides little in the way of precise measurement, but the skulls' blacked-out eyes and sharp fangs lend a sinister aura to the image. Finally, at the bottom, an enlarged view of what is identified as a colony of *Pasteurella pestis* provides a dramatic conclusion to the scientists' investigation: visual evidence of American bacteriological perfidy.

Gannan was not only the centerpiece of the PRC's investigations but also the center of attention for the World Peace Council's committee of international scientists. The foreign commission stayed in the PRC for a month and a half, from late June to mid-August 1952. The committee spent most of its time in Beijing listening to reports from Chinese scientists, with side trips to the cities of Shenyang and Pyongyang. But for two days in mid-July, the committee visited Gannan, traveling by plane, train, and then for hours by car to finally reach the remote state farm where the voles had fallen that spring. There they interviewed residents, viewed the site of the vole attack,

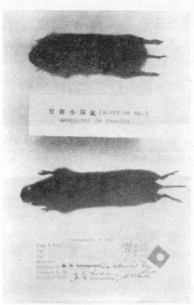

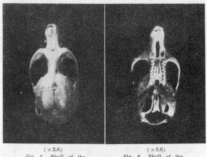

(×2.6)
Fig. 5. Skull of the vole found at Kan-Nan: Dorsal view.

(×2.6)
Fig. 6. Skull of the vole found at Kan-Nan: Ventral view.

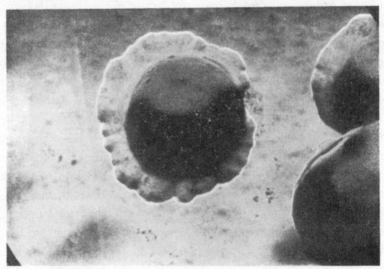

Fig. 4. External appearance of the vole found at Kan-Nan (upper) in comparison with *Microtus gregalis* (Pallas) (lower).

(×2.6)
Fig. 7. Skull and mandible of the vole (*Microtus sp.*) found at Kan-Nan: Lateral view. Mandible, below.

Fig. 12. Colonies of *Pasteurella pestis* after incubation for 48 hours.

Fig. 7.1 Anatomical and bacteriological investigations of the Gannan voles. *Report of the International Scientific Commission for the Investigation of the Facts Concerning Bacterial Warfare in China and Korea* (Peking, 1952): 257–59.

and examined the scientific field equipment used to conduct investigations. In his notes from the field, Joseph Needham acknowledged that many of the committee's encounters at Gannan included staged elements orchestrated for the benefit of the foreigners, although he considered them a part of a familiar Chinese tradition of "welcome and friendship." At Gannan, the group was received with what Needham described as a "soiree, quite wonderful" that included Mongolian and Manchu dances, folk operas, and a musical performance by a "[very] pretty female tractor driver." Equally orchestrated, Needham realized, was the "mise en scene for one person" performed in a mobile bacteriology lab set up in a Gannan village, where a technician, "dressed up in mask, rubber gloves, etc.," examined slides under a microscope "in the intense heat of the Manchurian summer" a good three months after the attack of the voles had occurred.[18]

Even though Needham noted some strange aspects of the Gannan situation in his private writings, the commission's public report does not express just how truly strange the Gannan investigations actually were. Even though the report included detailed information on the background of the Chinese scientists involved in the PRC's germ warfare investigations, no one on the international commission commented on the remarkable background of the Northeastern Plague Prevention Institute's personnel. First, all of the experts involved in the Gannan investigations—Ji Shuli, Xin Jun, Zhang Jiefan, and the institute's director, Cui Jizheng—actually lived and worked in the northeast, a rare quality among the dozens of scientists who were part of the PRC government's germ warfare investigations, the majority of whom were stationed in Beijing. But of greatest interest is the educational and employment history of these men. From their brief resumés included at the end of the *Report*, we learn that all of these plague experts mobilized by the PRC government in 1952 had been trained in scientific institutions in Manchukuo during the Japanese occupation: either at the Manchuria College of Medicine in Shenyang or the Harbin Medical College. The former had been part of the Japanese South Manchuria Railway Company's network of medical institutions, the latter was the educational arm of Wu Lien-teh's North Manchuria Plague Prevention Service, which had been taken over by the Japanese after 1931. Even more intriguing is the fact that all the Chinese plague prevention experts involved in the 1952 investigations had received advanced training in Japan during the war, at universities in Nagoya, Hokkaido, and Tokyo. Incredibly, the institute's director, Cui Jizheng, and vice director, Zhang Jiefan, had both been trained in bacteriology at Japan's Army Medical School in the

1930s—the very same program (and the same era) that had given rise to Ishii Shirō, the creator and leader of Unit 731, and the institution from which Ishii had drawn many of his colleagues.[19]

The PRC's case that the United States used biological weapons during the Korean War rested heavily on the idea that the United States had inherited not only Unit 731's data but also Japan's role as China's top imperialist enemy. Connections to Japan's role were explicitly made through the structure of the *Report*. The appendixes containing the Gannan investigations (M, N, O and P) were placed strategically after two reports by Chen Wengui, a Chinese physician who had investigated the Japanese plague-germ attack on Changde, Hunan in 1943—one of the best-documented occurrences of biological warfare launched by Unit 731 during World War II.[20] Appendix K provides Chen's original report, written in 1943. In appendix L, Dr. Chen provided a statement confirming that the techniques used by the Japanese in 1943 had appeared again on the Korean front in 1952, leading Chen to conclude that "the use of bacterial weapons on a large scale by the American invaders . . . has followed exactly the route opened by Japanese fascist war criminals in waging bacteriological warfare against humanity."[21] The specter of Unit 731 clearly haunted the Chinese investigations of the Korean war allegations. No one acknowledged, however, that Japanese science and its modes of knowing the nature of Manchuria served as the very foundation of the PRC's investigations.[22]

Many questions remain about the events of 1952. Unfortunately, given the recent deterioration of Sino-American relations and the PRC government's concurrent restrictions on access to archives, now is not the right time to probe the veracity of the Korean War germ warfare allegations or the possibility that Japanese-trained Chinese bacteriologists participated in the activities of Unit 731. Nevertheless, the materials published in the international committee's report do allow us to reflect on how the events of 1952 fit into the larger challenge of creating natural knowledge on an Asian borderland.

As a whole, the PRC investigations can be seen as an act of claiming Manchuria: establishing control of its territory for China by demonstrating knowledge about its nature. This scientific reclamation was one element in a long process. By the beginning of the twentieth century, Chinese scientists had probed Manchuria's nature in a variety of fields, as we have seen with the bacteriological work of the North Manchurian Plague Prevention Service and the Chinese Geological Survey's paleontological investigations in Jehol. But instability and the presence of political and military competitors, particularly imperial Japan, placed Chinese knowledge making in a precarious position.

With the Mukden Incident of 1931 and the establishment of Manchukuo, the Japanese military conscripted Manchuria's numerous scientific institutions, from medicine and public health to the geological sciences, into a network that mobilized the nature of Manchuria for total war. The bacteriological enterprises of Unit 731 were the epitome of this combination of military dominance, political machination, and scientific acumen that turned the pursuit of knowledge of Manchuria's nature into tragedy for the region's inhabitants.

The Korean War germ warfare investigations can be seen as an attempt to use science to overturn this process, to turn a devastated plagueland into a land of confidence and control. Chinese scientists, many of whom had received their training in foreign scientific institutions, applied their world-standard techniques to acquire knowledge of a region that had once been under the sway of enemy empires. Through their precise observations, these men demonstrated that the nature of Manchuria was not singularly diseased, and the region was no longer a contested borderland. Instead, the terrifying germs of epidemics were brought in by enemies from beyond the borders of what was definitively "China's northeast."

✳

The end of the Japanese occupation and the establishment of an anti-imperialist "New China" had not meant the end of disease in China. As demonstrated in the work of Xiaoping Fang, Miriam Gross and others, the PRC government constantly dealt with old health problems, including cholera pandemics and intractable endemic diseases such as schistosomiasis.[23] Under conditions of civil war, Manchuria (particularly Changchun) was the site of massive plague outbreaks in 1947 and 1948 as the People's Liberation Army battled the Nationalist army for control of Manchuria's largest cities with sometimes devastating consequences.[24] Another plague outbreak in 1951, well after the establishment of the People's Republic, devastated parts of the Manchurian-Mongolian grassland and led to tens of thousands of deaths.[25] Even after the creation of New China, plague had remained entrenched in the very places and circumstances that had made Manchuria an epicenter of the disease throughout the first half of the twentieth century.

The germ warfare allegations and their investigation may be seen as a highly political public relations attempt at denying a reality of Manchuria's nature under conditions of modernity. Plague still flared in its old haunts, and New China's scientific professionals, like the scientific professionals of all regimes that preceded it, were fighting a desperate battle to know and control the webs that linked Manchuria's nonhuman and human worlds. Through participation in the 1952 investigations, Chinese scientists with backgrounds related to a legacy of imperialism—men with degrees from Western powers, especially the arch enemy United States, and even men who trained with the Japanese war criminals of Unit 731—publicly proclaimed their allegiance to the PRC's New China, even as they dealt with the problems remaining from the old.

Nevertheless, through their creation of natural knowledge in the midst of war, PRC scientists began to mark Manchuria as a place of hope: no longer a plagueland, but a land whose nature could support the future aspirations of "New China." In the years that followed, it would take a new "people's science," combined with the labor of the people themselves, to accomplish this seemingly impossible task.

RECLAIMED

Technology and Embodied
Knowledge on the Sanjiang Plain

Stand at any point and look at the land. Look at what those fields, those
streams, those woods produce, even today. Think it through as labour.
—RAYMOND WILLIAMS, *The Country and the City*

In February 1954, Zhang Yuliang and other scientists of the Forestry Soil Research Institute of the PRC Academy of Sciences made a study of the plants
and soil of the prairie in the middle of the Three Rivers Plain (Sanjiang
pingyuan). As he traversed the frozen ground in a jeep and on foot, Zhang
noted that the land had a slight pitch: elevations were higher to the southeast
near the Wanda Mountains, but as one proceeded northwest toward the interior, the elevation decreased approaching the Songhua River. Although this
tilted prairie seemed flat as a tabletop, it was actually shot through with large
depressions that accumulated water and turned into vast swamps in the rainy
season. Multiple smaller rivers—the Qixing, Naoli, Hamatong, and many
others—wriggled through the territory like meandering snakes, their slow-
moving waters creating spongy marshes on all sides. The land was dominated
by wetland ecosystems, which Zhang, following the academic trend of the
time (and because Soviet experts conducted the survey with him) described
using Russian words, their Cyrillic letters dropped awkwardly into his report's
Chinese text: болот (marsh), луговые степи и луга (grassland steppe and
meadows), and кустарниковые растения (shrub plants).[1] Zhang observed
that the majority of the land was covered in tall prairie grasses, their rhizome
root systems forming a tough underground net, their feathery reeds extending in all directions as far as the eye could see. These were the same north-

ern grasslands that had so enraptured the Russian botanist Karl Maximowicz when he stumbled upon their rich diversity as he explored the Amurlands in the mid-nineteenth century. But instead of finding the "silky shimmering reds" and "iridescent blues" of summertime wildflowers, Zhang encountered a barren yellow-brown winter landscape dusted with snow: Communist Party authorities in Beijing had insisted that the survey of the land be completed that winter. Conducting a plant survey in the wintertime, Zhang noted, was not ideal.

The land that Zhang Yuliang examined in the winter of 1954 had experienced many changes over the past decades. This area at the intersection of the Ussuri, Songhua, and Amur Rivers had been the homeland of the Nanai (Hezhe/Goldi) people who provided the Qing court with sable furs and whose Dersu Uzala had guided Vladimir Arsenyev as he explored the region. By the 1950s, however, the Nanai population had dwindled precipitously and were in the process of being categorized as the smallest of all fifty-five official "national minorities" of the People's Republic.[2] For a century, Han Chinese farmers had made agricultural inroads in the region, and surveyors encountered the settlers' fields and villages on the prairie. In spite of this development, millions of acres of wetlands and prairies still remained in this region. Because of the predominance of uncultivated land, this corner of northeastern China was commonly called the Great Northern Wasteland (*Beidahuang*).[3]

Zhang was aware that Mao Zedong himself had called upon the Great Northern Wasteland to become the nation's "grain factory" (*liangshi gongchang*), the site of large-scale mechanized farming that would produce food for New China under the leadership of the Chinese Communist Party (CCP). CCP leaders, together with advisers from the Soviet Union, had selected this site fifty miles east of Jiamusi as a perfect place to establish a model "state farm" (*guoying nongchang*): a scientifically advanced, rationally planned, industrial-model farm that would use the latest technology to achieve unprecedented yields. In their vision, this huge stretch of flat land would be transformed by armies of Soviet machines—tractors, seeders, harvesting combines—manned by thousands of Chinese workers and rationally managed by the state. With this combination of technology, labor, and CCP guidance, the land of Manchuria would be transformed from the Great Northern Wasteland (*Beidahuang*) into the Great Northern Granary (*Beidacang*). In Zhang Yuliang's expert opinion, however, this part of Manchuria was more water than land. While he couched his conclusions in terms of questions that "require more research," for Zhang the facts were clear. In spite of what zealous revolutionaries might say, scientific investigation revealed that the Three Rivers Plain was too waterlogged for mechanized farming.[4]

The Three Rivers Plain was once China's largest freshwater wetland, a vast alluvial delta formed by the confluence of the Amur, Songhua, and Ussuri Rivers. Remarkably, this delta in the extreme northern reaches of Manchuria is now a major grain-producing region, home to some of the most productive farmland in all of China. While the region contains several PRC national nature reserves, the vast wetlands Zhang Yuliang encountered in 1954 have all but disappeared as a result of the success of state-directed projects to reclaim agricultural land. Beginning in the late 1940s and continuing throughout the Mao years, the CCP transformed the land through state-created farms manned by millions of demobilized soldiers, prisoners (political and otherwise), and urban youth "volunteers." The dramatic transformation of the Sanjiang landscape from the Great Northern Wasteland to the Great Northern Granary was achieved through the massive expenditure of human labor, deployed through a specific form of social organization—the state-managed farm.

For central planners in New China, the reclamation of Sanjiang's land represented a reclaiming of Manchuria from a place of alienation, war, and tragedy to a place of revitalization, a place where the dream of a prosperous, strong China could be launched. Manchuria was once "the land where the dragon arose," a place of numinous energy that pulsed beneath the ground to sustain the Manchu-dominated empire. We have seen how in the nineteenth and twentieth centuries, competing Asian empires mined the territory for the subterranean energy stored in fossil fuels. Many had also tapped the energy contained within the soil itself: fertile soils that could produce seas of soybeans, wheat, and rice. During the 1930s and 1940s, this agricultural abundance was stripped from the land and sent to feed the Japanese homeland and its imperial armies. After 1949, the socialist modernizers of New China sought to reclaim northern Manchuria's soil to feed the People's Republic. Manchuria, once a place of imperialist domination, could become a place of national salvation. Reclaiming the land for the nation also meant "reclaiming" the wetlands from nature: a task that locked science and labor in a struggle with the watery environment. This struggle to make the land of the Three Rivers "productive" transformed both the land and the lives of those who labored on it.

This chapter examines the coproduction of both nature and knowledge through human labor where water met the land in Manchuria. The chapter follows two intertwined threads: the spatial aspects of environmental change and the connection of environmental changes to human meaning making through work. As David Blackbourn has observed in his monumental study of the transformation of Germany's riverine landscapes, the environmental and the cultural represent "two halves of a single history," and the "cultural construct framed by observers" ought to be told together with the "reality of rock, soil, vegetation, and water."[5] This goal is particularly important when discussing the Three Rivers Plain. While almost all locales in China have been altered over the course of millennia in the human search for farmland, agricultural reclamation in Manchuria during the second half of the twentieth century is one of the most dramatic examples of landscape transformation in PRC history. And while there are many reclaimed-land stories in China, the Three Rivers Plain is perhaps the most famous of them all. The story of land reclamation on the Sanjiang Plain not only forms the foundation for knowing Manchuria; it is central to knowing China itself.

The transformation of the Great Northern Wasteland is one of the most abiding positive cultural narratives of the Mao era (1949–1976). Well into the twenty-first century, the successes and sacrifices encountered by PRC citizens who "opened up the wasteland" (*kaihuang*) of Beidahuang in the 1950s through the 1970s have been celebrated in memoirs, novels, popular songs, films, and television series.[6] There is a genre of literature known as "Beidahuang literature," with its own journals and practitioners, as well as a distinctive form of "Beidahuang art." This enormous cultural output has multiple origins: PRC state media that strives to magnify the heroic successes of state-guided development, the writing of established intellectuals who were exiled to these state-managed farms during the political purges of the 1950s, and the memories of over a million "sent-down youth" from China's major cities who labored in the Great Northern Wasteland during the Cultural Revolution (1966–1976).

The narrative generated about Beidahuang is fairly consistent: the region was barren, unproductive, and devoid of human settlement before the arrival of workers under the Red Flag. Inspired by communist ideals and honest youthful naiveté, individuals sent to the region labored mightily against an extremely harsh environment to open the land. This was tough, exhausting, tedious manual labor, done at the cost of great personal tragedy, but the results were remarkable: the "barren" wetlands were transformed into some of the most productive fields in the nation. The combination of the pristine environment of Manchuria and the indomitable spirit of the Chinese people helped realize the national dream of self-sufficiency and modernity. Indeed, this *Beidahuang jingsheng* (spirit of the northern wilderness) is often lauded as a crystallization of the spirit of the Chinese people themselves: honest and simple, unafraid of hard work, accustomed to sacrifice, and as a result, destined for greatness.[7]

This chapter considers the Sanjiang Plain as a Manchurian landscape that opens a window into the knowledge created by labor. In his classic 1973 study *The Country and the City*, Raymond Williams sought to debunk nostalgic myths of an unchanging, picture-perfect English countryside. He pointedly urged readers to "stand at any point and look at the land. Look at what those fields, those streams, those woods produce, even today. Think it through as labour."[8] This exhortation, which forms the epigraph to this chapter, reminds us that even the most placid of agricultural landscapes have all been altered through intense human effort.

Such dramatic environmental "changes in the land" wrought by human labor have also created dramatic changes in human understandings of the nature of the land.[9] Thomas Andrews combines these two ideas in his concept of workscape, which he defines as "a place shaped by the interplay of human labor and natural processes. . . . [N]ot simply land, but also air and water, bodies and organisms, as well as the language people use to understand the world. The workscape concept treats people as laboring beings who have changed and been changed in turn by a natural world that remains always under construction. Wherever people work, in short, the boundaries between nature and culture melt away."[10] The Three Rivers Plain is a workscape extraordinaire: a place where every vista is the result of constant construction, even as the scenes to the eye give the impression of benign natural beauty. It is also a place where the "boundaries between nature and culture melt away," an ideal vantage point that allows us to be present at the exact point where human senses, environmental change, and meaning making meet.

This chapter seeks to establish embodied knowledge as the most important way of knowing Manchuria. Many environmental historians, following the lead of Richard White, have shown that work has been "a primary method through which humans have achieved knowledge of the natural world."[11] In contrast to scientific or aesthetic observation, working people, including lumberjacks, miners, sugarcane cutters, and farmers, engage with the non-human world through "backbreaking, enervating, heavy work"—digging, harvesting, hauling, trekking—and through this labor, gain intimate knowledge of nature.[12] This knowledge is extremely complex, manifested through intense bodily sensation and muscle memory developed through skilled grappling with the environment. The record of this work might not be found in the archive but is instead imprinted through pain upon bodies themselves, many of which bear what Thomas Rogers has called "the deepest wounds."[13] Some scholars have emphasized the way that work created linkages between man and nature, while others have emphasized the way that human labor, particularly when pushed to exploitative extremes on rivers, on plantations, or in mines, has resulted in extensive damage both to the environment and to human bodies.[14] Whether the emphasis is on holism or alienation, a focus on labor and environment brings our attention to the coproduction both of human experience and of nature itself: "How working people have experienced and transformed the natural world, as well as how they have been transformed by it."[15]

This chapter turns to the experiences of those who worked in the Sanjiang Plain in the 1960s and 1970s in order to understand how labor translated into knowledge of the Manchurian environment. The opening of the Great Northern Wasteland is often remembered today as a protean struggle that pitted the raw strength of man against a brutally harsh nature. The image of land reclamation performed during the Mao years is typically that of thousands of individual workers wielding pickaxes, shovels, and burden poles, workers "swarming like ants" in huge groups, "armies" that attacked the land as part of forced labor mass mobilization.[16] For centuries, the work of clearing and farming the land was a fundamental way that humans encountered Manchuria's nonhuman world. What is unique about the postwar period, however, is that many of those who labored left detailed written records of their experiences and impressions. Thousands of memoirs of state-farm laborers from the 1960s and 1970s are available from national and local presses, ranging from the literary creations of exiled intellectuals to the more straightforward recollections of sent-down "educated youth" (*zhiqing*) and soldiers. A surprising number of these recollections contain descriptions of specific forms of labor, along with musings about the ways that this labor shaped the way that humans understood nature. For the hundreds of thousands of young Chinese sent to northern Manchuria, knowing Manchuria has come through the experience of working its soil.

Work done by millions of laborers generated knowledge about Manchuria, but that labor was in turn shaped and mediated by science and technology. Northern Manchuria was the site of the PRC's most intensive effort to modernize and mechanize agriculture, and machines were central to dreams of transforming the land of the Three Rivers Plain. The historian of science

Sigrid Schmalzer has recently focused our attention on how important science was to the agriculture of the Mao years. Her influential study *Red Revolution, Green Revolution* focuses on science from the bottom up: relatively small-scale, intimate projects of breeding new seed strains or developing biological pest control, work conducted with the materials at hand in local experimental stations by farmers and specialists on the village level.[17] The Sanjiang Plain represents another side of Mao's Red and Green revolutions: large-scale, top-down, and industrial. Scientists from the newly established PRC took part in the work of surveying, planning, and executing the new state farms in the 1950s, and some state farms received an influx of "cutting edge" agricultural technology from the Soviet Union. In the 1960s and 1970s, a significant segment of the PRC's fledgling machinery industry output was directed toward state farms in the northeast. Indeed, while the Sanjiang Plain is undoubtedly a "workscape," a place shaped by the interplay of human labor and natural processes, it is at the same time a remarkable example of what Sara Pritchard has called an "envirotechnical landscape," a place that is made through the confluence of land, water, and the technological dreams of state developers.[18] On the Sanjiang Plain, smoke-belching tractors and whirring combines roamed the fields, while diesel-powered conveyor belts and threshers turned bucolic barns into factory floors. A consideration of the formation of knowledge at the point where human and environment intersect thus needs to consider the presence of a third party: the machine.

The chapter begins with a spatial and temporal introduction to the PRC's primary vehicle for transforming northern Manchuria—the state farm. Pinpointing the location of land reclamation projects over time reveals significant historical continuities between PRC state farms and land reclamation projects attempted by the Qing, the Chinese republic, and imperial Japan. Such a spatial approach clearly demonstrates that the goal of modernized agriculture on the Three Rivers Plain was a dream that encompassed multiple decades, multiple political regimes, and multiple actors. Only once we have situated state farms within the space of Manchuria can we begin to understand the environments that made land reclamation the site of struggle between land, water, human, and machine.

Defining Sanjiang

The Sanjiang Plain is located in the farthest northeastern corner of the People's Republic of China in Heilongjiang province (fig. 8.1). *Sanjiang* means "Three

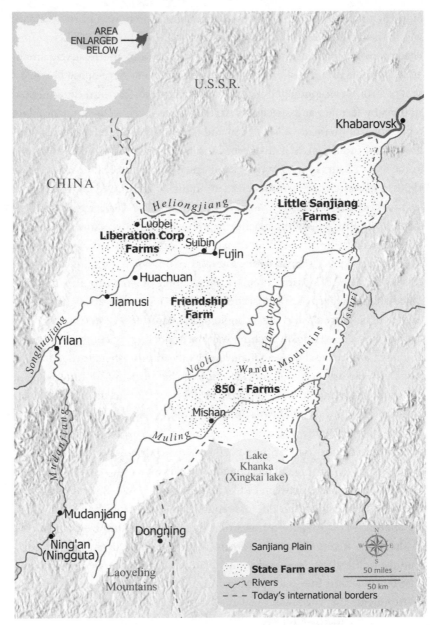

Fig. 8.1 The Sanjiang Plain, showing location of state farms.
Map by Jeff Blossom. See plate 20 for color version.

Rivers," and its triangle-shaped outline is formed by the confluence of northeast China's three largest rivers—the Amur (Chinese, Heilongjiang) River to
the north, the Ussuri (Chinese, Wusuli) River to the East, and the Songhua
River (Songhuajiang, known in Manchu and sometimes in English as the
Sungari) to the west. Part of the drainage basin of the Muling River (to the
east of Mudanjiang) is also included in the Sanjiang Plain, forming a central
axis that juts out from the bottom of the triangle. Today, the Sanjiang Plain
covers twenty-three counties, including the counties that make up the cities
of Jiamusi, Hegang, Shuangji, Qitaihe, and Jixi. Altogether, the region covers
approximately 109,000 square kilometers, or about 42,085 square miles. This
makes the Sanjiang Plain about the same size as the state of Virginia, or considering its borderland location at the northeasternmost corner of China, it
is perhaps better compared to the state of Maine. In terms of climate, a comparison with Minnesota or Michigan's Upper Peninsula works fairly well: the
Sanjiang Plain spans between the forty-fourth and forty-eighth northern par-
allels, and average temperatures range from –13 degrees Fahrenheit in winter
to 77 degrees Fahrenheit in summer.[19]

The Sanjiang Plain is considered synonymous with wetlands, but like
the whole of Manchuria itself, the region encompasses a wide variety of terrain and ecologies. In the southern and central parts of the region, the forested ranges of the Laoye, Zhangguangcai, and Wanda Mountains form an
arc of elevated terrain reaching as high as three thousand feet, creating an
Appalachian-like landscape of smooth-topped low mountains. The three major rivers, fed by their multiple tributaries, all flow (or wriggle) in a northeasterly direction through a broad flat plain that averages an elevation of about
160 feet above sea level. In the center of the Sanjiang Plain, the landscape
is dominated by wet grasslands. The landscape in the furthest northeast ex-
tremity of the region, known as the Small Sanjiang (essentially the top of the
Sanjiang triangle), is a low-lying delta formed by the confluence of the three
rivers. Here the landscape forms a watery world, dominated by looping meanders and massive marshes.

We have already visited this place in chapter 4: the Sanjiang Plain encompasses the prairie region where Karl Maximowicz encountered fields of
wildflowers in 1855. The region around Xingkai Lake (Lake Khanka) and
the Ussuri River in the Sanjiang Plain is where Dersu Uzala saved the life
of Arsenyev. Yamagata Miyuki did his wartime birdwatching here. Even the
early Qing exile Wu Zhaoqian was a denizen of the Sanjiang region: Ningguta
was positioned where the northernmost reaches of the Changbaishan range

begin to fade into the watery land of the river delta. On the border between Russia, China, and North Korea, encompassing the region's dense rainforests and expansive prairies, its White Mountains and its Black Waters, the Sanjiang Plain is a fitting site to end our exploration of the borderland natures of Manchuria.

This liminal space is also a paradox of nature. Although the Sanjiang Plain is one of China's coldest and most inhospitable regions, it is at the same time home to some of China's richest soil: the famed *heitu*, or "black earth" of Heilongjiang. The technical term for this black earth comes from the Russian, *chernozem*, a combination of the terms for black (*cherny*) and land (*zemlya*). The soil is naturally high in organic content, ammonia, and phosphorus, and even in an unimproved state can initially produce extremely high yields.[20] *Chernozem* exists in a few places on the globe: in the North American plains stretching from Canada to the central United States, and in a wide band in the Eurasian continent extending from Eastern Europe through the central Russian steppe and into Siberia. Beginning in the early nineteenth century, European scientists marveled at the ability of this soil to generate agricultural wealth, calling it for Russia "what coal is to England."[21] The head of the St. Petersburg Botanical Gardens (and Karl Maximowicz's boss), Franz Josef Ruprecht, even compared the riches of *chernozem* to that of "gold and diamond mines."[22]

Qing observers had long noted the black earth of the empire's northeast, but as the historian of Chinese agriculture Peter Lavelle has recently demonstrated, it wasn't until the crises of the late nineteenth century that the Qing court was jolted into realizing that "the preservation of their empire required the transformation of its borderlands into replicas of [Han] Chinese territory."[23] This meant rapid agricultural development of a region once conceptualized as a Manchu reserve. Once these more northerly regions were open to Han agriculturalists, the most rapid settlement took place in *heitu*-rich areas that afforded good riverine transportation, especially along the Songhua River. In these rich soils, Chinese farmers with few resources and relatively basic tools could still produce strong yields that made their enterprises economically viable. The flat lands of northern Manchuria not only attracted Han family farmers; modernizers from around the world attempted to introduce mechanized farming to the region in the hopes that the land could provide grain for modern armies and turn profits for investors. Many, however, discovered that the riverine environment foiled their machine-fueled dreams of exploiting the black earth.

The Elusive Dream of the Machine—Reclamation before 1949

Official PRC histories for the Sanjiang area tend to emphasize the radical unproductive or uncultivated (*huang*) aspect of this part of the Beidahuang before the triumphant advent of massive state-run farms. In these narratives, the land was fallow marsh where "the smoke from peoples' hearths was rarely seen" (*ren yan xi shao*). While there were certainly large stretches of land that had yet to be cultivated when the PRC was established in 1949, these lands were neither "wide open" nor "untouched"; for centuries they had been part of a complex web of land ownership and resource exploitation particular to Manchuria. As we have seen, the Qing court defended the northeast corner of their empire as a forested resource base for Manchus, but military outposts required food supplies. Both Ningguta, the place where Wu Zhaoqian languished in exile, and Yilan, the place where Qing officials feasted with indigenous hunters bearing fur tribute, were within the Sanjiang Plain. Given the demands of upkeep for Qing garrisons, land around these settlements had been under cultivation by Manchu banner estates from at least the mid-sixteenth century. As we have seen in chapters 2 and 4, much of the labor on these estates was provided by unfree Han Chinese laborers, prisoners and exiles who tilled the land under harsh conditions that at their worst could resemble serfdom. Records from the Qing detail searches for runaway laborers described by the appearance of their clothes and even the color of their skin. Some Chinese laborers nevertheless managed to establish themselves as long-term resident-cultivators of the land, but the land itself remained under the control of the Manchu state.[24]

The late nineteenth and early twentieth centuries saw a radical transformation as the Qing court began intentionally to open banner-lands in Manchuria to free Han settlement in an attempt to bring the region more firmly into control against the expansion of the Russian and Japanese empires. As early as the 1860s, the imperial government granted use of land in the southern parts of Sanjiang to Han Chinese owners in return for their promise to settle the land. Thousands of refugees, both Han and Korean, moved to the region from the south and the east during times of famine and natural disasters.[25] During the New Policies Period, the state finally ended its monopoly on land ownership, and millions of acres of land were auctioned off to private owners. In spite of attempts at limiting the amount of land that any one individual or concern could purchase, wealthy investors (individuals as well as companies) purchased huge tracts of land in Jilin and Heilongjiang provinces.

Large landowners rented their holdings to individual farming families, many of whom were able to ultimately purchase their own smaller parcels.[26]

It is estimated that by 1920, over half a million acres had been opened up to cultivation on the Sanjiang Plain, with soybeans, wheat, and especially opium poppies being the preferred cash crops.[27] Cultivated areas were focused on areas close to transportation along the lower reaches of the Songhua River, especially around the port city of Jiamusi, and also in areas near the China Eastern Railway, for example, around Ning'an (Ningguta), Mudanjiang, and Suifenhe. Cultivated areas also emerged around locales that were settled as the result of mining activities, especially in the vicinity of Mishan County or Xingkai Lake (Lake Khanka) on the Russian border, where coal was exploited first by Russians and then by the Japanese.[28] The narrative of the PRC's state farms rising heroically up from empty wastelands tends to ignore this vital economic history of northern Manchuria.

Huachuan county, to the east of Jiamusi along the Songhua River, is one example of the rapid development that parts of the Three Rivers Plain underwent well before the establishment of the People's Republic. Originally administered as part of the banner system from Yilan, Huachuan was a focus of multiple development plans under the Qing, including gold mining and land reclamation, with over fifty-three thousand acres of land opened to Han Chinese farmers in the late nineteenth century.[29] Some place-names in the region, for example, "Hunan Camp" (*Hunan ying*) reflected the origin of the workers who were recruited by the Qing state from as far away as Hunan province over fifteen hundred miles away. The region became a county with its own administrative offices the same year China became a republic. As James Reardon Anderson has pointed out, the establishment of the "county" system indicated that enough Han migrant agriculturalists had settled in the region to merit the establishment of standard Han-area governance—a process that manifests the spatial pattern of settler societies in northern Manchuria.[30] By the mid-1920s, when local elites edited the county's first history, Huachuan was producing over sixteen million pounds of wheat and counted tens of thousands of grain-laden carts passing over its roads.[31] While the gazetteer's compilers proudly touted the region's economic development, they also noted that natural resources, especially forests, were being rapidly depleted. According to observers writing in 1928:

> For the 267 years of the Qing, this area was viewed as a no-man's land, its grasses and trees vast and boundless, its rushing unrestrained rivers like

a moat at the edge of the world. All the animals and plants that soared or crept within its confines were free to grow and flourish in accordance with their own natures. Since the establishment of our county, trees have been cut down to build shelters, the land has been opened up and turned into farms, and the forests have been cut almost to the point of extinction. . . . If this rate of loss continues over the next few years, how will we be able to attain any resources?[32]

Locals were of two minds about the best way to approach the land, with some expressing enthusiasm for rapid development while others questioned the virtue of a rapacious market: "some people call for utilizing all the riches that the land has to offer, draining the endless stores of Creation. There are others who are concerned about the situation and advocate only using as much as is needed for self-sufficiency."[33] While this narrative of environmental decline from a former time of pristine simplicity had been applied to the Qing empire's borderlands since the seventeenth century, the twentieth century had undoubtedly brought a dramatic shift. Observers of the time noted that this impressive agricultural development and accompanying environmental depletion was accomplished not by machines but by individual farmers laboring with basic, even primitive tools: ox-drawn plows, rough-hewn wooden hoes, and spindly scythes.[34] The abundant wealth of the Heilongjiang soil was extracted through the backbreaking labor of individual men, women, and children.

There were several attempts to introduce mechanized farming to the region at the turn of the twentieth century as part of a late Qing drive to transform borderlands into agricultural landscapes in order to modernize and strengthen the empire.[35] During this time, dozens of companies formed by Chinese and foreign investors bought up hundreds of thousands of acres in the Three Rivers Plain area, mostly in Suibin county (located in the triangle of land between the Songhua and Heilongjiang Rivers), and Dongning county (the former site of Yamagata Miyuki's wartime birdwatching endeavors located to the east of Mudanjiang). Some of these groups experimented with large-scale corporate farming with tractors and modern equipment worked by managers, technicians, and hired farmhands. According to the PRC historian Sun Zhanwen, almost half of the land in Suibin county was controlled by five companies run by Chinese "warlords and government officials," a term that in this scholarship inevitably suggests corruption and oppression. Nevertheless, it was this "bureaucratic capitalism" that pioneered the use of foreign-made steam-powered tractors and harvesters in the far northlands in the early twentieth century.[36]

Americans were among those who brought machines to the Sanjiang Plain. Seeking to maintain its "open door" policy and offset increasing Japanese influence in the region after the 1904–1905 Russo-Japanese War, Washington aggressively pursued routes to insert American capital and technology in Manchuria at the beginning of the twentieth century.[37] In 1909, American political and diplomatic figures, including onetime State Department consul at Shenyang Willard Straight (1880–1918) and the US ambassador to China Charles Richard Crane (1858–1939) established the Manchurian Development Company that opened a fifteen-thousand-acre farm in Suibin county with the support of Chinese government officials of the early republic.[38] The farm was overseen by another American investor and China adventurer, William Morgan Palmer (1887–1925). Palmer, a Harvard graduate of the class of 1910, had taken up positions as (a very youthful) adviser to the new republican government of China and used his connections to court Chinese investment in the "American ranch."[39] On this American ranch between the Songhua and the Amur rivers just a few miles from the border with Russia, Palmer built a large number of houses, stables, and storehouses, managed hundreds of Chinese and Korean farmworkers, and employed American university professors as technical advisers. Palmer also oversaw the importation of a "great quantity" of advanced American agricultural machinery, and employed these machines to open miles of land. A rare photograph from Palmer's collection shows some of this machinery operating on the Sanjiang Plain: a tall tanklike tractor with a canopy to shade the driver from the sun, pulling a long line of five heavy-duty disc harrows (plows) behind it (fig. 8.2). The photograph shows this American technology opening up wide swaths of Manchurian soil while a lone worker stands balanced on the plows, looking off into the far distance.

The ranch, however, soon encountered insurmountable problems. In particular, the swampy nature of the land presented challenges to the machines: Palmer noted that a great deal of constant "dyking, ditching, and pumping" was necessary to protect the fields that had been "laboriously opened by the large tractors and worked with great difficulty in the wet soil."[40] Finally, political instability following the 1911 Revolution forced the abandonment of the farm. Admitting defeat, the Americans sold their imported tractors and plows to Russian merchants in Harbin. The Americans departed, but the Chinese and Korean farmers they had hired remained on the land, working their rented tracts of land without the benefit of the American's expensive but useless machinery.[41]

Fig. 8.2 American tractor and cultivators at the "American Ranch" on the Sanjiang Plain, c. 1911. William Morgan Palmer Collection, #P956, Harvard-Yenching Library.

The American efforts at mechanized farming had little sustained legacy, but Japanese agricultural projects in Manchuria had a far greater impact on the subsequent development of the region. From the turn of the century until 1945, Manchuria was a site of utopian schemes for the Japanese state. Establishment of Manchukuo in 1932 ushered in a new era of enthusiasm for north Manchuria's potential on the part of Japan's imperial planners. The Japanese formed their own Manchurian Development Company (Manshū takushoku kōsha) that was far more extensive and stable than its small-scale American predecessor. In the minds of Japanese planners, scientific research and industry in Manchukuo's cities would unlock new technologies that could overcome the region's challenging environment, and the millions of acres of "uncultivated and empty" land in the northern part of Manchuria would become the new colonial homes for poor Japanese tenant farmers who suffered from overcrowded conditions on the home islands. From 1934 to 1945, the Japanese state oversaw the transfer of hundreds of thousands of farmers from inland Japanese villages to the "wide open plains" of northern Manchukuo. On paper, each farmer was promised almost fifty acres of unoccupied virgin land, along with subsidies and expertise to introduce mechanized farming techniques to the land. Propaganda featured happy farmers astride modern tractors, reaping spectacular harvests that were crucial fuel for the Japanese war effort.[42]

The reality was different. Much of the "virgin lands" given to Japanese colonial settlers were actually cleared and cultivated tracts wrested from the Chinese farmers who had worked the land for decades. Many of these forced transfers took place in the Sanjiang Plain. For example, the "model" Japanese colonial villages of Iyasaka and Chiburi were located in Huachuan county, a place that was, as we have seen, the site of intensive Han Chinese agricultural development in the first decades of the twentieth century.[43] Other villages were established in the Mishan area in eastern Heilongjiang on the border with the Soviet Union, an area that had been developed for its coal mining in the late Qing and was one of the most heavily populated spots in the Sanjiang region. As Louise Young has described, Japan's colonization of Manchuria's "empty land" was in fact a massive land grab, "an enormous land transfer accomplished through price manipulations, coerced sales, and forced evictions."[44] Some dislocated farmers were conscripted to build Japanese fortifications. Others were rounded up into what were essentially concentration camps, fenced-off villages where they were "protected from banditry" but also monitored for anti-Japanese activity.[45]

The Chinese suffered greatly under Japan's colonization schemes, but the Japanese settlers themselves also faced insurmountable difficulties. The Japanese Manchurian Development Company had promised to provide tractors to its colonial farms, but there was a severe shortage of machinery for the almost three hundred thousand Japanese farmers who settled in Manchuria.[46] Given the dearth of modern equipment, the Japanese families found it impossible to farm their fifty-acre land grants on their own, and often wound up employing the previous Chinese owners as hired hands to work the land.[47]

The Sanjiang Plain was not untouched, primitive territory—instead, the region was an important site of agricultural extraction under strained conditions of capitalist and imperialist contestation going back centuries. Dreams of employing modern farm machinery tried to take root in the early twentieth century, but it was a dream that could not be fulfilled for economic, environmental, and political reasons. Modes of development were contested, empires inflicted misery and made false promises, and visions of a modern Manchurian breadbasket remained beyond reach.

Land Reclamation under the CCP

Dreams of opening the land of the Sanjiang Plain through machines continued under the CCP. Established in Shanghai in 1921 but forced to leave urban areas by Chiang Kai-shek's Nationalist forces, for more than twenty years the CCP grew roots in rural China, first in the lush sub-tropical southeast (Jiangxi) and later in the dry sandy foothills of north-central China (Yan'an). Events at the end of World War II and the beginning of the Chinese Civil War made the northern Manchurian countryside a key terrain for CCP military strategies. While the Kuomintang occupied Manchuria's cities after World War II, the CCP established its base in the Manchurian countryside and used Beidahuang as an "anvil of victory" in its encirclement campaigns against the Nationalists.[48] Previous scholarship has focused on how the CCP engineered violent class struggle and land reform among village populations in the countryside after 1945 as a way to establish its political power and enlist rural populations to its cause. Decades of unequal land-ownership patterns in north Manchuria certainly created circumstances favorable to communist-led redistribution of land to the tiller. But buried within the narrative of *fan shen* ("turning over the self," or poor farmers being reborn by overthrowing power structures and receiving individual parcels of land) is a narrative of *fan tu*, or "turning over the soil": realizing the potential of Manchuria by opening up more land to cultivation with modern machines under CCP-guided planning, investment and management.

Official histories credit Mao Zedong and his economic advisers Chen Yun (1905–1995) and Li Fuchun (1900–1975) with brilliant and original insight about the value of the Sanjiang Plain. As early as 1945, they began to think of the northern wastelands as a potential "grain factory" (*liangshi gongchang*)—a place where vast stretches of land could be opened up with modern farming equipment to provide food for the People's Liberation Army as it launched the military conquest of all of China.[49] For all their foresight, it is clear that Chinese communist approaches to utilizing north Manchuria's resources contained many similarities to those of previous regimes. Like the Manchu court before it, the CCP saw the northeast as a special power base that was crucial for the regime's military needs. Like the Qing court who sent convicts and exiles to Manchuria to do farmwork on lands around banner outposts, the CCP used political dissidents, criminals, and soldiers as a captive labor force to work the land. Like all previous regimes, the CCP saw northern Manchuria as a destination for "excess populations"—not only as a site of open migration but also as a site for the engineered transfer of underemployed or potentially dangerous groups. Finally, like the investors of the early Chinese republic and state planners of imperial Japan, the CCP also saw northern Manchuria as a frontier waiting for the application of modern farming technologies, a flat and fertile land where science would produce utopian harvests.

These visions came together in the establishment of *guoying nongchang*, or "state farms." Unlike the communes that would emerge from the collectivization of China's preexisting villages in the 1950s, these farms were created ex novo through direct state planning and the introduction of labor from other locations. Official histories point to the Soviet Union's *sovkhoz* (short for *Sovetskoe khozyaystvo*, or "Soviet farm") as the inspiration for China's state farms, and the Soviet model, combined with direct Soviet assistance, was crucial to the establishment of China's state farms through the 1950s.[50]

A spatial approach to the emergence of state farms in the Sanjiang Plain makes clear, however, that New China's state farms had picked up where a hundred years of agricultural colonization efforts launched by the Qing, the Republic, and imperial Japan had left off. Fig. 8.1 shows the general locations of the different types of farms established by the CCP. The CCP's first attempt to establish a mechanized farm in the region started in 1948, when a team led by a cadre from Yan'an dug three abandoned Japanese tractors out of the mud of a former Japanese settler community in Shangzhi county near Harbin, a region originally developed during Qing reforms in the mid-nineteenth century. The project had to move three times before they found land that the tractors could successfully plow, finally settling in Ning'an county near the city of Mudanjiang: a site, ironically, not far from the old Manchu settlement of Ningguta where Wu Zhaoqian and his southern Chinese literati colleagues had been exiled in the mid-seventeenth century.[51] In 1948, the CCP also established the Hejian Waterworks Farm at the site of an irrigation complex left behind from a Japanese colonial farm in Huachuan county worked by Korean farmers.[52] These and other early CCP farms not only used sites and equipment left over from the Japanese: in some cases they even employed "left over" Japanese settlers as technical personnel to drive and service the unfamiliar farm equipment, which included German Hanomag, American Ford, and Japanese Kubota tractors that remained from a half century of Manchuria's land development schemes.[53]

Throughout the 1950s, the landscape of the far northeast became dotted by more and more state farms in patterns that replicated earlier settlements: counties on the lower reaches of the Songhua River around Jiamusi City and the area around Mishan close to Xingkai Lake and the Ussuri River. It is not surprising that such land had already been occupied, and the "headquarters" of state farms were established in or around small Chinese villages. While they were centered on preexisting settlements, state farms set ambitious boundaries that far exceeded the parameters of the previously cultivated land around these villages. One pamphlet on state farms from the 1950s somewhat fancifully but vividly expressed this in the words of an "old local villager," who "in the past, always looked out her window and saw a vast prairie that no one cultivated." With the arrival of the state farm, she was now "thrilled that the land would finally be put to use." While local villagers had opened as much land to cultivation as was possible with plows, draft animals, and the labor available in their village, for state farm planners, the slogan was *huangyuan duo da, nongchang jiu duo da*: "However big the wasteland is, that's how big the farm shall be."[54]

To provide the manpower for the expansion of agriculture in the Great Northern Wasteland, the state used the same strategies employed by previous regimes, populating farms with unfree labor. Among the first state farms were those designated as People's Liberation Army Officer Education and Guidance Corps (Jiefangjun guanjiao tuan), or "Liberation Corps Farms" (Jiefangtuan nongchang) for short. These were farms designed to hold captured Nationalist army troops and those accused of being KMT sympathizers and spies who were forced to work the land under the direct supervision of People's Liberation Army soldiers. Altogether, over fourteen thousand men from all across China were sentenced to farm the Sanjiang Plain in what were essentially prisoner-of-war camps. The fledgling PRC state also established over three dozen "labor reform" (*laogai*) farms under the direction of the Public Security Bureau (China's centralized state police). These *laogai* farms held recalcitrant criminals—thieves, vagabonds, and dissidents—who had failed to be transformed by the more benign correction of "thought reform."[55] The inmate-farmers came from all over China, and they worked very hard. Even though the guiding principle of land reclamation in Beidahuang was to introduce mechanized farming and to maximize productivity, in these state farms, mechanization took a back seat to politics and punishment: the guiding slogan for these farms was *gaizao di yi, shengchan di er*, or "reform through labor first, production second." For the first few years of their existence, the massive farms were worked entirely by hand.[56] Given the location of the farms, their machine-free (and stridently communist) nature was a particular irony: they were established in the wedge of land between the Songhua and Heilongjiang Rivers that had once been the site of the "American ranch" of William Morgan Palmer, the Harvard man who had tried (and failed) to introduce capitalist mechanized farming to northern Manchuria in the early twentieth century.

The second half of the 1950s saw tens of thousands of soldiers from southern China shifted en masse to PLA Railway Corps farms in the Sanjiang region. The first wave of arrivals in the mid-1950s were primarily demobilized soldiers from the Korean War. Moving these soldiers to the north killed two birds with one stone: it alleviated potential employment problems caused by the demobilization of the massive Korean War–era People's Volunteer Army and provided manpower to grow more grain for a growing postwar PRC population. The first wave of soldiers was sent to the triangle between the Songhua and Heilongjiang, where the Army River (Junchuan) and Forever Army (Yanjun) farms joined the labor reform and Liberation Corps farms already in the region. In 1955, tens of thousands of soldiers of the Railway Corps under the direction of the famed PLA general Wang Zhen (1908–1993) were sent to the Xingkai Lake and Mishan area, the site where the Nanai hunter Dersu Uzala had once saved the life of his "Kapitan" by building a shelter of reeds in the face of a ruthless blizzard. There in the low foothills of the Wanda Mountains, the soldiers constructed ten farms designated with numbers from 850 to 859. By the end of the PRC's first decade, the Sanjiang Plain was home to over one hundred thousand soldiers who had hailed from all over China but now found themselves living as "pioneers" of land reclamation in the far northeast.[57]

Farm 853, nestled at the foot of the Wanda Mountains about forty miles to the west of the Ussuri, is a well-documented example of a farm established by the PLA Railway Corps. According to the farm's official history, the area had previously been the site of a failed reclamation village in the late Qing. During the occupation, Japanese authorities established two Japanese settler communities here by forcibly removing existing Han farmers into a "concentration village" ironically christened "Yamato" (a name propagated by Japanese wartime nationalists to designate the people and "spirit" of Japan, but read in Chinese as "Great Peace," or *da he*). By the time the Railway Corps soldiers arrived in 1956, the hundred or so remaining residents of Great Peace had been liberated and undergone land reform (plots of land redistributed to the poorest villagers), but to the north of the village lay seven hundred square miles of uncultivated meadows and marshes stretching from the Wanda Mountains to the banks of the Naoli River. Before the arrival of the Railway Engineer Corps soldiers, two thousand convicts were transferred to the region to build a forty-mile-long road from the county seat to Great Peace. Once the road was done, 2,700 Railway Corp soldiers traveled down it to the new site of Farm 853, where they encountered a daunting set of tasks. Roads needed to be built, fields needed to be surveyed, irrigation plans made, and above all, ditches had to be dug to drain swamps and prevent floods during the rainy summer months. Soldiers even had to build their own living quarters, initially simple lean-tos that did little to keep out the harsh cold in a place where average temperatures in the winter hover around 0 degrees Fahrenheit.[58]

One quintessential image captures the extreme hardship of these early years of the Railway Corps farms: a group of men in baggy army uniforms attempting to till the land with an ancient plow. The plow is guided by one soldier who stands knee-deep in mud. Instead of machines or farm animals, the plow is pulled by six men, each tethered to the plow like oxen, their muscles bulging and steps faltering under the strain of pulling the heavy iron plow through the thick black earth.[59] The labor of these soldiers, lauded in PRC stories and film, is well known in China as a symbol of PLA heroism and self-sacrifice.[60] While these experiences are used as symbols of the "Beidahuang spirit," there is no denying that work on early state farms entailed the extraction of human muscle energy under conditions of severe deprivation. Without machines, labor on the Sanjiang Plain inevitably meant suffering.

Even if they were designed as places for labor-as-punishment or staffed by soldiers already committed to bodily sacrifice, human labor alone could not fulfill the spectacular expectations for farming in the Sanjiang Plain: machines were needed to achieve the planners' dreams. A small number of new tractors from the Soviet Union, at first driven slowly at wintertime across the frozen Amur River, helped bolster the productivity of these early farms. Not surprisingly, the presence of machines made a huge difference in the amount of land a farm was able to clear. Farms that were not high on the mechanization priority list, such as the labor reform and Liberation Corps farms, had abysmal rates of cultivation. In the first year of the Number One Liberation Corps Farm in Baoquanling, one hundred people working in shifts using only hoes and shovels were able to open only a half an acre of land per day. The Qixingpao labor reform farm near Baoqing was only able to open a total of 26.5 acres in its first year. After the miserable performance, the Qixingpao farm was assigned nine tractors, and in its second year opened a total of 6,600 acres.[61] Clearly, machines were absolutely essential to the dream of "however big the wasteland, that's how big the farm will be."

But even the presence of machines could not guarantee the permanence of reclaimed lands: the environment would often prove resilient against human attempts at curbing it. During high-tide "production pushes" from farm managers (called reclamation "waves"), land was frequently "blindly opened" (*mangmu kaihuang*) without adequate surveying or soil testing. Such was the case, apparently, for the first farms established during the civil war with the

Nationalists in the late 1940s: an official gazetteer admits that of over twenty-one thousand acres that were opened by CCP-managed farms in Heilongjiang prior to 1949, more than half wound up returning to "waste." Similarly, the famous Army Railroad Corps in the Baoqing/Mishan area (the "850 Farms") claimed to have opened over half a million acres by 1956, but much of the land was low-lying wetland and wound up returning to marsh. An inability to master water control (*shuili*) resulted in a great disjuncture between *ken* (opening up land for agriculture) and *jian* (literally "construction"). Simply clearing a wetland of its naturally occurring flora—even if done with machines—was not the same as making the land productive.[62]

The Triumph of the Machine: The Sino-Soviet Friendship Farm

The transformation of the Sanjiang environment could not be accomplished by merely supplementing human labor—it required a mass infusion of advanced agricultural technology. The PRC's guiding force in the early years, the Soviet Union, would be the source of this necessary ingredient. The showcase of mechanized, modern state farm on the Sanjiang Plain was the Youyi (Friendship) Farm, established in 1954–1955 with massive assistance from the Soviet Union. Youyi was established in Jixian county, to the east of Huachuan county in flat prairie land between the Laoyeling Mountains and the Songhua River, only sixty miles from the Soviet border. According to official histories (and numerous pamphlets, newspaper articles, and even children's books from the 1950s featuring the marvels of the Friendship Farm), the farm was originally the idea of a Soviet agricultural delegation that pitched the idea of a massive Soviet-style, fully mechanized grain production farm to Mao during a visit to China in the autumn of 1954. This would be a shining example of the transformation of the Chinese countryside from unproductive wasteland to hyperproductive farm, an "agricultural factory" that would, because of its massive scale, terrain, and management, be capable of grain production unprecedented in its amount and efficiency. Under the direct management of the Central State Council, the Friendship Farm received high priority resources, manpower, and channel clearing, allowing agricultural visionaries to realize their plans as quickly as possible.[63]

Before the dream could be realized on the barren plain, infrastructure had to be set up from scratch. Soviet and Chinese personnel worked together to survey the site, erecting boundaries defining the farm's seven-hundred-square-mile perimeter, testing the soil, and establishing the location of headquarters and fields. Work teams constructed roads, laid electricity and telephone lines, and built a rail line from Jiamusi with a train stop exclusively for use by the farm. The central government sent dozens of recent graduates from China's agricultural universities, along with four hundred tractor drivers and mechanics from all over China, to provide the initial management and technical force. Unlike the army and labor reform farms where workers had to build their own primitive shelters, construction of office buildings and residential quarters was tasked to the Mudanjiang municipal construction bureau. Propaganda photographs from the time show a Potemkin-like town with handsome three-story Western-style brick buildings ordered on a grid of symmetrical paved roads, set on a plain otherwise absolutely empty to the far horizon. About 1,500 workers and their dependents made the barracks of the Friendship Farm their home.[64]

To support the farm, the Soviet Union provided massive amounts of equipment. During the winter of 1955, flatbed rail cars rolled into the newly built Youyi train station with over 2,000 pieces of Russian farm machinery, including tractors, trucks, automated seed broadcasters, and combine harvesters, along with numerous smaller cultivator/harrowing machines, winnowing machines, and conveyor belts.[65] This full set of machinery would make Youyi unique among China's state-run farms. While many farms possessed some mechanized equipment, particularly tractors and plows to open land, at the Friendship Farm, the entire process, from plowing to harvesting and storing, was to be thoroughly mechanized. This mechanization process was lauded and recorded in admiring detail in PRC media during the farm's first year.

Five years after the "liberation" of China, the land of northern Manchuria would finally be "liberated": Soviet machinery could plow where traditional methods had failed. According to reports, in the first thirty days after the machines arrived from the Soviet Union, the Friendship Farm was able to open up twenty-two thousand hectares. The dense, deep grassland soil was extremely thick and matted with roots. Local farmers claimed that to open up the soil to cultivation took a plow pulled by eight horses, and even then, the plow could only cut into the soil to a depth of a few inches, hardly enough for proper planting. Soviet technology had the answer. The Stalinets-80 tractor (known in Chinese material as the C-80, C for the Chelyabinsk factory where they originated) had an eighty-horsepower engine and tanklike continuous-track wheels that gave the machine an enhanced ability to tread over mucky terrain (fig. 8.3).[66]

One C-80 tractor pulling a five-prong moldboard plow could open up

Fig. 8.3 Soviet tractor and cultivators at the Friendship Farm, c. 1955. Reproduced from Ru Yan, *Guoying Youyi nongchang* (Beijing: Zhongguo qingnian chubanshe, 1956), 13.

fifteen acres of land in one ten-hour shift, the equivalent of ten days of work for one traditional peasant driving eight horses and pulling one two-pronged plow. The smaller Volgograd DT-75 tractor could open up seven acres in ten hours, a feat that would take peasants with a horse team five to six days to accomplish. The bright headlights and loud noises generated by the tractors also made plowing at night a possibility—tractors drove away dangerous nocturnal animals and "made the darkness disappear." Friendship Farm's ninety-eight tractors working all together for one twenty-four-hour period would be the equivalent of ten thousand people working for one day in the traditional manner. With these machines, breathless reporters observed, the "luxuriant grassland became black soil in an instant." Propaganda images from the early years of the Friendship Farm prominently feature this rich black earth (*heitu*). In a photograph from 1957 (plate 12), the *heitu* is carved into deep furrows through the might of the Soviet tractor (in the background, shining bright red against the blue horizon) and also through the wisdom of the Soviet expert (standing in the foreground, nattily dressed in a reddish-hued Western suit and surrounded by admiring Chinese colleagues).[67]

Machines continued through each step of the grain production process.

Soviet seed selecting systems determined size and quality of seeds, selecting several tons of superior seed in one hour. These seeds were then loaded
into planting machines that were pulled, three in a line, behind one tractor,
automatically sowing forty-eight lines at a time in precisely hewn rows only
two inches apart. One tractor could plant over forty acres in one shift. Harvest at Youyi Farm was accomplished with 94 C-6 semiautomated combines.
Pulled behind a tractor, one C-6 could harvest fifty acres of wheat in one shift.
The Friendship Farm also had several top-of-the-line Soviet fully automated
combines—nicknamed *tianjian gongchang*, or field factories. These combines
were faster than tractor-pulled combines and could not only harvest but also
winnow fifty acres of wheat in one shift. Media claimed that these one hundred combines working the fields of the Friendship Farm could harvest ten
thousand acres of wheat in one twenty-four-hour period, the equivalent of
the work of forty thousand "traditional" farmers. Grain processing was fully
automated as well. With electronic wheat-drying machines removing reliance
on weather, harvest could be managed in rainy conditions. Conveyor belts
moved the wheat to modern silo storage space, completing the process.[68]

As recounted in official media, the Friendship Farm seemed to have accomplished the dream of agricultural planners in China since the turn of the
century—turning the Great Northern Wastelands into the mechanized "grain
factory" of China. Massive machines allowed the reclamation of land that had
resisted large-scale cultivation for centuries. Automation reduced waste, bol-
stered efficiency, slashed labor needs, and resulted in unimaginable harvests,
all under CCP (and Soviet) management.

While the Friendship Farm was presented as the quintessential mechanized factory farm, there were obvious cracks in this shiny modern image.
In particular, radical discrepancies appeared in the amount of land "cleared"
versus the amount of land cultivated. Farms were vast, and in addition to cultivated fields, the stated size of farms also included roads, reservoirs, offices,
barracks, warehouses, roads, livestock pastures, and large areas of "empty"
land, land either deemed untillable because of marsh or simply land that
could not be worked due to limitations in machinery and/or labor force. The
size of the farm meant the geographical extent of the farm's boundaries and
not the area that was actually put to direct agricultural use. What land could
be sown was not determined by Soviet technology but was, rather, dictated by
the environment of the Sanjiang Plain itself.

Chinese scientists had anticipated these problems even before the first
tractor-drawn plow had turned over the soil. As we saw at the beginning of

this chapter, in the winter of 1954, a survey of the plants and terrain of the proposed site of the Friendship Farm was conducted by Zhang Yuliang, a botanist of the Chinese Academy of Science and a protégé of Liu Shen'e (1897–1975), one of New China's most distinguished plant scientists.[69] Zhang's survey was compiled with obvious deference to the Soviet experts involved in the farm project: he established that his survey was done according to protocols developed by Soviet scientists, and he used Russian words for technical terms describing the ecology and nature of the soil. Nevertheless, Zhang's survey came to a conclusion directly opposed to the claims of the farm's Soviet advisers: the land of the proposed Friendship Farm would not support China's dream of mechanized farming.

According to Zhang, the Friendship Farm encompassed three types of land: wetland or marsh, wet grassland, and slightly elevated shrub-bearing terrain. He admitted that the soil in areas of higher elevation was indeed the remarkably fertile chernozem: the famous black soil (*heitu*) of Manchuria. The presence of willows or fields of perennial flowering plants signaled *heitu* beneath, but these were a rarity at the Friendship Farm. The majority of the land was wetland covered with an intractable cover of prairie grasses, including the tenacious *Calamagrostis angustifolia* and *Calamagrostis purpurea* ("Scandinavian small-reed grass") that grew through the spread of a tough complex of rhizomes beneath the ground. Within the prairie wetlands, spots could appear dry in one season but would become waterlogged and unusable in another—the culprit being a remarkably high water table that rose from beneath and inundated the ground with any significant rainfall. Poorly drained depressions throughout the farm area gave rise to alkaline flats, capable of growing nothing but saltgrass (called "sheep grass," or *yang cao*, in Chinese). While Qing administrators and indigenous peoples of the region had once found multiple human uses for the region's reeds and grasses—to fatten war horses or as building material for life-saving shelters—from the perspective of socialist agricultural modernity, they were the very things that defined "wasteland." Zhang's report clearly concludes that the majority of land at the Friendship Farm was too waterlogged for mechanized agriculture, and was suited only for growing reeds and grasses that could be used as livestock fodder.[70] Given the emphasis that Beijing's political elite put on the success of this Sino-Soviet project, Zhang's scientific warnings went unheeded.

The opening of the Friendship Farm saw the opening of the next phase of the battle between the land of the Sanjiang Plain and the machines sent to tame it. In the first year of the Friendship Farm, heavy rains made it difficult to utilize the Soviet machinery in the efficient way envisioned by central planners. Tractors sank in unexpected marshes leaving only their top-mounted exhaust chimneys visible, or got hopelessly stuck in thick mud on waterlogged land. Chinese managers blamed the Soviets for choosing an area with a high water table and poor drainage. In response, the Soviets stated that they had conducted a thorough survey in the previous year, but it was during a time of drought, and they blamed the Chinese government for having no previous surveys or consistent long-term climate statistics for the area (a very unfair criticism, considering that the area had been occupied by Japan for more than fifteen years). Some critics even advocated abandoning the entire Friendship Farm project altogether, but the project was too politically sensitive to give up.[71]

The experience of the Friendship Farm demonstrated the centrality of drainage and water control to mechanized land reclamation in the Great Northern Wasteland. The waterlogged nature of the Sanjiang Plain had thwarted the tractor dreams of developers in northern Manchuria for generations, and the PRC goal of transforming the land into "grain factories" required an epic struggle between marsh and machine. The agricultural history of the region abounds in tales of these battles, and perhaps none is as famous as the story of Yanwodao. In 1957, the leaders of the Army Railway Corps 853 Farm at the base of the Wanda Mountains were determined to expand their cultivated area to include what appeared to be very fertile land in the middle of the Naoli River, on an island known as Yanwodao, or Crane's Nest Island. The island was actually in the middle of a "floating marsh," a land formation common in delta wetlands where groundwater is close to the surface, making the topsoil very unstable. A huge diversity of plants, thick grasses, and even small trees can grow out of the highly organic topsoil mat, giving the area the appearance of solid ground. This "ground" can even be walked upon but will easily give way to water if anything heavy traverses it. As the team from the 853 Farm made for the island across what appeared to be a solid natural bridge, they drove straight into a floating marsh. Six tractors sank in water up to their exhaust chimneys, stranding the team in the middle of the river. The machines were finally saved through the heroic efforts of a man named Ren Zengxue, who dove into the muck to secure chains around the axles of the tires for the tractors to be pulled out with large trucks that were usually used to haul grain after the harvest.[72]

Tales such as the story of Yanwodao are seen as manifestations of the spirit of the Great Northern Wilderness. Such stories also highlight other realities. The dream of opening vast amounts of wasteland—and the belief that it could be done with machines—clashed with the realities of the terrain; and in the struggle between water and machines, human bodies bore the brunt of the battle.

With new Soviet-made equipment increasingly out of reach as the Sino-Soviet relationship deteriorated through the late 1950s and early 1960s, the PRC began producing its own heavy farm machinery after almost a decade of industrial planning. The first Chinese-made Dongfeng (East Wind) tractors rolled (or rather, crawled) off the assembly line on their tanklike treads in 1958. By the mid-1960s, thousands of Dongfeng-52 and Dongfeng-75 wheeled tractors were being distributed to communes across China. As many as one-third of the tractors in the PRC were reserved for use by large state farms, with many going to open the wastelands of the northeast.[73] Eventually, as the water gave way to land, the tractor and its associated technologies became more and more common on the state farms. These machines would transform not only the Sanjiang Plain but also the way that humans experienced the environment.

Man-Machine: Sent-Down Youth on the Sanjiang Plain and Knowing Manchuria

We have seen how hundreds of thousands of convicts, prisoners of war, and soldiers labored on the Sanjiang Plain, but the clearest and most numerous voices that express personal experience of land reclamation work come from the urban youth who were sent to Beidahuang farms beginning in the 1950s, and particularly during the Cultural Revolution (1966–1976).[74] These "educated youth" (*zhishi qingnian*, or *zhiqing*, pronounced "jrr ching") were literate, and, in most cases, only temporary denizens of the Great Northern Wasteland. Most returned to cities where they assumed jobs, started families, and carried their memories of farm life into urban society. A small number went on to generate popular memoirs, films, novels, and art about their experiences. Even those who inhabited more mundane walks of life after their return have contributed essays, interviews, and oral histories for local and regional history projects over the ensuing years. Their recollections offer detailed accounts of forms of labor, allowing us to gain insight into the activities that actually constituted land reclamation and understand exactly how humans interfaced with the nonhuman environment.[75]

Photographs of sent-down youth in Beidahuang from the era—both those

taken by state media as propaganda, and those personal photographs reproduced in "down to the countryside" memory volumes, show young people
engaged, or about to engage, in a wide variety of labor.[76] Dressed in army
uniforms as members of the Cultural Revolution's military-style "Production and Construction Corps," fresh-faced young men and women with hoes
over their shoulders march out onto fields in lines behind the flags of their
"platoons." The lines are remarkably long, snaking along perfectly flat fields
and grasslands that stretch to the far horizon. Depending on the season, the
fields are covered in snow or luxurious with grass, but the vision of untapped
vastness is the same. Numerous photographs show the work of harvesting,
with youth (unrecognizable as such), bent over in fields or shin-deep in water,
cutting stalks with small sickles. Many other images show perhaps the most
"primitive" work of all—groups of people breaking rocks with pickaxes, moving massive boulders with ropes, sticks, and bare hands. The photographs
show great exertion, but exertion to what end is not always clear. Placing these
images of work into a sequence that reconstitutes the actual process of land
reclamation is a difficult task.

But one thing is quite striking: machines appear in almost all of the photographs. A snaking line of youth with scythes advances on the prairie, but
immediately beside them is a line of tanklike tractors. People are scattered
through a wheat field harvesting a few stalks at a time, but behind them in the
distance are massive automated combines, cutting down waves of wheat and
shooting endless streams of hulled kernels into a massive hopper. While these
images are intentionally composed to demonstrate the modernization of
farming in the north, a glimpse at *zhiqing* memoirs confirms the interweaving
of manual and mechanized labor. The extent to which one interacted with machines depended on a number of factors: what kind of farm one was assigned
to (a local commune, a labor reform camp, or a well-resourced state farm),
the natural environment and terrain in which the farm was situated, and even
such things as one's age and gender.[77] Examining *zhiqing* memoirs with an
eye for depictions of work allows us to go beyond the "hectares opened" and
"hectares cultivated" statistics of official histories and better understand how
human beings made sense of the nature of Manchuria through physical exertion that was mediated through machines.

Sent-down youth memoirs confirm the common presence of tractors in
Beidahuang—indeed, it becomes apparent that the vast scale of land reclama-
tion work in the 1960s and early 1970s was made possible only with tractors.
Molding massive fields into evenly spaced, miles-long furrows for straight-

Fig. 8.4 Using machines to weed rows of soybeans, c. 1970. Reproduced from
Heilongjiang sheng shengchan jianshe dui, zhengzhi bu, *Zhishi qingnian zai Beidahuang*
(Beijing: Renmin meishu chubanshe, 1973), fig. 54.

line crops was primarily accomplished behind the wheel (fig. 8.4). While these
images are meant to impress the viewer with the tractor's marvelous ability to
shape the land, the tractor also had a profound ability to shape the lives of the
humans who interacted with it. Having a tractor not only meant that a farm
would be able to fulfill its goal of "opening" the land—it was also a potent
symbol of modernity. Several memoirs speak of the excitement the arrival of
a tractor could bring. One former *zhiqing* remembers the day in October 1971
when his team received delivery of a shiny, fire-red seventy-five-horsepower
East Wind tractor. As soon as it was driven into the headquarters, all the teen-

agers excitedly gathered around it and clamored for a chance to drive it. He was selected by farm leaders to learn how to drive the tractor—an honor that in the retelling seemed to be the highlight of this young man's "down to the countryside" experience. Shunning the expertise of the "rustic" older army soldier who was the farm's tractor instructor, this teenage son of intellectuals claimed to have learned how to operate the tractor solely by studying the written manuals.[78]

If we looked into the personal photo albums of this young man from this time, chances are we would find a portrait of him standing proudly next to his tractor. Indeed, the ubiquitousness of the "tractor selfie" in *zhiqing* photos from the time demonstrates the attraction these modern machines had for both male and female youth in Beidahuang. Although it is not clear how common an occurrence it was, women were allowed to join the tractor corps. Those who wrote of their experience claim they were consciously following in the footsteps of the celebrated pioneering "tractor girls" of the early 1950s such as Harbin native Liang Jun, whose image graced the one yuan note in circulation at the time. One female *zhiqing* tractor driver relates proudly cutting her long hair and donning the required short jacket, cap, white gloves, and white neckerchief, her elite identity as a worker clearly advertised by her attire. The tractor had become a potent symbol of gender equality in socialist China.[79]

While the tractor created camaraderie and social identity for those lucky enough to operate them, the machines also profoundly shaped the lives and experiences of their operators, creating grueling work hours, producing stifling hierarchies, and exposing bodies to life-debilitating harm. Driving a tractor did not necessarily lessen a work burden, but brought a different, more disciplined, and potentially more dangerous kind of work to the fields of Manchuria. The machines required a highly structured discipline of time. *Zhiqing* tractor drivers recollected rising earlier than the manual laborers on the farm to tend to the maintenance and repair of the machines before the working day began. The tractor operators worked with their machines in clearly assigned shifts, and shifts could be extended past sundown—the powerful headlights on tractors meant that the tractors could work throughout the night, with some drivers taking shifts while other drivers caught some sleep in the fields.[80] Since most other farm equipment was pulled by tractors, tractor drivers became deeply enmeshed in the machinery of the mechanized farm: their memoirs frequently reference the model numbers and the technical terms for different types of plows, harrows, seeders, and harvesters pulled by the tractors. Tractors clearly manifested the factorylike aspects of factory farms.

Driving a tractor enmeshed humans with machines, but because tractors were the major implement of opening the land, they could also put humans face-to-face with the most dangerous and unpredictable aspects of the environment. As the 1960s and 1970s progressed, land reclamation activities in already-established farms took place farther and farther away from farm headquarters, and new production teams created with new arrivals were sometimes tasked with opening new lands at the poorly known periphery of a farm's borders. The 1970s also saw the establishment of new state farms in truly unruly territory as expansion pushed farther northeast and deeper into the "small Sanjiang" delta at the exact confluence of the Ussuri, Amur, and Songhua Rivers. Tractor drivers recall having to drive dozens of miles away from "civilization" on dangerous reclamation "expeditions" to uncharted areas. To clear roads, the teenagers felled massive trees using explosives and heavy iron chains to extract the stumps. Tractor crews would sometimes become lost or run out of gas at night, leaving drivers to light fires against the mosquitoes and wolves until the main production team came to find them. Some drivers recalled suddenly encountering unexpected "floating" marshes and desperately attempting to rescue their sunken tractors (which were, after all, their only means of transport back to their production teams miles away). Several recalled the deaths of fellow tractor driver youth who were crushed under the tanklike treads of their East Wind tractors. One *zhiqing* commented on the irony of his experience in Beidahuang. Surrounded by nature, he nevertheless grimly summed up his experience by saying that he had "given his entire life to the machine."[81]

Only a relative few *zhiqing* had "fully automated" work such as that experienced by tractor drivers, but many *zhiqing* in Beidahuang labored through some interface with machinery or in an environment shaped by machines. The work of clearing land and farming land existed on a spectrum of mechanization. Farms made due with a variety of implements and a wide range of machines. Different stages of the land reclamation process and different types of crops also required the use of different types of machines. These in turn had different impacts on human bodies, a visceral experience that was often differentiated by gender.

Seeding was one automated process that nevertheless relied on human agility and muscle. As they had been since the nineteenth century, wheat and soybeans were the two crops most often planted in the Great Northern Wasteland. Under agricultural modernity this continued, as both were straight-line crops that facilitated the mechanization of all stages of the agriculture process. Once topsoil was turned using five-bottom tractor-drawn plows, seeding would proceed. Some farms were equipped with tractor-drawn seeding machines with adjustable row distances. This was a remarkably efficient and labor-saving mode of farming, requiring only one person to drive the tractor and another to sit on the seeder to maintain the proper angle and spacing. Still, *zhiqing* recalled that great physical strength and agility were required to operate the seeding machine. Seed, like grain, was stored in huge sacks, and operators had to swing dozens of heavy bags up and over into the seeder tank before each run. The seeder operator had to observe that seed was coming out smoothly from each seed feeding tube. Clogs had to be manually freed up by the seeder operator, who would clamber across the central steel beam to clear the errant tube while the machine was operating.[82]

Weeding, however, was one process that was relatively machine-free. While plowing and seeding were automated, maintaining the young crops as they grew seemed to be the one aspect of land reclamation that remained stubbornly manual. In the absence of chemical herbicides, weeding the soil around the growing seedlings was accomplished by hand using simple single-head draw hoes. This tedious and physically exhausting labor was typically assigned to women.[83] Several female *zhiqing* wrote vivid recollections of this dreaded but unavoidable work. Women would work in teams, with one woman responsible for weeding one row or furrow. Standing at one end of a massive machine-created field (such as the one pictured in fig. 8.4) but with no machines to aid their work, these women came face-to-face with the Great Northern Wasteland. A single furrow could quite literally be miles long. As phrased by one *zhiqing*, "it is no exaggeration to say that in the Beidahuang, however far the horizon, that's how long the furrow was. Standing on one end of a furrow, there was no way you could see the end of it." Soybean fields were particularly difficult to navigate on foot, as the plants were planted along very deep furrows, with a high ridge holding the plant (called *longtai*) beside a deep, muddy ditch for maintaining moisture (*longgou*).[84] At the beginning of the workday, each person began to weed her own furrow, slowly progressing toward the horizon. All started at the same time, but one's progress, speed and distance were constantly challenged by others who might weed their rows a bit faster. Women recalled that because the ending point was off in the distant horizon, there was no way to figure a "halfway point" for rest. Sometimes it would take two days to finish weeding the same furrow. Even if a worker finished a row during her shift, she was not really finished, because there was an "infinite" number of rows, and when one row was finished, another awaited. These women had one of the jobs that put them in the most direct contact with nature—at the end of each day, they were quite literally covered with the black soil of Beidahuang. Yet some expressed that this work, done in such intimate contact with the earth, made them feel as if they themselves had become machines. If, as one *zhiqing* put it, "knowing the expansive vastness of Beidahuang's black earth was mostly done through hoeing," then that knowledge was acquired at the cost of becoming "human automatons."[85]

Harvest began a frenetic point in the agricultural process that clearly manifest the factorylike mode of mass production on state farms. This mass production varied in its degree of mechanization. Some farms had no harvesting machines and harvests were done entirely by hand. Some "semimechanized" farms required workers to constantly interface with small machines that supplemented (and shaped) human brawn, but did not replace it. Still other farms had massive, state-of-the-art, fully mechanized and self-propelled combines (called *kang-bai-yin*) that could harvest and winnow grain at a staggering rate.

Assignment to a less-mechanized farm meant that workers would become intimately familiar with one of the primary symbols of communism, the sickle, a tool that most urban youth had only seen wielded by happy peasants in propaganda films or gracing the corner of the CCP flag. When we look at a photograph of workers harvesting vast fields of golden wheat with sickles, we may see a peaceful pastoral scene (fig. 8.5), but these images more than likely indicated a desperate shortage of farm equipment. Even on a fully mechanized farm, the sickle might become the "preferred" method of harvesting if the land became too marshy with late summer rains to support the weight of machines. One *zhiqing* recalled having the misfortune to arrive in one such Beidahuang farm just as the harvest was beginning. He had done farmwork as a child in the countryside outside of Tianjin and thought he could handle farming, but was stunned at the massive size of the waterlogged wheat field he had to harvest with just a sickle. Standing in knee-deep water, unable to see the location of individual stalks, and unfamiliar with how to hold and manipulate the sickle, the *zhiqing* silently proceeded to harvest the wheat. His hands rapidly became covered in bloody blisters, his feet painfully swollen. Some sent-down youth even slashed their hands with their sickle just to be able to escape the backbreaking work demanded by the tool.[86]

Farms that did this kind of manual harvesting still employed threshing machines to separate the grains from the stalks brought in from the fields.

Fig. 8.5 Sent-down youth harvesting wheat with sickles, c. 1970. Reproduced from Heilongjiang sheng shengchan jianshe dui, zhengzhi bu, *Zhishi qingnian zai Beidahuang* (Beijing: Renmin meishu chubanshe, 1973), fig. 61.

Threshing machines used a diesel-powered motor that turned threshing blades using a leather belt. One *zhiqing* described in detail the factorylike process of threshing wheat. Six diesel-powered machines stood in a line, with each machine worked by six groups of people. First, a team of several dozen people brought the stalks from the field and dumped them near the machine. The second team of several dozen people gathered the stalks and positioned them in front of the machine. The third team fed the stalks directly into the blades. The fourth team collected the kernels using a wooden shovel and piled them to one side. A fifth team shoveled the kernels into bags, and the sixth team placed the bags onto a truck to send them to the storage silo. For his first day on the job, this *zhiqing* was assigned to feed the stalks into the thresher. He didn't understand why other workers had shown up to work with their faces almost completely covered in towels. As soon as the machines started up, he immediately understood: as he put the stalks into the machine, the kernels started pelting his face like a rain of fine shrapnel. Working in round-the-clock shifts, kernels and chaff dust flying, diesel fumes swirling, and the loud sound of the belt and the blades churning, for many *zhiqing* working the harvest was like working in a frantic, tense, and machine-driven factory.[87]

While harvests reaped through manual labor often led to mechanized grain processing, even a fully mechanized harvest in the end relied on manual labor. The quintessential image of the modern Beidahuang farm is a fleet of self-propelled combine harvesters moving through a vast field of golden wheat. These fully mechanized high-capacity combine harvesters produced mountains of golden wheat that were the ultimate symbol of the highly productive Beidahuang farm, the actualization of land reclamation's dream. Yet as one *zhiqing* wryly observed, after the combines had finished their work (and the photo ops with smiling leaders were done), "someone had to take those 'great mountains of grain' and put them into bags." This work typically involved endless days of shoveling bag after bag, carrying bag after bag on the back, heaving the bags up into truck beds, and driving them dozens of miles away on rutted roads to the central state-run depot. In the end, the great encounter with nature through agriculture on the Sanjiang Plain ended like any other factory process—with the packaging of a mass-produced product.[88]

This consideration of opening up and farming land in the Great Northern Wilderness reveals the central role of machines in mediating an understanding of nature. From clearing the land, plowing, harrowing, seeding, and harvesting, all processes had some interface with machines. Even within processes that were not highly mechanized—those forms of labor, such as hoeing, that seemed to put humans in direct, deep engagement with the environment— the specter of the machine could not be avoided. Because machines dictated the scale, organization, and rhythms of the farm, even manual labor made workers felt as though they themselves had been transformed into machines. Manchuria is often seen as a place that gave the Chinese people a pure, direct experience of an unspoiled nature, but it must also be seen as a place where nature was turned into a factory. This dream, in turn, has important significance for China's image of itself and its trajectory in the twenty-first century.

Conclusion

The 1985 report of the Friendship Farm's Scientific Experimental Station was clear: even though the farm had started as a model of agricultural mechanization, over its twenty-nine years of existence it faced a constant struggle with the environment of the Sanjiang Plain. Youyi was still a massive, mechanized farm manned by thousands of workers who cultivated over four hundred square miles of land, but the "black soil" had lost significant amounts of its organic content and was being drained of its nitrogen and potassium.[89] Since its establishment in 1954, by the mid-1980s the farm had encountered eight years of serious flooding and seven years of serious drought.[90] Even in years marked by flooding, the wheat that constituted 40 percent of the farm's crops faced "seasonal drought" conditions: since rains didn't come until much later in the summer, spring seedlings suffered from a lack of water. The alternation of not enough water followed by too much water made the wheat crop yield fluctuate wildly.[91] Study after study revealed the harsh truth of farming on the Sanjiang Plain: if the soil was not "managed properly according to the conditions of its nature," then the entire farm would face "man-made disasters" (*renwei de huohai*).[92]

With the beginning of Deng Xiaoping's Four Modernizations, farming at Youyi received a new injection of technology that it hoped would help it avoid "man-made disasters." In 1978, an experimental station at the farm was able to import a set of sixty-two agricultural machines from the American John Deere company. Guided in part by the American William Hinton (1919– 2004), the author of *Fanshen* and a well-known supporter of the Communist Revolution, the experimental station quickly put to work the seven tractors, three combines, numerous seeders, irrigation machines, and dozens of other pieces of equipment and studied the results.[93]

If the introduction of Soviet farming technology revolutionized agriculture on the Sanjiang Plain in the 1950s, the introduction of Western technology in the late 1970s revolutionized agriculture once again. The team found all the equipment to be faster, more reliable, and more flexible than anything they had ever encountered, and concluded that "in all aspects, PRC-made machines cannot compare to US machines."[94] One John Deere tractor in one shift could plow over fifty acres, while Chinese tractors could only plow three to ten acres. While increases in productivity were clear with wheat and soybeans, the greatest increase was in corn cultivation: in the same amount of time that it took Chinese machines to sow 50 acres, the US machines could sow over 130 acres, and in the amount of time it took Friendship Farm workers to pick a half acre of corn (by hand), the John Deere corn harvester could bring in the crop from 75 acres.[95] With these machines the experimental station dramatically increased the amount of land it brought under cultivation, doubling the size of its small branch farm from 2,100 acres in 1978 to over 4,200 acres the next year. With this increased acreage, harvests almost tripled, from over 4 million pounds of grain in 1978 to over 11 million pounds in 1984.[96] The most stunning change was in productivity per worker. In the seven years from 1970 to 1977, the branch farm team's 242 workers produced an average of about 4.5 million pounds of grain per year. After the introduction of the imported equipment, the team produced 8.3 million pounds per year with only twenty workers. In the 1970s, each worker produced about 16,000 pounds of grain. By the 1980s, that figure had increased to 415,000 pounds per worker.[97] The experimental station's experiment had been a resounding success. John Deere had made humans superfluous.

The arrival of sixty-two pieces of American equipment to the Friendship Farm experimental station in 1978 was the beginning of what would become a massive flood of machines arriving on the Sanjiang Plain. Through the end of the twentieth and beginning of the twenty-first centuries, the region's farms imported thousands of pieces of advanced agricultural machinery from Europe, Japan, and especially the United States. By the 1990s, massive tractors and combines the size of small houses rumbled through ever-expanding farmland, reducing the need for labor in an increasingly mechanized and computer-automated approach to farming.[98] "Water control" was the linchpin of this process. Statistics on agricultural equipment used in the Hongxinglong agricultural management district, which includes the Friendship Farm, indicate the symbiotic relationship between drainage work and agricultural mechanization. Over the past decades, there has been a considerable increase in the number of large tractors in the region, from 3,000 in 1978 to over 11,000 in 2015. Fully automated combines increased from around 1,600 to over 5,600 during that time. The one form of technology that had the most remarkable increase, however, was "powered water-drainage machines" (*paiguan dongli jixie*)—that is, water pumps—which saw an almost sixteen-fold rise from about 1,000 in 1978 to almost 17,000 by 2015.[99]

This increase in pumps manifests, in terms of machine units, the most dramatic change in the region's environment that has been brought by post-Mao economic reforms: in the battle between man and water on the Sanjiang Plain, man has, at least for the moment, emerged victorious. *Shuili* or water control was always a key segment of land reclamation work, but it was among those aspects most resistant to mechanization. It was not until the dramatic increase in technology related to water control that the marsh that had been the machine's greatest enemy was finally beaten back. Studies have shown that since the 1950s, over 80 percent of the wetlands in the Sanjiang Plain have been eliminated, with the most significant destruction occurring after Deng's reforms.[100] Mao may have called for the creation of *liangshi gongchang* (grain factories) in northern Manchuria, but it was not until after his death that his dream was realized.

Ironically, this destruction of wetlands was accompanied by the very fast rise of wet-paddy rice cultivation on the Sanjiang Plain.[101] Once a form of agriculture practiced primarily by Korean immigrants to Manchuria, by the beginning of the twenty-first century, riziculture was becoming standard practice.[102] Today, thousands of pumps draw upon the immense store of water in the area's high water table and gush it onto the surface, flooding the reclaimed former-wetlands in order to grow rice. Experiments in improved and organic or "green" rice farming in the region have resulted in some of China's most coveted and expensive brands of rice, with brand-names like Black Soil (*Heitu*) that hark back to an idea of the unspoiled nature of Beidahuang as the Great Northern Wilderness—not a *wasteland* of useless marsh, but a pristine nature untouched by modern chemicals and corruption. In a climate of deep uncertainty about food safety in contemporary China, with scandals ranging from tainted milk powder to contaminated dumplings, the Sanjiang Plain has become a marketable symbol of environmental purity and wholesomeness in China's booming economy. This image of Beidahuang as a site that bestowed a generation of youth with a direct experience of untouched nature obscures the fact that the region was also "ground zero" for China's mechanized farming revolution. Also ironic is the marketing of Beidahuang rice as pure and natural, when the Sanjiang Plain is in fact a "enviro-technical landscape" par excellence, as well as the site of some of China's most radical environmental destruction.

✳

The slogans of *ren ding sheng tian* ("Man must conquer nature") and the tale of *Yu gong yi shan* ("The foolish old man moves mountains") were central symbols of the early PRC, an era when Mao tried to use human muscle and zeal to transform China. In the ancient parable of the "foolish old man," Yu Gong attempted to move two mountains one shovelful at a time, a task that he knew would be completed only by the labor of many future generations working tirelessly toward the goal. In 1945, Mao famously reinterpreted the two mountains as imperialism and feudalism: the chains of foreign interference combined with the chains of China's outdated traditional culture. Judith Shapiro and others have pointed out how this fable of incredible effort and struggle was utilized during the Mao years to justify massive state-run labor mobilization projects that transformed China's environment at the cost of tremendous human suffering.[103]

Perhaps the foolish old man could move a mountain, but man alone could not drain the Sanjiang Plain. The presence of agricultural machines in Manchuria is a story of the desire for a mechanized modernity held by Qing officials, American capitalists, Chinese reformers, Soviet advisers, and Japanese planners across the entire course of the twentieth century. While this dream had a long history, it gained new valences of significance in the People's Republic of China after 1949. The CCP had reclaimed the lands of the northeast for their new government after decades of imperial rivalries and wartime extractions. The former Qing "land where the dragon arose" had, under the deprivations of colonial occupation, become a "plagueland," a site of suffering and degradation. The CCP would employ the techniques of science and technology (sometimes, as we saw in the case of post-1949 plague prevention work, quite literally the same as those of imperial powers) but organized in a different manner. Where many had tried but failed, under Soviet guidance and with the hard work of Chinese "volunteers," Manchuria would be redeemed from its past as its land was reclaimed for the future of the Chinese state through the application of technology.

The story of environmental transformation under agricultural machines is one that is familiar from many places around the world. In one of the fundamental narratives of US environmental history, Daniel Worster pointed to the rise of the tractor, coupled with capital's drive for profit and the state's unrelenting vision of bountiful harvests, as the drivers of environmental change and environmental devastation in the American Midwest in the first half of the twentieth century. Federally subsidized tractors encouraged mass-scale, straight-line monocropping, which made the Midwest the most productive farmland in the world. Tractors also drove tenant farmers off the land, and in the environmentally fragile southern plains, led to the tragedy of the Dust Bowl.[104] A less well known but equally radical transformation of the environment took place further north (and much earlier) in the American Midwest when the vast wetlands that covered much of the Great Lakes states were drained and turned into farmland. The "black soil" of this region, fertile but prone to waterlogging, was transformed in the mid-nineteenth century through the labor of Irish ditchdiggers who carved drainage canals "all over the wet prairies from the Great Black Swamp of Ohio, along the Wabash in Indiana and Illinois, to Michigan, southeast Wisconsin, and central Iowa."[105] Today, this transformation is all but forgotten, and the thousands of square miles of quintessential rolling American farmland in this region, with its rich harvests of corn, wheat, and soybeans, appears as a permanent and natural fixture of the American landscape. Little thought is given to the labor that made it. The American imagination may laud the nation's "pioneers," but the idea of what they actually did is vague at best: there is no celebration of the ditchdiggers of the Black Swamp as central to the national character.

This chapter has attempted to see the Sanjiang Plain as a "workscape," to "think through the fields, streams, woods, [and marshes] as labor." This labor simultaneously transformed the landscape and produced knowledge of Manchuria's nature that was both deeply embedded in individual bodies and enduring on a national level. Direct and extensive experience of the land shaped the knowledge of all the actors discussed here: the Chinese, American, Japanese, and Russian reformers who saw the key to China's modernity in Manchuria's plains; the soil scientists and botanists of New China's research institutions who trudged across miles of frozen prairie to survey the land; and the millions of workers, many of them unfree, who found themselves in the far northern corner of Manchuria at the behest of (or through the designs of) state demands for increased production: the Han Chinese serfs who labored on Manchu banner-farms and eventually claimed the fields for themselves; the "reluctant pioneers" of the turn of the century who emigrated from Shandong (China proper), Honshu (Japan's main island), or Hwanghae (Korea); the soldiers who in the 1950s served as beasts of burden and pulled tractors out of the mud in order to meet the urgent demands of a modernizing state; and finally, the "educated youth" who answered Mao's call to go down to the countryside in the 1960s and 1970s and found themselves physically inserted into the mechanisms of industrial farming.

All these laboring people knew Manchuria through their work, but it is the knowledge acquired by the sent-down youth of the Cultural Revolution, now retirees in their sixties and seventies, that has claimed an oversized voice in establishing a way of knowing Manchuria for an entire nation. While China in the post-Mao era now mourns the personal tragedies endured by many of the sent-down youth, the simplicity of the work undertaken, the ruggedness of the environment, and the selfless motives of land reclamation in the Great Northern Wasteland are venerated in popular culture as the "Beidahuang spirit." Indeed, we might say that those engaged in the work of land reclamation are thought to have been imbued with a spirit reminiscent of Leopold's *Sand County* "split-rail values": urbanites experienced honest, direct engagement with the land and nature, unmediated by machines and modernity.[106] While those values of hard work, tenacity, ability to weather suffering, and closeness to the soil are associated with a simpler time before the arrival of market-based economic reforms in the 1980s, they are also deeply associated with Manchuria as a place. Through them, the northeast became a unique site that preserved and nurtured essential aspects of a timeless Chinese spirit, even as it was transformed into a showcase of Chinese modernity.

There are those, however, for whom the experience of nature via the machine was profoundly dehumanizing.[107] Sent-down youth memoirs are full of stories of harsh conditions: brutally long working hours that drained human vitality, dangerous conditions that could lead to maiming or gruesome death, and year after year of sheer drudgery. Youth were crushed under the treads of tank-tractors, drowned in rivers and burned in fires, or simply reduced to the level of "automatons" who "gave their lives to the machine." As Sigrid Schmalzer has pointed out, the "dramatic remaking of the landscape" during the Mao era has been "presented as a testament to the bold vision of Chinese socialism to transform heaven and earth," but that moving of heaven and earth "required human sacrifice."[108]

Even through suffering, however, knowledge is created. While the reclamation work done by sent-down youth shaped the natural environment of the Sanjiang Plain, the memories they bear within their minds and upon their bodies have shaped the national consciousness. The circulation of their stories in multiple venues, from widely consumed mass media to the intimate space of the home, constitutes knowledge of Manchuria. If Manchuria today is a known space, its nature cataloged and understood, its landscape indigenized and made a central part of the national narrative of China, it is because of the memories of those who labored there.

CONCLUSION

A VIEW FROM THE MOUNTAIN

Fellow hikers on my first ascent up the Long White Mountain in 2006 did not seem particularly happy to be there. The region had only recently caught on as a tourist attraction in China, and workers from Jilin had been bused in that weekend as part of a company holiday. The newly constructed enclosed staircase built on the northern slope allowed for better traction on the pumice surface of the volcano's cone, but the stairs were steep and the climb was rough going. "Who the #$% said that Changbaishan was so #$% scenic?" (in Chinese, literally "good to see"), one exhausted man asked of no one in particular as he struggled to catch his breath. "There's #$% nothing up here to see at all!"

I had to admit the man had a point. Unlike other mountains in China that were well-known sites and sights, Mount Changbai literally had "nothing to see." Mountains south of the Great Wall like Taishan in Shandong or Huangshan in Anhui had been sites of pilgrimage and tourism for millennia. The cliffs were etched with commentary on the scenery written by emperors and famous poets; splendid temples greeted travelers at both the base and at the summit. Even individual pine trees and grottos had names passed down from antiquity.

Unlike these mountains, Mount Changbai was not an inscribed landscape—it was devoid of temples and calligraphy, and had no visible history to offer the viewer. In the early 2000s, regional authorities had made some attempt at adding human history to the mountain, placing statues of Daoist deities along the trails and painting poetic names (in red calligraphy) on oddly shaped volcanic rocks. At the summit, instead of a temple, there was a granite monument inscribed with the calligraphy of Deng Xiaping announcing the Chinese name of the mountain: "Changbaishan."

Missing was any indication that Changbaishan was considered the sacred origin of China's dragon vein as was once theorized by the Kangxi emperor and *fengshui* masters of the Qing court. But that doesn't mean that dragons were entirely missing from the mountain. Near the summit, tourism promoters had installed a representation of the "Tianchi Monster," a cross between the Loch Ness Monster and the mysterious dragon of the waters from local lore. The ten-foot-tall plastic statue formed the backdrop to tourists' photographs as they stood along the spectacular shores of Heaven Lake.

Most of those on the hike to the summit that day were not Chinese, however. Large groups of tourists from South Korea had been bused up on their own tour buses or had emerged from the Korean-owned hotel near the Songhua waterfall. These visitors climbed the mountain they called Paektusan with a different outlook. Happy and lighthearted because they were on holiday, they nevertheless seemed to approach the mountain with a sense of reverence and attention. Several wandered the shore of Chŏnji, pausing to build stone cairns along the edge of the brilliant blue water. Many stood at the edge of the lake, gazing across at the opposite shore. While there were no people visible on the North Korean side on that day, the South Korean tourists cupped their hands on either side of their mouths and shouted across the lake, calling out to their compatriots and expressing hopes for the reunification of Korea. None, however, dared to walk around the lake to the other side.

My first foray up Mount Paektu/Changbai both encapsulated and inspired many of the themes of this book. The fact that a fraught national boundary splits this beautiful mountain in two is a dramatic and frustrating manifestation of Manchuria's historical status as a contested borderland. Rivers and mountains are not in and of themselves borders: plants, animals, and humans can spread around and about them. But for centuries, multiple political entities have laid competing claim to these territories and attempted to carve and contain an undulating nature within borders of empire and nation.

My experience of climbing Mount Paektu/Changbai revealed that making sense of nature is not a straightforward process, and making sense of the nature of Manchuria poses particular challenges. I had traveled to the mountain from Yanji, the county seat of the Yanbian Korean Autonomous Prefecture in eastern Jilin province, located fifteen miles north of the North Korean border and one hundred miles to the northeast of the mountain. As the unlicensed taxi I hired sped along winding mountain roads, I began to fear that I was being hijacked: no matter how I strained to look out of the car windows, Mount Paektu/Changbai was nowhere in sight. Even when I arrived at the embarkation point at the base, I *still* could not see the mountain. It was only after I had begun the ascent and progressed along a road clear-cut of trees that I finally saw the purple cone of Mount Changbai soaring up before me: a sight so dramatic and imposing that it made me gasp in wonder. At this moment, the difficulty experienced by those who tried to see the mountain in the past—Umuna, Mukedeng, Sŏ Myŏng-ŭng—suddenly made sense to me. Contrary to what Yi-Fu Tuan assumed, a mountain is not always an object that is "highly visible." Seeing a mountain, and indeed, seeing any aspect of the environment, depends on where one is located and how one moves across the land.

Seeing also depends on what one expects to see. The hiker who complained that Changbaishan had "nothing to see" used phrasings that were less elegant than those used by exiles of the early Qing, but his experience of nature was similar to that of Wu Zhaoqian and other Han elites who sojourned in the heartland of Manchu territory in the mid-seventeenth century. As Wu Zhaoqian began his progression northeast from his Beijing jail cell in 1659, he initially noted an environment that resonated with his cultural expectations: indeed, he saw a landscape covered with a layer of ghosts and historical ruins. But once he passed through what he called the Shady Ditch Gate of the Willow Palisade, Wu came face-to-face with what for him was a disorienting true frontier. Unmoored from cultural roots of nostalgia, Han exiles attempted to produce a natural knowledge that was divested of history and personal sentiment, a more "objective" way of knowing. While they still saw Manchuria as a landscape of defeat, with their writings they began to bring the frontier into the conceptual boundaries of their world.

My climb up Mount Paektu/Changbai also revealed that there are many different forms of natural knowledge. The Chinese tourists may have been disappointed, but the reverence of the South Korean tourists (much more so than the plastic "Nessie" statue at the shore of Heaven Lake) was a clear reminder that the mountain was once thought of as a sacred landscape: the source of the dragon vein that poured its energies throughout northeast Asia. Both the Qing and the Chosŏn courts launched major projects to establish the exact location of the mountain: projects that could not proceed without the assistance of local trackers and ritual specialists. This newly found empirical knowledge of Paektusan's geography was expressed in ways that bundled the mountain's sacrality. Using topographically exact maps, Qing rulers conceptualized Mount Paektu/Changbai as the origin of the dragon vein that buttressed the empire. Even as latitude and longitude, perspective and accuracy became a concern for Chosŏn maps, Mount Paektu remained floating magnificently in a sacred space, tethered to the geo-body of Korea through the cords of ki/qi that emanated from the mountain and girded the peninsula with its powerful energy. Indeed, in the eyes of early modern elites, these lines of numinous energy emerged from Manchuria and shot through northeast Asia like a network of nerves or like arteries, swelling with unstoppable flows like the rush of waters from a towering mountain, making Manchuria "the most excellent of lands." Indeed, it was this network of dragon veins that defined Manchuria itself.

A glance at a map of Manchuria from the early twentieth century, however, would reveal the emergence of a different sort of arterial network that defined the region: lines of steel that shot through the entire body of Manchuria, not flowing from one point of origin but spreading instead like a web, with tendrils linking remote grasslands and mountain forests to multiple urban nodes. This vast railway network linked the natural resources of Manchuria to the world. Today these lines provide transportation for millions of people who ride the trains without a thought of who originally established many of the tracks: the Russian empire, the fledgling Chinese republic, and the empire of Japan. This web of steel symbolized a new era of modernity and a new mode of contesting the nature of the Manchurian borderland, built to transport not the energy of kings but the fuel of modernity: coal.

For reasons of my nationality, I was unable to ride these rails directly south into North Korea, although lines traversing the border exist on either side of Mount Paektu. The dream of traveling the Paektudaegan, the great mountain ridge or line of ki that forms the spine of the unified Korean Peninsula, is an impossibility today. But anyone can ride the rail lines from the area around Mount Paektu/Changbai west-southwest in the direction of the flow of the Qing dragon vein, toward the original heartland of Manchu power. The rail lines bypass Xinbin, the Manchu Autonomous County that is the site of Hetu Ala and Yongling, the tomb of the distant ancestors of the Qing emperors. But a transfer to the more northerly Shenyang-Jilin railway will take you along the same pathway as the old Qing road between Jilin and Shenyang, the Manchu capital of Mukden. On this journey one can stop at "Yinggemen" (the Willow Pallisade's Origin Gate, or Wu Zhaoqian's Shady Ditch Gate), envision the bloody battlefields of Sarhu lying beneath the waters of a reservoir, or think of Kangxi killing tigers on Ironback Mountain during the eastern tour. Marked in blood from the beginning of the 1600s, the region was again to become a battlefield at the turn of the twentieth century as the Russian, Japanese, and

Qing empires fought vicious modern wars for control of this terrain and its natural resources.

The Jilin-Shenyang rail line has a stop at Fushun, the demolished and haunted Ming garrison that became a center of coal mining under the management of the Japanese-run South Manchuria Railway Company. Fushun was once thought of as being directly along the path of the dragon vein that ran from Mount Changbai through the imperial tombs to Mukden. Today, what was once the land of the dragon vein is now the site of Asia's largest open-pit mine. A century of excavation at Fushun has resulted in an increasingly precarious environment. Hollowed-out ground can no longer support the surface, opening up massive sinkholes that swallow minivans and flood neighborhoods. Mine-induced earthquakes rattle the region, further endangering miners below and residents above. While coal still exists beneath the ground, the land's energy has been exhausted.

Traveling one hundred miles on the rails due west brings us to another important mining city: Beipiao, situated in an arid mountainous area that used to be known as Jehol. Jehol was once home to the Qing court's summer palaces, a place where the Manchu emperors communed with Inner Asian allies and imbibed the power of Manchuria's dragon veins. Here, too, dragon veins yielded to coal veins in the early twentieth century. Today, in cities like Beipiao, mining is on the wane, and as in Fushun, the ground shakes and sinks from a century of intensive extraction.

But a hike into nearby backcountry hills (over pitted gravel roads and past herds of cows) reveals another way the underground nature of Manchuria is still producing valuable resources. Since the 1990s, farmers have been carefully combing through the layers and layers of rocks in the region's exposed outcroppings, meticulously cracking open pieces of thin shale in hopes that the discovery of a new fossil could lead to fame and fortune. Some enterprising farmers have even skillfully attached pieces of different fossils together to create new "species" that have been passed off to unsuspecting scientists as astounding missing links. Today, Jehol's farmers have mastered a scientific way of seeing. They too scrutinize minute structures to parse the diversity of nature into discrete categories, for reasons that go far beyond the objects themselves.[1]

Recent fossil discoveries have supported an explosion of paleontological museums in the region: the cities of Shenyang, Dalian, Jinzhou, and even the coal town of Beipiao all boast state-of-the-art natural history museums featuring the region's fossil wealth. The "crown jewels" of these fossils—the feathered dinosaurs of the Jehol biota—even have a place of honor in the nation's premier paleontological institution, the Beijing Museum of Natural History. Here, the *Lycoptera*, Kangxi's "stone fish," are so ubiquitous that they are on sale in the gift shop. More rare is the *Yixianoris grabaui*, the most complete specimen of an ancient bird ever found, and named after the German American paleontologist Amadeus Grabau. Nearby lies a perfectly preserved specimen of *Monjurosuchus splendens*: the "Shining Crocodile of Ancient Manchukuo," its body gently curving as if swimming through a placid pool. The original fossil was lost in World War II, visitors may learn, but another holotype was discovered in 2000, thus resolving a scientific "enigma."[2] No mention is made of the conditions of the fossil's original discovery at the height of the Japanese occupation, or of the fact that its very name embodies the hated puppet government of Manchukuo. Visitors remain unaware that these fossils reflect the fractured history of knowledge making in Manchuria during the age of empire.

From the coal-mining regions of Jehol it's a quick ride north to Tongliao, the city on the Manchurian-Mongolian grasslands that was an epicenter of plague in the early twentieth century. The PRC is now highly confident of its ability to control plague on these grasslands, as indicated by the remarkable transportation network that flourishes at Tongliao. Today, the city is one of the most connected places in the country, served by six different lines including high-speed rail that can whisk passengers to the megacities of Shenyang and Beijing in just a few hours. Years of constant monitoring, destruction of animal habitats, and pesticide use on the grasslands has made the occurrence of plague cases extremely rare. The emergence of several cases of pneumonic plague from Inner Mongolia, including two people who traveled all the way to Beijing before they were diagnosed, produced considerable concern in the fall of 2019, but no epidemic ensued. Networks of scientists who consider themselves heirs to the plague prevention work of Wu Lien-teh remain hypervigilant, but the western stretches of Manchuria where scientists pursued *Yersinia pestis* are no longer thought of as a plagueland.[3]

Natural occurrences of plague may be a distant memory, but the plague machinations of Unit 731 are intentionally maintained in China's collective consciousness. Opened in 2015, Harbin's state-of-the-art Exhibition Hall of Evidence of Crimes Committed by Unit 731 of the Japanese Imperial Army is located on the original site of Unit 731's headquarters. The museum's dramatic black-granite edifice houses multiple exhibits detailing Unit 731's conception, research, and wartime atrocities. In the labyrinth of rooms inside, black walls and low lighting create an aura of dread and horror, reflecting how science brought about Manchuria's darkest hour. Realistic dioramas reenact atrocities such as the Anda proving-ground plague-bomb experiments; vivisection surgeries are projected as shadows behind a diaphanous screen. But everyday laboratory items such as test tubes, chemical storage bottles, and microscopes are also displayed as instruments of terror used by nefarious scientists. One of the most striking exhibits contains pages from the "Report of Q," the autopsy results of the plague victims from the 1940 Nong'an and Changchun plague outbreak. Each individual page of the massive report is encased in a clear container and suspended in space to create a wall of evidence that towers over the viewer. Here, a record of the relentless pursuit of microbes into the human landscape stands as mute evidence of Manchuria's victimization at the hand of foreign science.[4]

Another museum in the center of Harbin promotes a very different narrative: the redemption of Manchuria's nature through the combination of technology and the Chinese Communist Party after Liberation. The Beidahuang Museum is part of a network of museums in northeastern China that commemorate the transformation of the Great Northern Wasteland into the Great Northern Granary from the 1940s to the present. The collections are amazingly rich, and include photographs, documents, and hundreds of intimate artifacts ranging from uniforms, mess kits, and washbasins to diaries and letters written by the sent-down youth of the Cultural Revolution. These youth, now urban retirees in their sixties and seventies, flock to the Great Northern Wilderness on summer holidays. The visitors return to the state-run farms where they once sojourned to reconnect with their own personal history, with the local residents, and with the land. They may pause to honor young friends who died, and marvel at the way that nature has been transformed in the reform era through the application of technology and machinery they could not even have imagined. All along the Amur, Songhua, and Ussuri Rivers in high summer, mile after mile of rice fields saturate the land in vivid green against the endless blue sky. The visitors can revel in the abundance of a land that their bodies helped to create.

In the land near where the Ussuri and Amur meet, tourists can also take motorboat tours of the last remaining wetlands on the Sanjiang Plain. In some of these nature reserves the landscape is marked with symbols of political exemplars who struggled against the "wasteland": life-size bronze statues depicting surveyors standing on frozen prairies, or soldiers frantically rescuing a sunken tractor from the marsh. In other reserves where environmental protection is emphasized, networks of platforms and viewing towers allow eco-tourists to quietly observe rare migrating water birds such as the red-crowned crane.[5] These last few acres of wetland on the Sanjiang Plain help one envision what the region might have been like when it was the dwelling place of Nanai fishermen and hunters, or when the Russian botanist Karl Maximowicz looked out over the plant life of the prairies along the Amur and felt an overwhelming sense of awe at the stunning diversity that surrounded him. But superficial similarities might be deceiving. Most of these nature reserves are surrounded by agricultural lands, and in some places cultivated fields of soybeans and rice are separated from protected wetlands by only a low earthen dike. Runoff of fertilizers and herbicides has altered the environment. We cannot be sure that the reeds were always this dense and green, or that the species we see today were there one hundred years ago.

The CCP's transformation of this region began one hundred years after the Russian botanist Karl Maximowicz created his systematic ordering of the flowers of the Amur. Today, we can look at the botanist's sample of *Maximowiczia amurensis*, dried and pasted to cardboard, its delicate flowers slit open and dissected, and understand it as an attempt to untangle a living plant from a network of local human meanings and uses. Maximowicz's knowledge was designed to project the plant as an object, via numbers, into a global order. Similarly, when Amadeus Grabau scrutinized the tiniest whorls in a petrified shell, it allowed him to visualize ancient landmasses that linked continents and obliterated nations. The atomizing analytical gaze of science scrutinized and disentangled, but it could also lead to soaring holistic visions of the world as one.

We cannot fault Maximowicz and his science alone for the disappearance of the Amur's diversity. While the *Florengrenze* (botanical boundaries) he discerned totally ignored the newly formed borders between Russia and China, other economic and political forces exerted changes on the land that wrote political borders into nature itself. Today, the contrast between flora of the southern (PRC) and northern (Russian) banks of the Amur is a stark reminder of the very real environmental significance of the region's borders. On the China side, the concerted effort of multiple regimes to develop agriculture over the course of a century and a half is etched onto the land, which appears as a dense patchwork of rectangular monocropped fields that push right up to the river. This pattern is particularly pronounced in the small triangle between the Songhua and the Amur, around Suibin county. Here, where Harvard men and Chinese investors once experimented with mechanized farming at the turn of the century and where Chinese interned in forced labor camps opened up land by hand in the 1950s, multiple leaseholders now produce organic crops for global markets. Across the river, agriculture struggles, but as a result, in many places, the flowers along the Amur still survive.[6]

And so we end where we began, with borders etched across the nature of Manchuria. This study has created a collaged image of a borderland that was never truly a country but was often imagined as a singular place. We have seen how this singular place was in fact made up of multiple environments, including alpine mountains, arid grasslands, and vast alluvial deltas. Its borders shifted, expanded, and contracted across the centuries as political powers sought to control the land's resources and energies. Indeed, any unity to a place that might be called Manchuria was the product of this quest to squeeze terrain into borders and thus lay claim to the wealth that its nature could produce.

As empires, kingdoms, and nations staked their claims, they created knowledge about Manchuria's environments. Certain forms of natural knowledge seemed to be wrapped up in exercises of displacement and destruction. Karl Maximowicz's project of classifying the diverse plants along the Amur River was facilitated by local knowledge, but his universal system was designed to make indigenous knowledge obsolete. The Russian botanist's dream for the future involved eliminating old-growth forest and plowing under the flowers of the prairie. Obsessive attention to the most minute of visual differences could lead to enormous wealth: the tiny swirls on a fossil shell might indicate the location of a mineral motherlode; or at the very least, such attention might generate a new species, another confident step on the inevitable road of bringing all of nature into the realm of the known. The Kangxi emperor was not immune from apprehending nature through domination and violence: he marked his progress across the historical paths his ancestors trod with the blood of stags and tigers, and tried to monopolize the vital energies of the earth for his own ruling house. Still, there was a certain humility to Kangxi's approach to nature. Like other elites in the East Asian intellectual milieu of his day, he allowed that there would always be a lively excess to nature that could not be totally grasped through analytical human intelligence: a sentiment frequently expressed through the standard practical-learning phrase "the principle of things can not be exhausted." While moderns sought more and more precise gauges with which to parse nature, early modern observers held that there was more to the world than meets the eye.

The fit between styles of creating natural knowledge and specific political projects was never perfect. Those who had once served the Ming created sacred landscapes for the Qing. Some observers examined tiny fossils in order to plunder coal, while others did the same in order to imagine the outline of conjoined continents. Even bitter enemies sometimes used the same techniques to make sense of the complex linkages between environment, animal and man in Manchuria, a pursuit that alternately led to both tragedy and redemption. Our examination of natural knowledge creation in the war-ravaged environment of Manchuria has shown that even the most analytical of "regimes of attention" do not have to be linked to domination or motivated by a desire to maximize resource extraction. As we joined a paleontologist in silent concentration at his laboratory bench or followed a local forager as he navigated a mountain landscape, we understand how many modes of human sensing can be quiet, mindful celebrations of the myriad things, exercises at once intimate, generous, and border-transcending.

This book has attempted to place historical actors within specific environments. I have followed their footsteps, gauged their reactions to specific landscapes, and attempted to see what they could see. It has placed knowledge making as the direct product of physical embeddedness and sensorial entanglements between humans and the environment. It suggests that natural knowledge—that of the worker, the poet, and the scientist—is created through intense bodily engagement: whether hoeing a row of soybeans miles long or disemboweling thousands of rodents and humans; whether carrying a deer home after a hunt or counting the ovaries concealed within the base of a flower. The environment is always an actor that both shapes and limits human knowledge. As they attempted to know nature with an eye toward domination, humans seeking power struggled to cram an unruly nature into the borders of nations, often with devastating consequences. But human knowledge making at a local level could also transcend the physical limits of space and time, like the shamans who created maps of long-gone landscapes, flying with spirits over hundreds of miles from golden palaces atop an invisible mountain.

Some observers in the past were able to combine a desire for precision and a sense of the sacred. It is my hope that through this vision of Manchuria as multiple, as bordered and boundless, that the reader might achieve for themselves a more vivid sense of this important place. It is also my hope that this "collaged vision" of knowledge making on an Asian borderland might inspire a sense of wonder at the possibility of an embodied yet unbordered human engagement with the world.

ACKNOWLEDGMENTS

A great many people helped with this project over the years. The first to thank are the librarians. In its earliest stages, Martin Heijdra of Princeton's East Asian Library knew this project better than I did. For more than a decade, Yuh-Fen Benda, Vanderbilt University's East Asia librarian, has tirelessly tracked down key materials for me. Without her expert help, I could not have written this book. Vanderbilt's Interlibrary Loan staff allowed me to access sources from around the globe while I was teaching in Nashville. Librarians and staff at many other institutions provided assistance, including those at the Harvard Yenching Library, the Shanghai Library, the National Library of Medicine, the Libraries of Hong Kong University, the Liaoning Provincial Archives, the Dalian Library, and the historians of the Friendship Farm.

Many people helped with logistics, connections, information, and encouragement, including Angela Leung, Wang Peihuan, Mark Elliott, Nianshen Song, Prasenjit Duara, Seonmin Kim, Jeff Wasserstrom, Marta Hanson, Sakura Christmas, Rob Perrins, Karl Gerth, Ning Wang, Ian Miller, Cheng Xiurong, Christos Lynteris, Guan Min'ge, Gao Wangling, and Wen-hsin Yeh. Liu Haiyan helped dig up sources here and abroad and worked his troubleshooting magic at critical junctures. I am particularly grateful to Bin Xu for opening doors (and guiding me through them) in the Great Northern Wilderness.

Sue Naquin, Angela Creager, Ben Elman, David Howell, He Bian, Dan Usner, Gerald Figal, Guojun Wang, Tracy Miller, Bill Caferro, Frank Wisclo, Emily Greble, Helmut Smith, Rob Campany, We Jung Yi, Ole Molvig, and Yoshi Igarashi were among colleagues at Princeton and Vanderbilt who provided sources, suggested elegant turns of phrase, and tirelessly answered questions in their areas of expertise.

The project was improved through talks and workshops given at several universities, including UC Berkeley, the University of British Columbia, Yale, Harvard, the University of Washington, Princeton, Oxford, Johns Hopkins, and Columbia. Shellen Wu, Lukas Riepell, Suman Seth, Helen Tilley, and D. Graham Burnett offered detailed criticism on individual chapters. Sue Naquin supported this project from its inception, and near its end, she read the entire manuscript and kicked it into shape. Much of the manuscript was also read by my writing group partners at Vanderbilt, Arleen Tuchman and Sarah Igo. For the past five years, they have been keen interlocutors, staunch advocates, and supportive friends. The once-a-week dose of energy I received from them at our meetings made the creation of this book possible.

Working on Manchuria means working with different languages and different ways of seeing. So Eun Ahn, Shiori Suemasa, Polina Dimova, Seung-Hee Jeon, Katherine McKenna, Cao Qican, and Jun Min Shin provided crucial translation assistance. Several key individuals also helped me translate the language of space. Mickey Casad, the executive director of the Vanderbilt Center for Digital Humanities, introduced me to important campus resources. Map work done by Samantha Turley and especially by Gabriela Ore helped train my mind to visualize Manchuria's terrain. The cartographer Jeff Blossom translated my visual meanderings into the book's robust set of maps. Phil Nagy painstakingly ensured the quality of the book's many images, while Cliff Anderson helped navigate permissions. Chris Benda's copyediting expertise caught numerous errors that my eyes simply could not see.

This project has received funding from the American Philosophical Society, the National Science Foundation, and the John Simon Guggenheim Foundation. Vanderbilt University generously provided money for travel, for resources, and, most importantly, for time to write.

At the University of Chicago Press, editor Karen Darling escorted the book through the publication process with precision and cheer. Tristan Bates, Deidre Kennedy, Caterina MacLean, and Katherine Faydash answered my many questions with patience and grace. I am incredibly grateful to the two anonymous readers for the press who read the manuscript with critical yet benevolent eyes. Their extensive insights transformed the book's framework and tightened its arguments. The failings that remain are entirely my own.

My family has been more than patient over the years. To Eddy, Safa, and my mom, I can say this book is now done, and I promise the next one won't take as long.

GLOSSARY

Romanization of terms in Chinese, Japanese, and Korean, followed by Chinese characters. Names of well-known people and places commonly found in standard references are omitted.

Aesun 愛順
ai 艾
Amnok 鴨綠
Anda 安達
Andō Kōji 安藤洪次
Bahai 巴海
Baichengzi 百城子
baihao 白蒿
baihe 百合
baishan heishui 白山黑水
baizi 稗子
Baoquanling 寶泉嶺
Beidacang 北大倉
Beidahuang jingsheng 北大荒精神
Beidahuang 北大荒
Beipiao 北票
Bencao gangmu 本草纲目
Bencao tujing 本草圖經
Bishu shanzhuang 避暑山莊
bu neng ming zhe 不能名者

Bukui jilue 卜魁紀略
Chǒnji 天池
Changbaishan 長白山
Chen Shixiang 陈世骧
Chen Yun 陳雲
Chengde 承德
Chifeng 赤峰
Chihaya 千早
chihōbyō 地方病
chishao 赤芍
Chǒng Sanggi 鄭尚驥
Chongzhengdian 崇政殿
cihuang 雌黃
Da Ming yi tongzhi 大明一統誌
Dai Fanglang 戴芳瀾
Daling (River) 大凌河
dan 石
Dasheng Ula 打牲嗚啦
Dazhengdian 大政殿
densenbyō 伝染病
Ding Wenjiang 丁文江
Dong yue 東嶽
Dongfeng tuolaji 東風拖拉機
Endō Ryūji 遠藤隆次
fan 礬
Fang Gongqian 方拱乾
fengshui 風水
Fengyang 鳳陽
fūdobyō 風土病
Fuling 福陵
Funing 撫寧
Fushun 撫順
fu 賦
gaizao diyi, shengchan di'er 改造第一, 生產第二
Gao Shiqi 高士奇
Gaoli 高麗
ge zhi 格致
gewu zhizhi 格物致知

guanwai 關外

guoying nongchang 國營農場

guzi 穀子

haibao 海豹

Haicheng 海城

Hamatong 蛤蟆通

hao 蒿

Heilongjiang wai ji 黑龍江外記

heitu 黑土

Helouju 何陋居

Hezhe/Hezhen (Nanai people) 赫哲/赫真

Hong Se-t'ae 洪世泰

Hsinkyō 新京

Hu Xiansu 胡先驌

Huachuan 樺川

huang 荒

huangmao 黃茅

huangqin 黃芩

Huangyu quanlan tu 皇輿全覽圖

huangyuan duo da, nongchang jiu duo da 荒原多大農場就多大

huashi 化石

Hucong dongxun rilu 扈從東巡日錄

Hulan 呼蘭

Hun (River) 渾河

Hunan ying 湖南營

Hwanung 桓雄

Hyesan 惠山

Ishii Shirō 石井四郎

Iyasaka 彌榮村

ji 箕

Jiamusi 佳木斯

Jiefangjun guanjiaotuan 解放軍管教團

Jinzhou 錦州

Jiufotang 九佛堂

jue yu 絕域

Kaema 蓋馬

kai jiang 開疆

kai huang 開荒

Kaiyuan 開源
kang bai yin 康拜因
kang 亢
Kantō (Kwantung) 關東
kanyan 看研
Kapsan 甲山
ken 墾
ki 氣
Koguryŏ 高句麗
Koryŏ sa 高麗史
Koryŏ 高麗
Kuandian 寬甸
Kumgang 金剛
Kunlun 崑崙
laogai 勞改
Laoyeling 老爺嶺
Li Dejin 李德金
Li Fuchun 李富春
Li Shizhen 李世珍
Liang Jun 梁軍
liangshi gongchang 糧食工廠
Liaodong 遼東
Liaoxi 遼西
Liaoyang 遼陽
ling 靈
lingdangmai 鈴鐺麥
liuhao 柳蒿
long xing zhi di 龍興之地
longgou 壟溝
longmai 龍脈
longtai 壟台
Lüshun 旅順
makou yu 馬口魚
man cao 漫草
Man-Mō Gakujutsu Chōsa Kenkyūdan 滿蒙學術調查研究團
Manshū sangyō chōsa shiryō 滿洲產業調查資料
Manshū takushoku kōsha 滿州拓殖公社
Minami Manshū tetsudō kabushiki kaisha (SMRC) 南滿州鉄道株式会社

Mishan 密山

mizi 米子

mobei yabulu 漠北押不盧

Mudanjiang 牡丹江

Mukedeng 穆克登

Muling 穆棱

murata 丸太

Naoli (River) 挠力河

Nen (River) 嫩江

Nihonbashi 日本橋

Ning'an 寧安

Ningguta 寧古塔

Nong'an 農安

Paektu taegan 白頭大幹

Paektusan gi 白頭山記

Paektusan 白頭山

Pingfang 平房

Qian Qianyi 錢謙益

Qianshan 千山

qiaomai 蕎麥

qinghao 青蒿

Qiqihar 齊齊哈爾

Qixing 七星河

Qixingpao 七星泡

Qiyunshan 啟運山

Qu Yuan 屈原

Rehe 熱河

ren ding sheng tian 人定勝天

ren yan xi shao 人煙稀少

Sabsu 薩布素

saiwai 塞外

Samjiyon 三池淵

Samsu 三水

Sanjiang pingyuan 三江平原

Sanjiaodi 三角地

Shang Kexi 尚可喜

Shangzhi 尚志

Shaodian 少典

Shengjing tongzhi 盛京通志
Shengjing 盛京
shenling 神靈
Shennong bencao jing 神農本草經
Shi clan (Jušeri) 石氏
shi yu 石魚
Shilou 石樓
shixue 實學
shu 秫
Shuangshan 雙山
shui long 水龍
shuili 水利
Siping 四平
sirhak 實學
Sŏ Myŏng-ŭng 徐命膺
Su Song 苏颂
Suibin 綏濱
Suifenhe 綏芬河
Suzi River 蘇子河
Taebaek 太白
Tajima Tadako 田島忠子
Takahashi Masahiko 高橋正彦
Tan'gun 檀君
Tan Xichou (H. C. T'an) 譚錫疇
Tensenko 天津子
Tianchi 天池
tianjian gongchang 田間工廠
Tieling 鐵嶺
Tokunaga Shigeyasu 德永重康
Tongliao 通遼
Toubi ji 投筆集
tumai 土脈
Unhung 雲興
Wanda 萬達
Wang Ang 汪昂
Wang Bo 王勃
Wang He 王和
wang ji 望際

wang qi 王氣

Wang Zhen 王震

wanwu 萬物

wei 尾

wo chao 我朝

wowo 窩窩

Wu Lien-teh (Wu Liande) 伍連德

Wu Qijun 吳其濬

Wu Zhaoqian 吳兆騫

wuweizi 五味子

Xiao Baishan 小白山

Xiao Han cun 小韩村

xiaomai 小麥

xing (nature) 性

Xingkai hu 興凱湖

Xinjing 新京

Xiqing 西清

xiu 宿

Xu Xiake 徐霞客

Yamagata Miyuki 山縣深雪

Yamato (Da he) 大和

yangcao 羊草

Yanwodao 雁窩島

Yi Sŏnggye 李成桂

Yi 李

Yilan 依蘭

yimucao 益母草

yinchen 茵陳

Yinghe 英和

Yingkou 營口

Yin'gou guan 陰溝關

Yiwulü 醫巫閭

Yŏngjo 英祖

Yongling 永陵

Youchao 有巢

youji 遊記

Youyi nongchang 友誼農場

Yu Gong yi shan 愚公移山

Yu Paektusan'gi 遊白頭山記
Yuan Hong 袁宏
yue 嶽
Zhang Jinyan 張縉顏
Zhang Yuliang 張玉良
Zhangguangcai 張廣才
zhangmao 章茅
Zhaoling 昭陵
zhen 陣
zhishi qingnian (zhiqing) 知識青年
Zhiwu mingshi tukao 植物名實圖考
Zhu Tinghu (T. O. Chu) 朱庭祜
ziran 自然

NOTES

INTRODUCTION

1. International Scientific Commission for the Investigation of the Facts Concerning Bacterial Warfare in Korea and China, *Report of the International Scientific Commission* (hereafter, *Report*), 27–29. Details of the Gannan incident are discussed in chapter 7.

2. For the background to the report and its accusations (which many Western scholars dismiss as a hoax), see Rogaski, "Nature, Annihilation, and Modernity."

3. Gannan voles, *Report*, 217–58; Xiuyan leaves, *Report*, 185; autopsies, *Report*, 449–70.

4. Lattimore, *Manchuria: Cradle of Conflict*; Lattimore, "The Unknown Frontier of Manchuria."

5. Li, *Kangxi jixia gewu bian yi zhu*, fire mountains, 82; permafrost, 9; petrified wood, 23; stone fish, 85; sea creatures into deer, 84; dense forest, 80.

6. An incomplete sampling includes Young, *Japan's Total Empire*; Mitter, *The Manchurian Myth*; Elliott, "The Limits of Tartary"; Duara, *Sovereignty and Authenticity*; Isett, *State, Peasant, and Merchant*; Rawski, *Early Modern China and Northeast Asia*.

7. Important exceptions include Bello, *Across Forest, Steppe, and Mountain*; Schlessinger, *A World Trimmed in Fur*; N. Smith, *Empire and Environment in the Making of Manchuria*.

8. On regimes of attention, see Pamela H. Smith, "Nodes of Convergence, Material Complexes, and Entangled Itineraries," in Smith, *Entangled Itineraries*.

9. On the construction of Manchu identity, and particularly the incorporation of multiple northeast Asian indigenous groups as "New Manchus," see Crossley, *A Translucent Mirror*; Elliott, *The Manchu Way*; Bello, *Across Forest, Steppe, and Mountain*; Kim, *Ethnic Chrysalis*.

10. On border making, see Chengzhi (Kicengge), "The Illusion of the Nerchinsk

Treaty Boundary-stone"; Perdue, "Boundaries, Maps, and Movement"; Perdue, *China Marches West*; Song, *Making Borders in Modern East Asia*; Kim, *Ginseng and Borderland*.

11. Kwong, *War and Geopolitics in Interwar Manchuria*; Suleski, *Civil Government in Warlord China*.

12. Iriye, *The Origins of the Second World War in Asia and the Pacific*.

13. Young, *Japan's Total Empire*; Matsusaka, *The Making of Japanese Manchuria, 1904–1932*; Duara, *Sovereignty and Authenticity*.

14. Levine, *Anvil of Victory*.

15. Ding and Elliott, "How to Write Chinese History in the Twenty-First Century."

16. Elliott, "The Limits of Tartary," 603.

17. Naquin, "The Material Manifestations of Regional Culture," 376.

18. Brown, *A Biography of No Place*, 3.

19. For influential examples of cross-border studies of this region, see Rawski, *Early Modern China and Northeast Asia*; Perdue, "Crossing Borders in Imperial China" and *China Marches West*.

20. Lattimore, *Manchuria: Cradle of Conflict*, 290–302.

21. These discussions emerge in Crossley's critique of the term "transfrontiersmen." See Crossley, *A Translucent Mirror*, 32–33, 48, 84–88; for "slough of hybridity," 32n64.

22. Crossley, 33.

23. Hinchliffe, "Inhabiting—Landscapes and Natures," 215. On the coproduction of nature and social orders, see Jasanoff, *States of Knowledge*; Latour, *The Politics of Nature*.

24. The classic example is Adas, *Machines as the Measure of Men*.

25. Raj, *Relocating Modern Science*.

26. This phenomenon has been bountifully illustrated by multiple scholars, including Cams, *Companions in Geography*; Chengzhi (Kicengge), *Daichin Gurun to sono jidai*; Perdue, "Boundaries, Maps, and Movement."

27. As observed by Suman Seth, "the local is socially and historically produced through dynamic interaction . . . and not a space where indigenous sensibilities reside in any simple sense." Seth, "Putting Knowledge in Its Place," 378. On the phenomenon of "nested" imperialisms, see Rogaski, "Knowing a Sentient Mountain."

28. For inspiring studies that take into account complex borderlands and multiple empires, see White, *The Middle Ground*; Gomez, *The Experiential Caribbean*; Strang, *Frontiers of Science*; Coen, *Climate in Motion*.

29. Fan, "Science in Cultural Borderlands," 215.

30. Livingstone, *Putting Science in Its Place*; Latour, *Reassembling the Social*; Kohler, *Landscapes and Labscapes*.

31. Hersey and Vetter, "Shared Ground"; Jørgensen, Jørgensen, and Pritchard, *New Natures*; Pritchard, *Confluence*; Mitman, "Living in a Material World"; Nash, *Inescapable Ecologies*.

32. "A category that encompasses things like animals, plants, planets, minerals, lands, waters, and peoples." Strang, *Frontiers of Science*, 7.

33. Jordheim and Shaw, "Opening Doors," 4; Delbourgo, "The Knowing World."

34. For an exploration of these themes in an Asian context, see Ghosh, *The Great Derangement*; Duara, *The Crisis of Global Modernity*; Ian Miller's introductory essay in Miller, Thomas, and Walker, *Japan at Nature's Edge*.

35. Marcon, "The Critical Promises of the History of Knowledge," 28.

36. Turnbull, "'Travelling Knowledge: Narratives, Assemblages, Encounters," 238–39.

37. White, "The Spatial Turn in History."

38. For these debates, see Steinberg, "Down to Earth"; Nash, "The Agency of Nature or the Nature of Agency?"

39. For a useful introduction to digital humanities for history, see Hillier and Knowles, *Placing History*.

40. On science and styles of seeing, see Daston and Galison, *Objectivity*; Latour, "Visualization and Cognition: Thinking with Eyes and Hands"; Bleichmar, *Visible Empire*; Fukuoka, *The Premise of Fidelity*.

41. Ingold, *The Perception of the Environment*; Eyferth, *Eating Rice from Bamboo Roots*; Geurts, *Culture and the Senses*; Smith, *The Body of the Artisan*.

42. Bello, *Across Forest, Steppe, and Mountain*; Schlesinger, *A World Trimmed with Fur*.

43. Victor Seow, *Carbon Technocracy*, chap. 1.

44. Carter, *The Road to Botany Bay*, xxii.

45. Carter, 294.

CHAPTER 1

1. Ingold and Lee, "Fieldwork on Foot: Perceiving, Routing, Socializing," 73. On movement as the primary action defining space, see White, "What Is Spatial History?"

2. Perdue, "Chinese Exploration"; Stewart, "The Exploration of Central Asia."

3. Schafer, *The Vermilion Bird*.

4. Daston and Park, "The Age of the New," 6; Greenblatt, *Marvelous Possessions*; Cook, *Matters of Exchange*.

5. Prieto, "Classification, Memory, and Subjectivity."

6. Grafton, *New Worlds, Ancient Texts*.

7. Daston and Park, "The Age of the New," 8.

8. Owen, *Remembrances*; Wu, *A Story of Ruins*.

9. Schlesinger, *A World Trimmed with Fur*, 10; Bello, *Across Forest, Steppe, and Mountain*.

10. Zhang, *Ningguta shanshui ji*, 5.

11. Bello, *Across Forest, Steppe, and Mountain*, 30.

12. On Qing observers moving across cultural borders, see Perdue, "Crossing Borders in Imperial China."

13. Details of Wu's life drawn from Li Xingsheng, *Jiangnan caizi*.

14. Elman, *A Cultural History of Civil Examinations in Late Imperial China*.

15. Wakeman, *The Great Enterprise*.

16. Qin, *Lake Taihu, China*.

17. Shih, *Chinese Rural Society in Transition: A Case Study of the Lake Tai Area*.

18. Li Xingsheng, *Jiangnan caizi*, 20–27.

19. Li Xingsheng, *Wu Zhaoqian nian pu*, 27.

20. Knechtges, *The Han Rhapsody*.

21. Wu Zhaoqian, "Chun fu" [Spring Rhapsody], *Qiu jia ji* (1965), 8:3686.

22. On the aesthetics of the late Ming literati, see Brook, *The Confusions of Pleasure*; Clunas, *Superfluous Things*.

23. Wu Zhaoqian, "Chun fu," *Qiu jia ji* (1965), 8:3686.

24. Li Xingsheng, *Jiangnan caizi*, 23.

25. Strassberg, *Inscribed Landscapes*, 139–50.

26. Li Xingsheng, *Jiangnan caizi*, 26.

27. Williams, "The Pity of Spring," 138.

28. Strassberg, *Inscribed Landscapes*, 6.

29. Li Xingsheng, *Jiangnan caizi*, 35–59.

30. Li Xingsheng, *Dongbei liuren shi*, 133–47; Elman, *Civil Examinations and Meritocracy in Late Imperial China*, 88.

31. Li Xingsheng, *Jiangnan caizi*, 60–82; Yim, "Traumatic Memory, Literature and Religion in Wu Zhaoqian's Early Exile," 123–25.

32. Li Xingsheng suggests that Wu had actually been very willing to work with the Manchu government and was simply paralyzed by fear. Such a perspective on Wu's actions in the 1650s certainly helps support Li's assertion that Wu became a patriotic supporter of Qing rule in the face of Russian incursions in the northeast during the 1670s.

33. On Wu Weiye, one of the foremost poets of the Ming-Qing transition, see Li, "History and Memory in Wu Weiye's Poetry"; Huang, "History, Romance, and Identity."

34. Wu Weiye, "Bei ge zeng Wu Jizi."

35. Chan, "Beyond Border and Boudoir."

36. Yim, *The Poet-Historian Qian Qianyi*, 88–89.

37. Chaves, "The Yellow Mountain Poems of Ch'ien Ch'ien-i," 487; Chaves, *Every Rock a Universe*.

38. Oh Youngchan and Byington, "Scholarly Studies on the Han Commanderies in Korea"; Byington, *The History and Archaeology of the Koguryŏ Kingdom*.

39. For an overview of this complex history, see Rawski, *Early Modern China and Northeast Asia*, 21–60.

40. *Funing xian zhi*, 27–28, 125.

41. "Xiao fa Funing ti nilü qiang" [Written upon the wall of the inn upon leaving Funing at dawn], Wu Zhaoqian, *Qiu jia ji, Guilai caotang chidu*, 22.

42. Rojas, *The Great Wall*; Waldron, *The Great Wall of China*.

43. "Shanhaiguan," Wu Zhaoqian, *Qiu jia ji, Guilai caotang chidu*, 22–23.

44. "Chu guan" [On leaving the pass], Wu Zhaoqian, *Qiu jia ji, Guilai caotang chidu*, 23.

45. On Lady Meng Jiang, see Rojas, *The Great Wall*, 81–86.

46. Jung, "Liaoxi."

47. One such observation was made by the European Jesuit Ferdinand Verbiest during the eastern tour of 1683. See chapter 2.

48. "Guangning dao zhong zuo" [Written on the road to Guangning] and "Deng Guangning cheng wang cheng wai fo she, yin tong zhu gong que guo" [Climbing the city wall at Guangning, seeing a Buddhist monastery outside of the walls, having passed it on the way back with the gentlemen]. Wu Zhaoqian, *Qiu jia ji, Guilai caotang chidu*, 26–27.

49. Wakeman, *The Great Enterprise*, 67.

50. "Shahe dao zhong" [On the road to Shahe], Wu Zhaoqian, *Qiu jia ji, Guilai caotang chidu*, 27; "Ci qian wei" in Wu Zhaoqian, *Qiu jia ji, Guilai caotang chidu* 26.

51. Pamela Crossley has cautioned us to avoid thinking of Liaodong as "balanced between" dominant categories (Manchu, Han) and urges us instead to think of Liaodong identity as a "coherent one with a history and a discrete geographical contour." Crossley, *A Translucent Mirror*, 50.

52. Robinson, "Chinese Border Garrisons."

53. Sun Yang, "Song Wu Hancha zhi Ninggu" [Seeing off Wu Zhaoqian on his way to Ningguta], in Zhang Yuxing, *Qingdai dongbei liuren shi xuan zhu*, 149.

54. On this mountain, see chapter 3.

55. Wakeman, *The Great Enterprise*, 59–62.

56. Swope, *The Military Collapse of China's Ming Dynasty, 1618–44*, 19–24.

57. For a spatial rendering of the Tang-Koguryŏ wars, see Injae et al., *Korean History in Maps*, 29.

58. "Gaoli ying" [The Korean citadel], Wu Zhaoqian, *Qiu jia ji, Guilai caotang chidu*, 32. On Du Fu's poetry, see Stephen Owen, *The Great Age of Chinese Poetry*.

59. All together the Willow Palisade consisted of three sections: a western section marking Mongol and Han settlement, an eastern section demarking Han and Manchu settlement, and a northern extension marking Mongol and Manchu territories. See Edmonds, *Northern Frontiers of Qing China and Tokugawa Japan*, 55–70.

60. Reardon-Anderson, *Reluctant Pioneers*, 23.

61. Today, a village at this site sixty miles east of Fushun along the Hun River still bears the name "Yingge Gate Town" (Yinggemen zhen). On the concentration of Manchu place-names around what was the Willow Palisade, see Zhu, "Using Toponyms to Analyze the Endangered Manchu Language in Northeast China."

62. "Yingou guan" [At "Shady Ditch" Pass], Wu Zhaoqian, *Qiu jia ji, Guilai caotang chidu*, 32.

63. Fang Gongqian, *Ningguta zhi*, 1b.

64. "Wu ri zu shui Nianmahe" [Stuck for five days at the Nianma River], Wu Zhaoqian, *Qiu jia ji, Guilai caotang chidu*, 32.

65. Xiong Yi, *Zhongguo he liu*, 98–101.

66. On Xuantu, see Twitchett, Fairbank, and Loewe, *The Cambridge History of China*, 449–50; Pai, *Constructing "Korean" Origins*, 455.

67. "Du Huntongjiang" [Crossing the Songhua River], in Zhang Yuxing, ed., *Qingdai dongbei liuren shi*, 242.

68. "Jiaohe shan zhong ye xing" [Traveling at night in the Jiaohe River mountains], Wu Zhaoqian, *Qiu jia ji, Guilai caotang chidu*, 35.

69. Or "Good Hunting" mountains, since "Julgen sain" is a phrase uttered after a

safe return from a hunting trip. Mu Huajun, "Juzhu zai Zhangguangcailing de Manzu 'bala' ren."

70. Jilin sheng difang zhi bianxiu weiyuanhui, *Jilin sheng zhi: Ziran dili zhi*, 80–130

71. "Da Wuji" [The great forest], Wu Zhaoqian, *Qiu jia ji, Guilai caotang chidu*, 35.

72. Lee, *The Manchurian Frontier in Ch'ing History*, 9–10.

73. On Hailin's "Snow Town," see http://15tianqi.cn/hailinxuexiangtianqi/.

74. "Shang fumu shu" [Letter to my parents], Wu Zhaoqian, *Qiu jia ji* (1993), 293.

75. On Fang's life, see Li Xingsheng's introduction to Fang's collected poems, Fang Gongqian, *Helouju ji*.

76. *The Analects*, 9:13. The translation of this passage by Legge reads: "The Master was wishing to go and live among the nine wild tribes of the east. Someone said, "They are rude. How can you do such a thing?" The Master said, "If a superior man dwelt among them, what rudeness would there be?" Legge, *The Chinese Classics*, 1:171.

77. Fang Gongqian, *Ningguta zhi*, 1b.

78. For an introduction to Chinese geographical history sources, see Wilkinson, *Chinese History*, 131–69. On Xu Xiake, see Julian Ward, *Xu Xiake: The Art of Travel Writing*.

79. Fang Gongqian, *Ningguta zhi*, 1a.

80. Fang Gongqian, 1b.

81. The list of exiles cited in Qian Wei's preface to *Ningguta shanshui ji* begins with Cai Yong, an official who was exiled to the north during the Han dynasty (c. 2nd century CE). On Liu Zongyuan, see Strassberg, *Inscribed Landscapes*, 139–50.

82. See Li Xingsheng's introduction to *Ningguta shanshui ji*, 1–5.

83. Zhang Jinyan, *Ningguta shanshui ji, yu wai ji*, 5.

84. Zhang Jinyan, 6.

85. Zhang Jinyan, 5.

86. Schaffer, *The Vermilion Bird*, 42.

87. Zhang Jinyan, *Ningguta shanshui ji, yu wai ji*, 1–2.

88. Zhang Jinyan, 14.

89. Zhang Jinyan, 1.

90. Zhang Jinyan, 48.

91. Zhang Jinyan, 49.

92. On *shixue* as "substantive learning," see Peterson, *Bitter Gourd*.

93. Elman, *From Philosophy to Philology*.

94. Emma Teng and Pat Giersch have pointed out that for Qing Taiwan and Yunnan, Han encounters with borderlands inspired empirical observation, but empiricism remained entangled with classical learning. Teng, *Taiwan's Imagined Geography*, 48–49; Giersch, *Asian Borderlands*, 68–69.

95. Wu Zhaoqian, *Qiu jia ji* (1965), 7:14a.

96. Literally translated, the Chinese for *song* would read something like "it was *songed*," making the adjective *song* (towering) into a verb. Zhang Jinyan, *Ningguta shanshui ji, yu wai ji*, 5.

97. The idea of a *zaowuzhe* emerges from texts associated with the Daoist tradition, particularly the *Zhuangzi*. Sinologists have coined multiple terms to distinguish this *zaowuzhe* from a monotheistic concept of a creator God. Ames translates *zaowuzhe* as "transformative creative power"; Elvin uses "the fashioner of things." Yates uses the term "maker of the phenomenal world" and cautions that "the Chinese never gave human beings a privileged position as being created in the deity's image and as possessing the right or duty to name and therefore control and exploit the rest of creation" (Yates in Cua, *Encyclopedia of Chinese Philosophy*, 657). While Zhang's approach to nature does not dwell on a creator deity, it is clear that he considered the act of naming to be an important human task. Cua, *Encyclopedia of Chinese Philosophy*; Elvin, *Another History*; Harbsmeier, "Towards a Conceptual History of Some Concepts of Nature in Classical Chinese: Zi Ran 自然 and Zi Ran Zhi Li 自然之理."

98. He Tangzuan, Zhang Jinyan, et al., *Ming shan sheng gai ji* (Ming Chongzhen reign, 1628–1644); Zha Zhilong, *Dai shi* (1586) [1654 reprint]. Both available through the National Library of China—Harvard-Yenching Library Chinese rare-book digitization project. Thanks to Sue Naquin for pointing out the potential complexity of Zhang's role in these publications.

99. Dott, *Identity Reflections*.

100. Zhang Jinyan, *Ningguta shanshui ji, yu wai ji*, 26. On "plants that cannot be named" see Métailié, "Plantes et noms, plantes sans nom dans le 'Zhiwu Mingshi Tukao.'"

101. Zhang Jinyan, 15, 16, 23.

102. Zhang Jinyan, 12.

103. Zhang's perspective on the *qi* of different regions echoes the thinking of the influential Ming-Qing scholar Wang Fuzhi, who hypothesized how the energy of different regions produced both landscapes and human beings with different qualities. See David Bello's evocative discussion of *qi*-based "Hanspace" in *Across Forest, Steppe, and Mountain*, 23–36.

104. Robson, *Power of Place*, 17–56.

105. Zhang Jinyan, *Ningguta shanshui ji, yu wai ji*, 48.

106. Whiteman, *Where Dragon Veins Meet*, 68.

107. Daston and Galison, *Objectivity*, 29.

108. Daston and Galison, 36. For a critique of Eurocentrism in the history of objectivity, see Delbourgo, "The Knowing World."

109. Schama, *Landscape and Memory*; Olwig, *Landscape, Nature, and the Body Politic*.

110. Hummel, *Eminent Chinese of the Ch'ing Period (1644–1912)*; Baihai, 14–15; Sabsu, 630–31.

111. "Suhuan" grew up to write (under the name Wu Zhenchen) what is considered the standard memoir of exile life, *Ningguta jilüe*.

112. To understand the threat in Manchuria in the context of the larger Qing geography, see Perdue, *China Marches West*.

113. "Feng zeng fu shuai Sa Gong" [A poem penned for colonel Sabsu] Li Xingsheng, *Dongbei liuren shi*, 269.

CHAPTER 2

1. Gao Shiqi gives the date as the fifteenth day of the second lunar month of the twenty-first year of the Kangxi reign. Gao Shiqi, *Hu cong dong xun ri lu* (*HCDXRL*), 88.

2. The first eastern tour had traveled about one hundred miles north of Mukden. Tie Yuxin and Wang Peihuan, *Qing di dong xun*, 13–14.

3. This description from Verbiest, "Journey into Tartary," 108. Verbiest estimated that the entire retinue included more than seventy thousand people. This would make the second eastern tour considerably larger than the three thousand or so members of Qianlong's southern tours described in Chang, *A Court on Horseback*, 117. Some scholars call Verbiest's estimate "unrealistic" and "baseless" and estimate the tour size at around two thousand to three thousand people. Tie and Wang, *Qing di dong xun*, 52–56.

4. Wang, "Kangxi dong xun shi shi gou bu."

5. Verbiest, "Journey into Tartary," 108; Gao Shiqi, *HCDXRL*, 124.

6. Unlike Wu Zhaoqian's poetry, which mentions only a few specific locales, members of the eastern tour (as well as official chronicles) list every stop along the entire route. To link the tour route to specific places on current-day maps, I rely on the equivalents provided in appendix 4 of Tie and Wang, *Qing di dong xun*. To help visualize Qing sensibilities about the relationship between the route and the terrain, I have used the illustrated Beijing to Hetu Ala route map dated to the Qianlong reign (*Mukden-i dedun uden-i nirugan*, or Map of the overnight and daytime rest stations in Mukden) reproduced in Andrej Rudnev, "A Manchu Itinerary."

7. The first chapter, "In Motion," which paints a vivid picture of the young Kangxi's love of nature, the hunt, and warfare, includes several of Kangxi's observations of the Manchurian environment penned during the eastern tour. Spence, *Emperor of China*.

8. Chang, *A Court on Horseback*, 1, 28.

9. Elliott, 609–12.

10. Berger, *Empire of Emptiness*, 40.

11. Crossley, "The Rulerships of China," 1483.

12. Gao Shiqi, *HCDXRL*, 84–87. Gao includes poems from such famous early Qing literati as Wang Shizhen, Xu Qianxue, Wang Youlin, and Mao Qiling. The lines quoted here are drawn from poems by the Hanlin academicians Zhang Ying, Shen Quan, and Wang Shizhen.

13. The Long White Mountain (Changbaishan), Penghai, "Black Waters" (Heilongjiang), and Pine Flower River (Songhuajiang) are all locations in the northeast. Muyeshan (in today's Inner Mongolia) was held to be the origin of the (non-Han) Khitan Liao dynasty.

14. On Gao Shiqi, see Hummel, *Eminent Chinese of the Ch'ing Period*, 413–15.

15. Gao Shiqi's account was originally published in the early eighteenth century in Gao's collected works, *Qingyintang quanji* [Complete collected works from the hall of pure elegies]. This chapter relies primarily on a modern punctuated version pub-

lished in 1986. Details are cross-checked with available versions of the original, including a Qing-period *Qingyintang quanji* available on line via the Harvard-Yenching library (https://iiif.lib.harvard.edu/manifests/view/drs:51789410$1i).

16. Witek, *Ferdinand Verbiest*.

17. Chu, "Scientific Dispute in the Imperial Court"; Jami, *The Emperor's New Mathematics*.

18. On Verbiest's relationship with Kangxi, see Spence, *To Change China*, 3–32.

19. Golvers, *Letters of a Peking Jesuit*, 448.

20. Verbiest's account was republished several times. This chapter relies on the English version published by the Hakluyt Society in 1854, *History of the Two Tartar Conquerors of China, Including the Two Journeys into Tartary of Father Ferdinand Verbiest in the Suite of the Emperor Kang-Hi* (hereafter "Journeys into Tartary"), cross-referenced with Golver's transcription of a Latin copy of Verbiest's August 1682 report held in the Vatican ("Iter P[atris] Ferdinandi Verbiest in provinciam Leaotum et Tartariam Orientalem institutum 1682, die 23 Martii," referenced here as "Golvers"). See also Pang, "An Evaluation of F. Verbiest's Account of His Journey to Manchuria."

21. Golvers, 447.

22. Golvers, 448.

23. Gao Shiqi, *HCDXRL*, 89–90.

24. Gao Shiqi, 90.

25. Gao Shiqi, 92–93.

26. To visualize the connection between the eastern tour and the battles of the Qing conquest, I have utilized the "Annotated Map of Battle Sites in the Regions of Shengjing, Jilin, and Heilongjiang" (*Shengjing, Jilin, Heilongjiang deng chu biaozhu zhanji yutu*, hereafter the *Zhanji yutu*), a large map of Nurhaci and Hong Taiji's military engagements throughout the northeast compiled in 1776 as part of the Qianlong emperor's cultivation of Manchu history and identity. The map indicates the main Qing imperial roads and relay stops, as well as the Willow Palisade and fortifications. All geographical features, including rivers, mountains, settlements, and passes, are labeled in Manchu and Chinese, along with bilingual markers for battles (Manchu-Qing victories only) indicating the date of the battle and the identity of the defeated enemy (Ming generals, other Jurchen tribes, etc). The version used here is a twenty-five-sheet reproduction published during the Japanese occupation of Manchuria. Wada, *Seikyō Kitsurin Kokuryūkō tōsho hyōchū senseki yozu ni tsuite*.

27. Swope, *The Military Collapse of China's Ming Dynasty, 1618–44*, 56–59, 158–67.

28. Gao Shiqi, *HCDXRL*, 94.

29. Gao Shiqi, 95.

30. Elliott, *The Manchu Way*, 57.

31. Gao Shiqi, *HCDXRL*, 93.

32. Chang, *A Court on Horseback*, 45–46.

33. Locations of hunts determined by cross-referencing Gao Shiqi's record with the *Qing Shilu* (Veritable Record) for the Kangxi reign. Chronicles for the eastern

tour (along with numbers of tigers bagged) can be found in the *Shilu* for Kangxi 21, 101:1–102:12.

34. On tigers, see Marks, *Tigers, Rice, Silk and Silt*; also Coggins, *The Tiger and the Pangolin*; Matthiessen, *Tigers in the Snow*; Han, "The Forest Stand Structures in Northeastern China; Their Potential Effects on the Suitability of Forests for Animals, Plants, and Other Values; and the Possible Relationships to Amur Tiger (*Panthera tigris altaica*) Conservation."

35. Verbiest, "Journeys into Tartary," 109.

36. Verbiest, 70–71. "Magnetic declination" is the angle between true north and the north as indicated by the needle of a compass (the earth's magnetic field). Because of movements in the earth's magnetic field, this value changes over time. In 1682, the magnetic north and true north at Shenyang were indeed very closely aligned. See the "Historical Magnetic Declination Map," National Oceanic and Atmospheric Administration National Centers for Environmental Information (https://maps.ngdc.noaa .gov/viewers/historical_declination/).

37. Zhang, "The Introduction of European Astronomical Instruments."

38. It is not clear what species this "sea bull" was. Today the Yellow Sea and Sea of Korea (Japan) coasts of Korea are home to several species, including the harbor seal and spotted seal, as well as sea lions. Won and Smith, "History and Current Status of Mammals of the Korean Peninsula."

39. Verhaeren, *Catalogue de la bibliotheque du Pei-t'ang*. The Beitang cathedral was constructed about ten years after the eastern tour, but it is not unreasonable to think that Verbiest was in possession of some of that library's books before the library's construction.

40. Entries on seals or sea lions are included in Conrad Gessner's *Historiae animalium*, book 4, *De piscium & aquatilium animantium natura* (1587), 489; Aldrovandi, *De piscibus* (1638), 209; and Joannes Jonstonus, *Historiae naturalis de piscibus et cetis* (1647), plate 44. All available at https://www.biodiversitylibrary.org.

41. Verbiest, "Journeys into Tartary," 111. On this episode, see also Spence, *Emperor of China*, 94; Iannaccone, "Lo zoo dei gesuiti:"

42. On seal skins or furs in Qing Beijing, see Schlesinger, "The Qing Invention of Nature," 71, 85. On medicinal uses, see Sabine Wilms, "Seal Penis and Testes in Gynecology," *Happy Goat Productions*, https://www.happygoatproductions.com/blog/ 2018/12/4/seal-penises-and-testes-in-gynecology. On the extinction of the Japanese sea lion (once common off the coasts of both Japan and Korea), see W. Perrin et al, eds., *Encyclopedia of Marine Mammals* (Elsevier, 2008), 170–72.

43. Golvers, 449.

44. For general introductions to Shenyang history, see Guo, "Shenyang"; Sepe, "Back to the Roots."

45. On the name "Mukden," see Elliott, "Limits of Tartary," 605n5.

46. Nappi, *The Monkey and the Inkpot*, 50–68.

47. Snyder-Reinke, *Dry Spells*; Overmyer, *Local Religion in North China in the Twentieth Century*; Pomeranz, "Water to Iron, Widows to Warlords."

48. For examples of immortals riding dragons, see Campany, *Making Transcen-*

dents, 71–72 On dragons as means of visualizing internal energies in Taoist inner alchemy techniques, see Kohn, *The World Upside Down*.

49. Sterckx, *The Animal and the Daemon in Early China*, 180.

50. Brook, *The Troubled Empire*, 6–23.

51. Wang Liquan, *Zhonghua long wenhua de qiyuan yu yanbian*.

52. Zhou Jingnan, "Zijincheng li de long wen zhuang shi"; Yan, "Long zai zijincheng."

53. Tang Gengsheng, "Zhu Yuanzhang san fan jiaxiang Fengyang kaoshu."

54. For an in-depth exploration of how Manchu court clothing created a "statement of ethnic identity and universal domination," see Vollmer, *Dressed to Rule*.

55. This exchange happened the following year on an imperial trip north to the Mulan hunting grounds. Gao Shiqi, *Sai bei xiao chao*, 3b.

56. In his response to the emperor, Gao paraphrases the ancient dictionary, *Shuowen jiezi* (2nd c. CE), which glosses "dragon" as a beast that "can sometimes appear and can sometimes be hidden. In the spring, they ascend into the sky, in the fall, they hide in the watery depths." On early modern Chinese elite sensibilities on the existence and elusiveness of dragons, see Elvin, "The Man Who Saw Dragons: Science and Styles of Thinking in Xie Zhaozhe's Fivefold Miscellany"; Zhang, "From 'Dragonology' to 'Meteorology.'"

57. Verbiest, "Journeys into Tartary," 107.

58. On these motifs in Mukden palace architecture, see Chen Bochao and Piao Yushun, *Shengjing gongdian jianzhu*; Zhang, *Shenyang gugong jianzhu zhuangshi yanjiu*; Wang Mingqi and Wang Qingxian, "Shenyang gugong"; Itō Seizō, *Hōten kyūden kenchiku no kenkyū*.

59. Wang Mingqi and Wang Qingxian, "Shenyang gugong." The three-dimensional dragons so characteristic of the Mukden palace architecture resemble the freestanding architectural dragon motifs used to great effect in the Forbidden City's Raining Flowers Hall (Yuhua ge), an architecturally striking Tibetan Buddhist meditation hall built at the orders of the Qianlong emperor. See Berger, *Empire of Emptiness*, 97–103.

60. Carroll, *Between Heaven and Modernity*, 105.

61. Whiteman, "Kangxi's Auspicious Empire."

62. Wheatley, "The Ancient Chinese City as a Cosmological Symbol."

63. The *Shengjing tongzhi* was first published in 1684. For this study, I have used the digitized version of the Kangxi-era text available via the Kyoto University Rare Materials Digital Archive, at https://rmda.kulib.kyoto-u.ac.jp/en/item/rb00008733.

64. *Shengjing tongzhi xu*, 1b. On *wang qi*, see Clunas, *Screen of Kings*, 30.

65. *Shengjing tongzhi xu*, 3a.

66. On the celestial mansions system, see Needham, *Science and Civilisation in China*, 3:242–59.

67. On the recording of portents, see Wilkinson, *Chinese History*; Smith, *Mapping China and Managing the World*.

68. On the "cosmic demarcation" (*fenye*) system and the boundaries of the Chi-

nese empire, see Jiang, *The Mandate of Heaven and The Great Ming Code*, 100–141; Pankenier, *Astrology and Cosmology in Early China*.

69. *Shengjing tongzhi*, 7:3b–10b.

70. These portents were not recorded in the *Shengjing tongzhi*, most likely because they were not recorded in Chinese texts. On these portents as recorded in the "Old Manchu Chronicles," see passages translated by the Manchu Studies Group, at http://www.manchustudiesgroup.org/translations/lao-manwen-dang/fascicle-one.

71. Deane, "The Chinese Imperial Astronomical Bureau."

72. For an overview of the case and Verbiest's reaction, see Chu, "Against Prognostication."

73. Chu, "Against Prognostication," 444.

74. For documents containing descriptions of Yang and Du's errors in *fengshui* calculations (including verbatim transcripts of technically detailed interrogations), see An Shuangcheng, *Qing chu Xiyang chuanjiaoshi*, 219–35.

75. Sweeten, "Early Qing Imperial Tombs."

76. Fushun shi renmin zhengfu difangzhi bangongshi, *Qing Yongling zhi*; Quan Yuedong, "Qiantan gujin Qing Yongling zhi chayi."

77. Kangxi Veritable Records (*Shilu*), 101:16

78. Kangxi Veritable Records, 101:18. The *Shilu* simply states that the text was the same as that delivered at Fuling.

79. Hansen, *Changing Gods in Medieval China, 1127–1276*, ix, 27.

80. Bruun, *Fengshui in China*; Field, "In Search of Dragons."

81. Gao Shiqi, *HCDXRL*, 102.

82. Gao Shiqi, 100.

83. "San yue chu liu ri gao ji Fuling gong shu shi yun," in Xu Xin, *Nu'erhachi ling*, 57.

84. For discussion of these ancestors of Nurhaci, see Crossley, *A Translucent Mirror*, 135–222.

85. Gao Shiqi, *HCDXRL*, 103.

86. According to the *Zhanji yutu*, several forts were located at this crucial location, indicating the strategic importance of this river juncture.

87. Swope, *Military Collapse*, 19–23. Today, part of the battlefield is beneath the waters of the Dahuofang Reservoir. The partially flooded "Ironback Mountain" forms a little peninsula at the eastern end of the reservoir, offering a view of the Suzi River and the mausoleum complex of the twentieth-century warlord Zhang Zuolin, which is located on the opposite bank.

88. Gao Shiqi, *HCDXRL*, 104.

89. "Saerhu, bing xu," in Tie Yuxin and Wang Peihuan, *Qing di dong xun*, 44.

90. Crossley, *A Translucent Mirror*, 78–79; Rawski, *Early Modern China and Northeast Asia*, 52.

91. Crossley, *A Translucent Mirror*, 146–48. The relative locations of Hulan Hada, Fe Ala, and Hetu Ala can be seen on sheet 2.4 of the *Zhanji yutu*.

92. Sweeten, "Early Qing Imperial Tombs."

93. Crossley, *A Translucent Mirror*, 135.

94. Fushun shi renmin zhengfu difangzhi bangongshi, *Qing Yongling zhi*, 17.

95. Crossley, "Manzhou Yuanliu Kao and the Formalization of the Manchu Heritage"; Sweeten, "Early Qing Imperial Tombs."

96. Fushun shi renmin zhengfu difangzhi bangongshi, *Qing Yongling zhi.*

97. Tie Yuxin and Wang Peihuan, *Qing di dong xun*, 16.

98. Loewe and Shaughnessy, *The Cambridge History of Ancient China.*

99. Whiteman, "Kangxi's Auspicious Empire: Rhetorics of Geographic Integration in the Early Qing."

100. Gao Shiqi, HCDXRL, 105.

101. See "Kangxi di mo bi mi yu" [Secret edict from the pen of the Kangxi emperor] Kangxi 20/12/8. *Kangxi chao Manwen zhupi zouzhe quanyi*, 7. This edict is also mentioned in Chang, *A Court on Horseback*, 82.

102. Gao Shiqi, HCDXRL, 106.

103. Elliott, *The Manchu Way*, 65–71.

104. Gao Shiqi, HCDXRL, 107.

105. Gao Shiqi, 108.

106. Nurhaci had married one of his daughters to the Ula headman, Bujantai, but Bujantai had arrows shot at his bride as a punishment for an unrecorded crime. Official histories suggest this indignity was the reason for Nurhaci's subjugation of the Ula. Crossley, *Translucent Mirror*, 209.

107. Bello, *Across Forest, Steppe, and Mountain*, 97–100; Schlesinger, *A World Trimmed with Fur*, 65–68.

108. Gao Shiqi, HCDXRL, 109–10; Verbiest, 76; Elliott, "Limits of Tartary," 612.

109. On Kangxi's long-distance pursuit of Galdan across the Gobi Desert, see Perdue, *China Marches West*, 174–208; on Kangxi's climb of Wutaishan, see Köhle, "Why Did the Kangxi Emperor Go to Wutai Shan?"

110. Verbiest, "Journeys into Tartary," 117.

111. Verbiest, 119–120. Verbiest was not the only one in the seventeenth century to assume a connection between the far northeast of Asia and the northwestern reaches of North America. In 1634, an emissary of Samuel de Champlain to the Winnebago donned a Chinese silk robe, believing the Indians in the vicinity of Lake Huron would naturally be familiar with the "majesty of the Chinese court." Vollmer, *Dressed to Rule*, xi.

112. Gao Shiqi, HCDXRL, 110–11.

113. Gao Shiqi, 112.

114. Gao Shiqi, HCDXRL, 113. On the "New Manchus," see Elliott, *Manchu Way*, 84–86; Bello, *Forest, Mountain, and Steppe*, 87–90. On the foraging work expected of bannermen at Ula, see Schlessinger, *World Trimmed in Fur*, 55–92.

115. Gao Shiqi, HCDXRL, 115.

116. Krykhtin and Svirskii, "Endemic Sturgeons of the Amur River: Kaluga (*Huso dauricus*) and Amur Sturgeon (*Acipenser schrenckii*)," 221–39.

117. Verbiest, "Journeys into Tartary," 76.

118. Gao Shiqi, HCDXRL, 116.

119. Gao penned a poem, "Song of the Water Dragon" to commemorate the event. HCDXRL, 116.

120. Gao Shiqi, *HCDXRL*, 113.

121. For battles in this area, see the *Zhanji yutu* sheet 3.3.

122. Verbiest, "Journeys into Tartary," 78.

123. Verbiest, 79. This episode is delightfully recounted in Jonathan Spence, *Emperor of China: A Self-Portrait of Kang-hsi*, 59.

124. Jami, "Imperial Control and Western Learning."

125. Gao Shiqi, *HCDXRL*, 120.

126. Gao Shiqi, 122.

127. Kangxi Veritable Records (*Shilu*), 102:10.

128. Kangxi Veritable Records (*Shilu*), 101:22. Also quoted in Chang, *A Court on Horseback*, 83.

129. On Bumbutai, see Rawski, *The Last Emperors*, 135–36.

130. Verbiest, "Journeys into Tartary," 71–72.

131. Verbiest, 119.

132. Perdue, *China Marches West*.

133. On the role of Qing logistics and artillery in the victory at Albazin and the Treaty of Nerchinsk, see Andrade, *The Gunpowder Age*, 221–30; Perdue, *China Marches West*, 165–69.

134. On these Manchu mapping expeditions after the treaty of Nerchinsk, see the unique Manchu-based research of Chengzhi (Kicengge), "Manwen 'Wula deng chu difang tu' kao"; Chengzhi (Kicengge), "The Illusion of the Nerchinsk Treaty Boundary-Stone."

135. Cams, *Companions in Geography*.

136. This description of the dragon vein rushing down mountains and following in the path of waterways suggests a strong connection to the flooding potential of Manchuria's rivers. Indeed, the connection between geomantic force and water flows was implicit in this document, which was (in spite of its lofty rhetoric) a proposal from the Qing official Gao Qizhuo to improve the drainage around Nurhaci's tomb. See Xu Xin, *Nu'erhachi ling*, 60.

CHAPTER 3

1. This foundation story was featured at the beginning of histories that detailed the rise of the Manchus from scattered Jurchen tribes in northeastern Asia to the rulers of the Qing empire. Crossley, "An Introduction to the Qing Foundation Myth."

2. The edict is recorded in the Kangxi *Shilu* 69:3.

3. Mitter, *The Manchurian Myth*; Gao Shuqiao, *Baishan heishui de zunyan*.

4. Schmid, *Korea between Empires, 1895–1919*, 175–98.

5. Debarbieux, Rudaz, and Price, *The Mountain*; Sahlins, *Boundaries*.

6. Since the mountain is little known in the West, there is no one commonly accepted English name for it. What we call the mountain depends on which side of the border we approach it from—both physically and conceptually. In this chapter, I refer to it as the "White Mountain," since the word "white" was embedded in the Manchu, Chinese, and Korean names for the mountain. When referring to the contemporary

framing of the mountain as a geographical border object, I combine the Korean and Chinese names, and call the mountain "Mount Paektu/Changbai." When considering the mountain from a historical perspective, I use the name contained in the source text using the pronunciation of the spoken language used by the writer or the English translation of that particular name. For a summary of debates about the mountain and boundaries in northeast Asia, see Ahn, "China and Korea Clash over Mount Paektu/Changbai: Memory Wars Threaten Regional Accommodation"; Rawski, *Early Modern China and Northeast Asia*, 235–63.

7. Song, *Making Borders in Modern East Asia*; Schmid, "Tributary Relations and the Qing-Chosŏn Frontier on Mount Paektu."

8. Haeussler, "Descriptions of the Baekdusan and the Surrounding Area in Russian and German Travel Accounts," 166.

9. Cosgrove and Dora, *High Places*.

10. Tuan, *Space and Place*, 161.

11. Levine, "Questioning the View," 250.

12. Millward, "Coming onto the Map"; Perdue, "Boundaries, Maps, and Movement"; Perdue, *China Marches West*.

13. On the importance of non-Western participants in the creation of science, see Raj, *Relocating Modern Science*; Fan, *British Naturalists in Qing China*; Mueggler, *The Paper Road*.

14. Tang, "Changbaishan Volcanism in Northeast China Linked to Subduction-Induced Mantle Upwelling."

15. Wang Jiping, *Changbaishan zhi*, 70–96.

16. On the Millennium Eruption, see Xu et al., "Climatic Impact of the Millennium Eruption of Changbaishan"; Wei, Liu, and Gill, "Review of Eruptive Activity at Tianchi Volcano"; Horn, "Volatile Emissions."

17. Bertoni, "Charming Worms."

18. Xu, "Climatic Impact."

19. Scarth, *Vesuvius*; Renne, "40Ar/39Ar Dating into the Historical Realm."

20. Jin and Cui, "Changbaishan Tianchi huoshan pengfa lishi."

21. Cocco, *Watching Vesuvius*; Bernstein, "Whose Fuji?"

22. Chen Hui, "Changbaishan chongbai kao."

23. Cho, "The Significance in Perception of Baekdusan in Baekdu-Related Myths."

24. Li Huazi, "Ming-Qing shiqi Zhong-Chao dilizhi dui Changbaishan ji shuixi de jishu," 57.

25. Schmidt, *Inventing Exoticism*, 84.

26. Elliott, "The Limits of Tartary," 612–13; Crossley, *A Translucent Mirror*, 297; Stary, "Il 'vero' esploratore del Changbaishan"; Lee Hun, "Ch'ŏng ch'ogi Changpaek-san t'amsawa hwangjegwon."

27. Both Lee Hun and Giovani Stary provide lists of multiple versions of the Umuna narrative. I rely here on two versions of the narrative, one in Chinese included in Wang Shizhen's *Chi bei ou tan* of 1691, and Stary's Italian translation of the 1783 Manchu version from *Manjusai da sekiyen-i kimcin bithe* (Manzhou yuan

liu kao). Lee Hun's detailed analysis of the "Golmin šanyan alin be tuwanaha ejetun" from the *Golmin šanyan alin-i ejetun* (based on a rare Manchu copy of the *Chang-baishan zhi* held in the Bibliothèque nationale de France) also provides excellent guidance through the narrative. For a detailed comparison of different Chinese and Manchu texts, see Stary, "Il 'vero' esploratore del Changbaishan," 271–80. Multiple versions of the narrative have resulted in multiple renditions of the name of the expedition leader. Here I follow Elliott's spelling "Umuna." Stary emphatically states that the true Manchu name of the explorer was "Umne."

28. Crossley, "Manzhou Yuanliu Kao and the Formalization of the Manchu Heritage."

29. Kim, *Ginseng and Borderland*.

30. Jami, *The Emperor's New Mathematics*.

31. *Manzhou shilu* 1.6a. Nurhaci defeated the Ehe Neyen tribe at the foot of the Long White Mountain in 1593.

32. When the Chosŏn envoy Shin Chung-il crossed the terrain of these tribes in the late sixteenth century, he noted that many of the settlements were practically empty, their menfolk all having gone off to serve Nurhaci at his fort along the Suzi River. Shen Zhongyi (Shin Chung-il), *Jianzhou jicheng tulu*.

33. Bello, *Across Forest, Steppe, and Mountain*, 63–115.

34. Wang Shizhen, *Chi bei ou tan*, 1:89.

35. For a contemporary PRC exploration of this area, see Zhang Fuyou, *Xunfang Eheneyin*.

36. Lee Hun, "Ch'ŏng ch'ogi Changpaeksan t'amsawa hwangjegwon," equates the Ehe Neyin River with the Toudao River that flows from Changbaishan's western slope (and past what is now the Changbaishan International ski resort complex).

37. On Sabsu, see Hummel, *Eminent Chinese of the Ch'ing Period (1644–1912)*, 518–19.

38. Stary, "Il 'vero' esploratore del Changbaishan."

39. Wang Shizhen, *Chi bei ou tan*, 1:90.

40. Wang Jiping, *Changbaishan zhi*, 97–99.

41. Ellis, "The Distribution of Bars at Pompeii"; Knowles, "What Could Lee See?"

42. Jilin sheng difang zhi bianxiu weiyuanhui, *Jilin sheng zhi: ziran dili zhi*, 91–97.

43. Wang Shizhen, *Chi bei ou tan*, 1:91.

44. All following quotes from the "Changbaishan fu" in Wu Zhaoqian, *Qiu jia ji*, 1965, 8:3691–93.

45. Yuan Hong, "Han Guangwudi ji," in *Hou Han ji*.

46. Yang, An, and Turner, *Handbook of Chinese Mythology*, 235, 138.

47. On this attempt to release Wu from exile, see Li Xingsheng, *Jiangnan caizi*, 148–56.

48. According to Xu Ke's *Qing bai lei chao*, on his deathbed, Wu Zhaoqian said to his son: "I want to hunt the White Mountain deer with you, catch fish from the Songhua and bring them home, pick some fresh mushrooms from our yard, and give

them all to your mother so she can make a stew. Do you think we still can?" Quoted in Li Xingsheng, *Wu Zhaoqian nianpu*, 175.

49. Millward, "Coming onto the Map."

50. Perdue, "Boundaries, Maps, and Movement"; Perdue, *China Marches West*; Millward, "Coming onto the Map."

51. Schmid, "Tributary Relations and the Qing-Chosŏn Frontier on Mount Paektu"; Kim, *Ginseng and Borderland*; Song, *Making Borders in Modern East Asia*.

52. Robinson, "From Raiders to Traders."

53. For a detailed presentation of the anxieties the Koreans felt and the many subterfuges they used against the Qing attempts at reconnoitering the border, Schmid, "Tributary Relations and the Qing-Chosŏn Frontier on Mount Paektu."

54. Cams, *Companions in Geography*.

55. For an exhaustive examination of the mapping expeditions to the northeast based on Manchu sources, see Chengzhi (Kicengge), "'Huangyu quanlan tu' Dongbei dadi cehui kao: yi Manwen dangan wei zhongxin."

56. Cams, *Companions in Geography*, 118–22.

57. On Korean border crimes (typically involving fights over ginseng) as a stimulus for Qing border enforcement, see Kim, *Ginseng and Borderland*.

58. For Ripa's description of his Manchu friend, see Ripa, *Giornale (1705–1724)*, 2:26.

59. Ma Menglong, "Mudekeng chabian."

60. Schmid, "Tributary Relations and the Qing-Chosŏn Frontier on Mount Paektu"; Song, *Making Borders in Modern East Asia*.

61. Fuchs, *Der Jesuiten-Atlas der Kanghsi-Zeit*, 238. This latitude was surprisingly accurate: today's reading at the highest summit is N 42°00′20″, E 128°03′19″.

62. On mountain sacrifices, see Robson, *Power of Place*; Dott, *Identity Reflections*; Lewis, "The Feng and Shan Sacrifices of Emperor Wu of the Han."

63. *Wang ji* did not pertain only to Changbaishan—it could also be performed for other mountains as well as for rivers and oceans. Du Erwei, *Zhongguo gudai zongjiao xitong*, 142–45. Thanks to Sue Naquin for this important point.

64. Wang, *Changbaishan zhi*, 67.

65. The following description of the ritual from the "Chun qiu zhi ji: Changbaishan zhi li," *Jilin tongzhi* 36:7a–8a.

66. Whiteman, "Kangxi's Auspicious Empire."

67. "Taishan shan yong zi Changbaishan lai," in Li Di, *Kangxi jixia gewu bian yi zhu*, 90. This essay is analyzed in Whiteman, "Kangxi's Auspicious Empire," 39–43; and Dott, *Identity Reflections*, 161.

68. Li Di, *Kangxi jixia gewu bian yi zhu*, 90. Translation by author.

69. Morris-Suzuki, *To the Diamond Mountains*.

70. Kukt'o chiri chŏngbowŏn, *The National Atlas of Korea*, 9.

71. Horn, "Volatile Emissions during the Eruption of Baitoushan Volcano (China/North Korea) c. 969 AD"; Wei, Liu, and Gill, "Review of Eruptive Activity at Tianchi Volcano, Changbaishan, Northeast China."

72. Mason, *Spirit of the Mountains*.

73. Cho, "The Significance in Perception of Baekdusan in Baekdu-Related Myths."

74. Grayson, *Myths and Legends from Korea*, 30–51.

75. Breuker, *Establishing a Pluralist Society in Medieval Korea, 918–1170*, 73.

76. Kim, *Ginseng and Borderland*, 27–39; Robinson, "From Raiders to Traders."

77. Setton, *Chŏng Yagyong*.

78. Jung, "The People of Joseon's Perception of Baekdusan"; Lee, "Korean Intellectuals' Perceptions of Baekdusan"; Kwon, "Changes in Perception of Baekdusan."

79. Shin, "Travel to Baekdusan."

80. The following discussion all derives from Database of Korean Classics, Hong Se-t'ae, "Paektusan gi."

81. The narrative does not specify Aesun's nationality. Hong's narrative suggests that he is Korean speaking but treats him as different from both the Manchu and the Chosŏn officials. He could be seen as representative of borderland "transfrontiersmen" whose local identity was not defined by one group or another.

82. Ingold, *The Perception of the Environment*, 25.

83. Kim, *Questioning Science in East Asian Contexts*, 83–84.

84. Following narrative all drawn from Database of Korean Classics, Sŏ Myŏngŭng, "Yu Paektusan gi."

85. Min, "The Shadow of Anonymity."

86. Sŏ, "Yu Paektusan gi," 17b–18a.

87. Sŏ, 23a–b. Sŏ's understanding of regional geography was quite accurate: the latitude of Shenyang is 41.8 degrees north.

88. On literati distaste for manifestations of mountain shamanism, see Lee and Koo, "The Confucian Transformation of Mountain Space."

89. Sŏ, "Yu Paektusan gi," 25b–26a.

90. Sŏ, 26b–28b.

91. Kwon, "Disputes about the Holding of Rituals at Baekdusan."

92. *Chosŏn wangjo sillok*, Yŏngjo 43/7/9.

93. *Chosŏn wangjo sillok*, Yŏngjo 44/7/14.

94. Yi T'ae-jin and Yi Sang-t'ae, *Kwanbuk ŭpchi—Kapsanpu ŭpchi*, Hamgyŏng-do 6:43:234–35.

95. *Chongmyo ŭigwe*, https://iiif.lib.harvard.edu/manifests/view/drs:7551429$77i. Prayer to Paektusan at sequence 77.

96. For examples, see Han et al., *The Artistry of Early Korean Cartography*.

97. Ledyard, "Cartography in Korea"; Song, "Imagined Territory: Paektusan in Late Chosŏn Maps and Writings"; Yi Sŏ-haeng and Chŏng Ch'i-yŏng, *Ko chido wa sajin ŭro pon Paektusan*.

98. Jin, "Paektudaegan."

99. Ledyard, "Cartography in Korea," 310.

100. Ledyard, 305–13.

101. "Kapsan chido" National Library of Korea #3 Ancient Map 61–51.

102. Goodman et al., "Spectacular Environmentalisms."

103. On combinations of senses that link internal feeling and external stimuli, see Geurts, *Culture and the Senses*.

104. Wang Mingxia, "Huifa bu de lishi yu saman wenhua."

105. Shamanism in this area was revived as a public practice after the Cultural Revolution, making it difficult to extricate today's rituals from tourist spectacle, but several families from this area have retained written records of shaman rituals and shaman genealogies dating as far back as the early nineteenth century. See Song Heping and Meng Huiying, *Manzu saman wenben yanjiu*.

106. For the summons of the ancestral shamans and the summons of animal spirits, see Shi Guangwei and Liu Housheng, *Manzu saman tiaoshen yanjiu*, 178–81; Song Heping, *Manzu saman shenge yizhu*, 183. Translations by author.

107. "Song of Seeing Off the Spirits" *Song shen*, from Song Heping, *Manzu saman shenge yizhu*, 352.

CHAPTER 4

1. Maximowicz, *Primitiae*, 406.

2. Zatsepine, *Beyond the Amur*, 3.

3. Bassin, *Imperial Visions*, 143–73.

4. Needham and Lu, *Science and Civilisation in China, Botany*; Métailié, *Science and Civilisation in China, Traditional Botany*. On the "fruitless" comparison, see Métailié, 89n537.

5. Quote from Mueggler, *The Paper Road*, 16; see also Fan, *British Naturalists in Qing China*; Nappi, "Winter Worm, Summer Grass"; and Bian, *Know Your Remedies*.

6. Mueggler, *The Paper Road*, 24.

7. Simonov and Dahmer, *Amur-Heilong River Basin Reader*, http://amur-heilong .net/http/fullindex.html.

8. Hafner and Feichtenberger, *Amur—Asia's Amazon*.

9. While we may use these modern tribal labels, Jonathan Schlessinger has pointed out that the Amur River basin was historically home to "a bewildering constellation of people" who were combined and homogenized into identifiable tribes through the pressure of empires and nation-states. Schlesinger, *A World Trimmed with Fur*, 66. Definitive historical studies on the indigeonous peoples of this region include Matsuura, *Shinchō no Amūru seisaku to shōsū minzoku*; Kim, *Ethnic Chrysalis*; Janhunen, *Manchuria*; and Slezkine, *Arctic Mirrors*.

10. The Qing gradually established a ring of permanent military bases and supply lines in the Amur basin, first with an outpost on the Mudan River (Ningguta, 1652), followed by a riverine navy shipyard on the Sungari (Girin, 1676), then an advance outpost on the right bank of the middle Amur itself (Aigun, 1683), followed by a garrison on the more westerly tributary of the Nen River (Mergen, 1686). With this network established, the Qing were able to defeat the Russians at their fort on the upper Amur at Albazin in 1686. For this process, and the Treaty of Nerchinsk, see Peter Perdue, *China Marches West*.

11. The northernmost borders were established approximately 1,300 miles north of Beijing, a distance representing approximately the span from Miami to Boston. Matsuura, "The Qing Surveys of the Left Bank of the Amur after the Conclusion of the Treaty of Nerchinsk," 11; Chengzhi (Kicengge), "Manwen 'Wula deng

chu difang tu' kao"; Chengzhi (Kicengge), "The Illusion of the Nerchinsk Treaty Boundary-Stone."

12. On the relocation of the Amurian Sibe people to far-western Xinjiang, see Sárközi, *From the Mists of Martyrdom*.

13. Bello, "The Cultured Nature of Imperial Foraging in Manchuria"; Bello, *Across Forest, Steppe, and Mountain*.

14. Matsuura, *Shinchō no Amūru seisaku to shōsū minzoku*.

15. Schlesinger, *A World Trimmed with Fur*.

16. Kim, "Saints for Shamans?"

17. Frost, *Bering*; Vinkovetsky, *Russian America*.

18. Bassin, "The Russian Geographical Society."

19. Bassin, *Imperial Visions*; Corrado, "The 'End of the Earth.'"

20. Plans of a US mission to "open" Japan prompted Russia to send its own envoy from St. Petersburg to Tokyo in 1852. The frigate *Diana* was dispatched a year later as a backup ship for the mission. See Lensen, *Russia's Japan Expedition of 1852 to 1855*.

21. Vucinich, *Science in Russian Culture*; Shetler, *The Komarov Botanical Institute*; Le Lievre, "Carl Johann Maximowicz, 1827–91 Explorer and Plant Collector."

22. Maximowicz, *Primitiae*, 1–2.

23. Grainger, *The First Pacific War*.

24. Corrado, "A Land Divided."

25. Maximowicz, *Primitiae*, 3.

26. Details on Murav'ev's journey can be found in Baranov, "Recollections of an Officer." For the events leading up to the Treaty of Aigun, see Paine, *Imperial Rivals*, 25–70.

27. Browne, "Biogeography and Empire," 312.

28. Browne, 305.

29. By then, other naturalists of the Great Siberian Expedition of the Russian Geographical Society had arrived on the river, including Richard Karlovich Maack, Leopold von Schrenk, Gustav Radde, and Friedrich Schmidt. Bassin, "The Russian Geographical Society."

30. Maximowicz, *Primitiae*, 4.

31. These "journey narratives" combine observations from Maack, Schrenk, and others who traveled downriver with Maximowicz's own observations collected when he traveled up the Amur on his return journey to St. Petersburg.

32. Maximowicz, *Primitiae*, 350.

33. Maximowicz, 406. This area later became known as the Three Rivers Plain (Sanjiang pingyuan). See chapter 8, "Reclaimed."

34. Maximowicz, 406.

35. Maximowicz, 6.

36. Candolle, *Prodromus systematis naturalis regni vegetabilis*; Stevens, "Haüy and A.-P. Candolle"; Drouin, "Principles and Uses of Taxonomy in the Works of Augustin-Pyramus de Candolle."

37. Maximowicz, *Primitiae*, 2–341.

38. Bretschneider, *History of European Botanical Discoveries in China*, 583–88.

39. Maximowicz is as much a compiler or synthesizer as a "discoverer." Many of the plants enumerated in *Primitiae* are from the work of Turczaninow (abbreviated as "Turc.") in the Transbaikal and Middendorff ("Midd.") in northern Siberia. Multiple plant identifications are attributed to Maximowicz's boss, Franz Joseph Ruprecht ("Rupr."). Maximowicz, *Primitiae*, 420–25; Turczaninow, *Flora Baicalensi-Dahurica*; Tammiksaar and Stone, "Alexander von Middendorff."

40. For a modern botanical description of *Schisandra chinensis*, see *Flora of China* (http://www.efloras.org/florataxon.aspx?flora_id=2&taxon_id=200008486).

41. Isotype of *Maximowiczia amurensis* Rupr., Conservatoire et Jardin botaniques de la Ville de Genève (G), G00356482, JSTOR Global Plants database, htpp://plants.jstor/stable/pdf/10.5555/al.ap.specimen.g00356482.

42. Candolle, *Theorie elementaire*, quoted in Stevens, "Haüy and A.-P. Candolle," 60.

43. Gordon, *Scientific Babel*, 23–50.

44. Maximowicz, *Primitiae*, 31.

45. Botanical Latin terms interpreted with the aid of Stearn, *Botanical Latin*.

46. On the significance of "autopsy" in classification schemes, see Hoquet, "Botanical Authority," 528.

47. Maximowicz, *Primitiae*, 416.

48. On Asa Gray's biogeography, see Hung, "Finding Patterns in Nature"; Hung, "Plants That Remind Me of Home."

49. Browne, *The Secular Ark*.

50. Lomolino, Sax, and Brown, *Foundations of Biogeography*, 7.

51. Maximowicz, *Primitiae*, 431.

52. Maximowicz, 448.

53. Xiqing, *Heilongjiang wai ji*, 82.

54. Yinghe, *Bukui ji lüe*, 2:840.

55. Hummel, *Eminent Chinese of the Ch'ing Period (1644–1912)*, 601–3.

56. Zhang Boying, *Heilongjiang zhi gao*, 2540.

57. Hummel, *Eminent Chinese of the Ch'ing Period (1644–1912)*, 931–33.

58. The Nen is a major tributary of the Black Dragon River, though it makes a convoluted journey before it joins the main flow. The river emerges in the Greater Khingan Mountains at a location only thirty miles from the Amur's banks, but the slope of the mountains forces the Nen to take a southerly course away from the Amur. It continues south almost five hundred miles until it joins the Songhua River, which then travels five hundred miles northeast until merging with waters of the Black Dragon. The vast plain created by the Nen and the Songhua Rivers constitutes the most fertile land of the Amur basin. Liu, "Land Use/Cover Changes and Environmental Consequences in Song-Nen Plain, Northeast China."

59. Xiqing, *Heilongjiang wai ji*, 83–84.

60. In order, broomcorn millet, foxtail millet, barnyard millet, with some glutinous and nonglutinous varieties, Xiqing, 82; Bray, "Millet Cultivation in China."

61. Xiqing, *Heilongjiang wai ji*, 82–83.

62. Reardon-Anderson, *Reluctant Pioneers*, 68; Xu Shuming, "Qingdai dongbei diqu tudi kaiken shulue," 102–5.

63. Liaoning sheng danganguan, *Qingdai Sanxing fu dutong yamen Man-Han wen dang'an xuanbian*, 77–79.

64. Xiqing, *Heilongjiang wai ji*, 4:5a.

65. This is now the Zhalong Wetland Nature Preserve. See Yinghe, *Bukui ji lüe*, 2:840; Gallagher, "Zhalong Wetland, Heilongjiang."

66. The name Xiqing gave to the swamp in Chinese, *Hongyan hatan* (from the Manchu for "Great Fiery Mountain Path"), as well as its general location northeast of Qiqihar, suggests that this was the volcanic field known today as Wudalianchi. According to Xiqing, even the birds and wild animals avoided the area, and carrion-eating wolves prowled for the bodies of those who failed the journey. The only way to avoid the perils of Hongyan hatan was to hire a native Oroqen guide, and even then, few were successful. See Chen Hongzhou, "Volcanic Eruptive Processes and Characteristics of the Current Volcanoes in the Wudalianchi Volcano Clusters Known from Manchurian-Language Historical Archives."

67. Xiqing, *Heilonjiang wai ji*, 83. On the problem of fakes in herbal medicine, see Nappi, "Surface Tension: Objectifying Ginseng in Chinese Early Modernity."

68. The "rectification" of plant names is described by Georges Métailié as "a firm belief that the importance of correct names really does constitute the basis upon which . . . all thought on natural living objects was founded." Métailié, *Science and Civilisation in China, Traditional Botany*, 6.

69. Nappi, *The Monkey and the Inkpot*; Needham and Lu, Gwei-Djen, *Science and Civilisation in China, Botany*; Métailié, *Science and Civilisation in China, Traditional Botany*; Dennis, *Writing, Publishing, and Reading Local Gazetteers in Imperial China, 1100–1700*.

70. Bray, "Millet Cultivation in China," 206.

71. Xiqing, *Heilongjiang wai ji*, 82. Norman (*A Comprehensive Manchu-English Dictionary*) glosses *fisihe* as "broomcorn millet" (*Panicum miliaceum*).

72. First published in 1684, republished numerous times throughout the Qing, and, according to Unschuld, "the most frequently consulted text on the *materia medica* from the Qing era." Unschuld and Zheng, *Chinese Traditional Healing*, 1109.

73. For a schema of *bencao* categories, see Nappi, *The Monkey and the Inkpot*, appendix B, 156–58.

74. Li Shizhen, *Bencao gangmu*, 533–34.

75. On Wu, see Fan, *British Naturalists in Qing China*, 94, 109; Bian, "Governmental Science."

76. Maximowicz, *Primitiae*, 497–500.

77. This list appears separately in Ruprecht, "Die ersten botanischen Nachrichten uber das Amurland"; Blaesing, "The Tungusic Plant Names."

78. For detailed tally of individual tribute submissions for Jiaqing 9 (1804), see document no. 73 in Liaoning sheng dang'anguan, *Sanxing fu dutong yamen Manwen dang'an yibian*, 189–204.

79. Lee, *The Manchurian Frontier in Ch'ing History*, 45–47.

80. Xiqing, *Heilongjiang wai ji*, 85. On mandrake/*yabrūh*, see Simoons, *Plants of Life, Plants of Death*, 101–35.

81. Xiqing, 85.

82. Wright, *Artemisia*.

83. Xiqing, *Heilongjiang wai ji*, 85.

84. On plant names in the Book of Odes, see Xiqing, *Heilongjiang wai ji*, 85; Needham and Lu, Gwei-Djen, *Science and Civilisation in China, Botany*, 98, 110; Métailié, *Science and Civilisation in China, Traditional Botany*, 83–84; Nappi, *The Monkey and the Inkpot*, 98–103; Stary, "Zue evolution der mandschurischen Ubersetzungstechnik anhand einiger Beispiele aus dem 'Buch der Lieder.'"

85. Maximowicz, *Primitiae*, 157–62.

86. Maximowicz, 449–50.

87. On edible plant "vegetables," see Maximowicz, 452.

88. Maximowicz, 452.

89. Maximowicz 453.

90. On Nanai medicinal plants, see Maximowicz, *Primitiae*, 454; Mitsuhashi, "Medicinal Plants of the Ainu."

91. Maximowicz, *Primitiae*, 458.

92. Maximowicz, 459.

93. The scholarship on the social and economic impact of migration to Manchuria includes Isett, *State, Peasant, and Merchant*; Shan, *Taming China's Wilderness*; Gottschang and Lary, *Swallows and Settlers*; Kong Jingwei, *Qing dai Dongbei diqu jingji shi*; Xin Peilin, Zhang Fengming, and Gao Xiaoyan, *Heilongjiang kaifa shi*; Xu Shuming, "Qingdai dongbei diqu tudi kaiken shulue."

94. Zhang, "Vegetation of Northeast China"; Zhang, "Spatially Precise Reconstruction of Cropland Areas."

95. Arsenyev describes this encounter in the first chapter of his novel, giving 1902 as the date. Using Arsenyev's handwritten field notes, scholars have determined that the actual encounter with Dersu occurred in 1906. Scholars have also surmised that Arsenyev combined experiences with three different indigenous trackers and attributed them all to Dersu. Arsenyev's notes, however, do make it clear that Dersu Uzula did exist as a distinct individual. Arsen'ev, *Dersu the Trapper*; Arsenyev, *Across the Ussuri Kray*, xii.

96. Arsen'ev, *Dersu the Trapper*, 8.

97. Arsen'ev, 15, 188.

98. Bello, "The Cultured Nature of Imperial Foraging in Manchuria," 74.

99. Bello, 76, 89.

100. Bello, *Across Forest, Steppe, and Mountain*; Schlesinger, *A World Trimmed with Fur*.

101. Sanxing fudutong yamen grain accounting figures from Jiaqing 8. See document no. 105, Liaoning sheng dang'anguan, *Sanxing fu dutong yamen Manwen dang'an yibian*, 362–63. Following equivalents in Yi Baozhong, *Dongbei nongye jindaihua yanjiu*, I am taking one dan 石 to be the equivalent of three bushels. Following USDA estimates, one bushel of millet is fifty pounds.

102. Kim, "Marginal Constituencies," 199–204.

103. Lü Xiulian and Zhao Kunyu, *Hezhezu nüxing lishi wenhua yanjiu*, 31n1.

104. Kim, "Marginal Constituencies," 222n134.

105. Arsen'ev, *Dersu the Trapper*, 120; Maximowicz, *Primitiae*, 401. *Clematis mandshurica* is quite similar in appearance to *Lonicera maackii* (Rupr.) Maxim., or the Amur honeysuckle. Identified definitively by Maximowicz in 1859, the flowering shrub is familiar to gardeners in the eastern United States (and property owners in Tennessee) as the highly invasive "bush honeysuckle."

106. Maximowicz, *Primitiae*, 496.

107. Arsen'ev, *Dersu the Trapper*, 46–48. This event was reproduced to great dramatic effect in Akira Kurosawa's 1975 film, *Dersu Uzula*. This film, a joint production of the Soviet Union and Japan and shot on location in the Russian Far East, won the Academy Award for best foreign-language film in 1976.

108. Worster, "Transformations of the Earth," 1092.

109. Deloria, *Indians in Unexpected Places*.

110. Liu Min, "Shizhe Qingdai zhi minguo shiqi Hezhezu renkou ruijian de yuanyin," *Heilongjiang shehui kexue*, 98:5 (2006) 149–151. Currently, more Hezhe (Nanai) are counted on the Russian side of the Amur: population estimates of overall Nanai population in the PRC are around five thousand and around twelve thousand in Russia. On the division of Nanai between Russia and the PRC, see Ed Pulford, "Material States."

111. Schiebinger, *Plants and Empire*, 196 and 226.

112. Burnett, *Masters of All They Surveyed*, 189.

113. The nature of colonial knowledge as a process of depeopling is most clearly demonstrated by its reverse: decolonizing projects of indigenous peoples that seek to "repeople" their knowledge. On this process, see Geniusz, *Our Knowledge Is Not Primitive*. As Geniusz points out, decolonizing knowledge means the reinjection of people: beginning with the individual names and tribal affiliations of those who possessed and taught the knowledge, and then reembedding that knowledge within indigenous rituals, songs, and beliefs. This social reembedding is a prerequisite for recovering technical knowledge of plants. Many thanks to my colleague Dan Usner for bringing this book to my attention.

114. Bian, *Know Your Remedies*, 133–35.

115. Nappi, "Winter Worm, Summer Grass."

116. Schäfer, *The Crafting of the 10,000 Things*; Elman, *On Their Own Terms*.

117. Mosca, "Empire and the Circulation of Frontier Intelligence"; Mosca, *From Frontier Policy to Foreign Policy*.

CHAPTER 5

1. Grabau, "Mollusca of North China."

2. For an overview of these and other species of the Jehol biota, see Chen, *The Jehol Fossils*.

3. "Feathered dragons" and "birds as modern day dinosaurs," see Lukas Rieppel, *Assembling the Dinosaur*, 220. For recent Jehol biota finds, see Qiang et al., "Two Feathered Dinosaurs from Northeastern China"; Chen, Dong, and Zhen, "An Excep-

tionally Well-Preserved Theropod Dinosaur from the Yixian Formation of China";
Sloan, "Feathers for T. Rex?"; Wilford, "Feathered Dinosaur Fossils Are Unearthed in
China."

4. Whiteman, *Where Dragon Veins Meet.*

5. Seow, *Carbon Technocracy.*

6. These citations reference Woodward, *Catalogue of the Fossil Fishes*, 1–4; Grabau, "Mollusca of North China"; Endo, "A New Genus of Thecodontia." For this sort
of brief history by citation, see Chen et al., *The Jehol Fossils*, 1–4.

7. Mayor, *The First Fossil Hunters*; Cutler, *The Seashell on the Mountaintop*;
Gould, *Leonardo's Mountain of Clams and the Diet of Worms.*

8. Osterhammel, *The Transformation of the World*, 651–58.

9. E. E. Ahnert, "The Geological Society and Science in Asia." The speech is also
referenced in Shen, *Unearthing the Nation*. 97.

10. On Rehe, see Dunnell et al., *New Qing Imperial History.*

11. Whiteman, *Where the Dragon Veins Meet*, 68–78.

12. On fossils and dragon bones, see Needham, *Science and Civilisation in China*,
Vol. 3, 611–25; Schmalzer, *The People's Peking Man*, 35–36; Nappi, *The Monkey and
the Inkpot*, 59–64.

13. Kangxi's essay appears in the Jehol gazetteer edited by the Qianlong emperor's favorite courtier, Heshen, *Qin ding Rehe zhi*, 96:14a entry for "Yu'er shi" (Little
Fish Stones); Li, *Kangxi jixia gewu bian yi zhu*; Liang Xiaodong, "Kangxi, Qianlong
huangdi dui huashi de yanjiu ji huashi wenhua."

14. The Jehol gazetteer is more specific about the location of the fossils, placing
them "in the mountains in the northwestern part of Chaoyang county." Heshen, *Qin
ding Rehe zhi*, 96:14a.

15. Kangxi references several ancient works and locales here: *Commentary on the
Classic of Waters* (Shuijing zhu) is from Northern Wei dynasty, c. fifth century CE.
Miscellaneous Tales from the Sunny Slope of Mount You (Youyang zazu) was written in
the Tang dynasty by Duan Chengshi (d. 863 CE). Hengyang is a city in Hunan province in south-central China; Yiyang county is in Henan province in central China,
600 miles to the north of Hengyang. Liang Xiaodong, "Kangxi, Qianlong huangdi dui
huashi de yanjiu ji huashi wenhua."

16. Such explanations for fossils had been posited since the twelfth-century
philosopher Zhu Xi, who speculated that fossil shells on mountaintops indicated that
these areas had once been under water. Needham, *Science and Civilisation in China*,
3:619–20.

17. Li, *Kangxi jixia gewu bian yi zhu*. Mirages, 16; fire mountains, 82; permafrost,
9; dense forest, 80; petrified wood, 23; stone fish, 85; sea creatures into deer, 84.

18. For a succinct discussion of Latour's concept, see Jöns, "'Centre of
Calculation.'"

19. Outram, "New Spaces in Natural History"; Kohler, *Landscapes and Labscapes.*

20. Otis, *Müller's Lab.*

21. Johannes Müller, "Fossile Fische," in Middendorff, *Middendorff's Reise,
Band 1., Theil 1*, 262–63.

22. Middendorff.

23. "Père" David traveled widely throughout all parts of the Qing empire, and was later credited with "discovering" the giant panda, as well as the diminutive breed of deer that bears his name. See David, *Abbé David's Diary*; Fan, *British Naturalists in Qing China*, 99.

24. Kilpatrick, *Fathers of Botany*, 25–31.

25. On nineteenth-century French paleontology, see Rudwick, *Georges Cuvier*. On Sauvage, see "Henri-Émile Sauvage (1842–1917)," *Data.bnf.fr*, http://data.bnf.fr/ 10608766/henri-emile_sauvage/.

26. Sauvage, "Sur un Prolebias (Prolebias Davidi) des terrains tertiaries du nord de la Chine." "Prolebias" is a now-defunct genus name. The fish retains its identity as a separate species, although it is now considered a member of the *Lycoptera* genus (*Lycoptera davidi*).

27. While as an ichthyologist he reigned supreme, later in his career Woodward strayed from fish and erroneously identified the remains of "Piltdown Man" as the missing link between apes and man. The revelation of the hoax severely injured his legacy. On Woodward's work, see Smith, "Arthur Smith Woodward."

28. Woodward, *Catalogue of the Fossil Fishes*, 1–4.

29. "H M Becher | British Museum."

30. Hibbett et al., "Climate, Decay, and the Death of the Coal Forests."

31. US Geological Survey, "Geologic Time: Index Fossils."

32. The definitive work on Richthofen in China is Wu, *Empires of Coal*.

33. Richthofen, *China*, 2:115.

34. The fossils that Richthofen did bring back were from eastern Liaoning province, near the Korean border. Like other field explorers before him, Richthofen "farmed out" the identification to specialists in the metropole (Berlin): see Wilhem Dames, "Cambrische Trilobiten von Liao-tung," 1–33, and Emanuel Kayser, "Cambrische Brachiopoden von Liao-tung," 34–36 in Richthofen, *China*, vol. 5.

35. Wu, *Empires of Coal*, 52.

36. In the "Triple Intervention" of Russia, Germany, and France, Japan was forced to surrender control of the southern Liaodong Peninsula, which it had conquered in the war. Paine, *The Sino-Japanese War of 1894–1895*.

37. Elleman and Kotkin, *Manchurian Railways*; Tollmachoff, "The Coal of Soochan and Its Importance in Pacific Trade"; Goudkoff, "Economic Geography of the Coal Resources of Asiatic Russia."

38. Steinberg, *The Russo-Japanese War in Global Perspective*.

39. Shimazu, *Japanese Society at War*.

40. Statements by the Japanese general Tanaka Giichi, quoted in Matsusaka, *The Making of Japanese Manchuria, 1904–1932*, 181.

41. Russian geological surveyors had certainly been active in Manchuria since the 1860s and were among the first to publish geological studies of the region. For a review of early Russian geological activity in Manchuria, see Ahnert and Lauroushin, "Subdivisions of the Jurassic, Cretaceous and Tertiary Coal-Bearing Strata of Russian Maritime and Amur Provinces, and of Sakhalin Island."

42. Similarly, immediately upon occupying Taiwan in 1895, Japan dispatched geologists to the island. Within a year they completed a comprehensive "Mineral and Geological Map of Formosa," a survey of land and soil that helped Japan's scientific colonization of the island. Chishitsu Chōsajo, *Imperial Geological Survey of Japan*.

43. Matsusaka, *The Making of Japanese Manchuria, 1904–1932*, 46–51; Suzuki, "Tracts of Japanese Geology."

44. Kantō tōtokufu minseibu, *Manshū sangyō chōsa shiryō*.

45. Braun, "Producing Vertical Territory."

46. For this early history of the Fushun mines and the Qing's concern about harm to the dragon vein, see Seow, *Carbon Technocracy*, chap. 1. Seow points out that in the present, the dragon vein has, quite literally, collapsed beneath the ground of this area. Due to overmining, the surface buildings and infrastructure in the city of Fushun are currently at risk of collapse, and mining operations there have been effectively halted.

47. Wu, *Empires of Coal*, 13.

48. Grabau, "Mollusca of North China."

49. Wu, *Empires of Coal*, 14.

50. On this community of geologists, see Shen, *Unearthing the Nation*.

51. On Andersson and his involvement in the discovery of Peking Man, see Schmalzer, *The People's Peking Man*, 32–45.

52. For details on Grabau's early life, see Mazur, *A Romance in Natural History*, 1–23.

53. On the dilemma of German American support for Germany in World War I, see Luebke, *Bonds of Loyalty*.

54. Mazur, *A Romance in Natural History*, 183–254.

55. Shen, *Unearthing the Nation*, 73–107.

56. Ahnert, "The Geological Society and Science in Asia," 10.

57. Ahnert, 10.

58. Ji Peng and Wu Zhiling, "Ding Wenjiang jingying Beipiao meikuang."

59. Huang Jiqing, *Zhongguo xiandai dizhixuejia zhuan*, 65–75.

60. See Tan Xichou, "Beipiao meitian dizhi"; Tan Xichou, "Liaoning Rehe jian ji Chao-Chi tiedao yanxian dizhi kuangchan."

61. Grabau, "Mollusca of North China." Grabau also referred to fossil samples collected by T. O. Chu (Zhu Tinghu), another early member of the Geological Survey. Tan's engagement with Beipiao surveys was longer and more extensive than Zhu's. On Zhu's career, see Huang, *Zhongguo xiandai dizhixuejia zhuan*, 86–97.

62. Tan, "On the Existence of Cretaceous Coal."

63. Grabau, *Principles of Stratigraphy*.

64. Dott, "An Introduction to the the Ups and Downs of Eustacy."

65. See Johnson, "A. W. Grabau's Embryonic Sequence Stratigraphy and Eustatic Curve," 45.

66. International Geological Congress, *Report of the XVI Session*; Mazur, *A Romance in Natural History*.

67. On de Chardin's unusual career and philosophy, see King, *Spirit of Fire*.

68. Examination of the photo reveals a total of eight Japanese geologists. Inter-

national Geological Congress, *Report of the XVI Session*, vol. 1, insert between 54 and 55.

69. On the complex politics leading up to the Jehol invasion, see Coble, *Facing Japan*.

70. Endō, "A New Genus of Thecodontia."

71. Endō. In the 1930s and 1940s, Endō named several new fossil species in honor of Manchuria/Manchukuo, including *Manchuriophycus inexpectans*, *Redlicia manchuriensis*, *Manchuriella prisca*, and *Manchurochelys manchoukuoensis*, or the "Manchurian Turtle of Manchukuo." For a sampling of some of these names, see Endō Ryūji, *Manshū no chishitsu oyobi kōsan*. On the connection between scientific names and empire, see Ritvo, "Zoological Nomenclature and the Empire of Victorian Science."

72. Matsusaka, *The Making of Japanese Manchuria, 1904–1932*; O'Dwyer, *Significant Soil*; Young, *Japan's Total Empire*; Young, *The Research Activities of the South Manchurian Railway Company, 1907–1945*; Fogel, introduction to Itō, *Life along the South Manchurian Railway*.

73. For Endō's early career, see Kobayashi, *Geology of Japan*, 6; Low, *Japan on Display*, 74–75.

74. Japanese members of the GSC included the paleontologist Hayasaka Ichirō (1891–1977), the Tōhoku university professor Yabe Hisakatsu, Okamoto Yohachirō (1876–1860), a professor of paleontology in Taiwan and the founder of colonial Taiwan's "Society for Natural History" (bestowed with the distinguished category of "lifetime member" by the Society), and even the director of the SMRC Geological Survey, Murakami Hanzō. See the membership list at the back of the *Bulletin for the Geological Society of China* 17, no. 2 (1938).

75. Endō, *The Canadian and Ordovician Formations and Fossils of South Manchuria*; Endō Ryūji, *Genjin hakkutsu*, 5–8.

76. Peattie, *Ishiwara Kanji and Japan's Confrontation with the West*.

77. Moore, *Constructing East Asia*, 4–9; Yang, *Technology of Empire*; Mizuno, *Science for the Empire*, 43–68.

78. Matsuda, *Mantetsu chishitsu chōsajo shiki*, 62–122; Yajima, "Japanese Wartime Geology"; Liang Bo and Feng Wei, "Mantie dizhi diaochasuo."

79. Komatsu, "Manshū ni okeru Nihon no sekiyu tankō."

80. For a consideration of the images and discourse of this expedition, see Low, *Japan on Display*, 60–79.

81. Tokunaga, *Report of the First Scientific Expedition to Manchoukuo*, I:45.

82. Matsuda, *Mantetsu chishitsu chōsajo shiki*.

83. Chishitsu chōsajo, *Manshū nanseibu no chishitsu oyobi chishi*, 87–134, 85 (photo).

84. Endō, *Genjin hakkutsu*, 24.

85. On Standard Oil's brief foray into Jehol, see "Standard Abandons China Oil Fields."

86. Endō likened his adventures and hardships in Jehol to those of pioneers in the American West. *Genjin hakkutsu*, 24–33.

87. "Jiufotang" is still used today as the formal name for the formation. See Chen, *The Jehol Fossils*.

88. Endō, "A New Genus of Thecodontia."

89. On the Manchukuo National Museum, see Fujiyama, *Shin hakubutsukan taisei*; Nagoya hakubutsukan, *Shin hakubutsukan taisei*; Zhang Wenli, "Riwei zai dongbei diqu chuangjian de bowuguan ji qi pingjia"; Li, "The Role of the Museum in the Rise of Modern Nationalism in East Asia." Thanks to Yoshikuni Igarashi for bringing the *Shin hakubutsukan taisei* to my attention.

90. Rieppel, *Assembling the Dinosaur*.

91. Li, "The Role of the Museum in the Rise of Modern Nationalism in East Asia."

92. For images of the museum's exhibits, see Nagoya hakubutsukan, *Shin hakubutsukan taisei*.

93. Endō, "A New Genus of Thecodontia." It is not clear how Gilmore provided assistance. On Gilmore, a curator at the Smithsonian and one of the "last major figures of the America's golden age of dinosaur hunting," see the Smithsonian National Museum of Natural History, "Dinosaur Hunter," http://paleobiology.si.edu/history/gilmore.html.

94. Endō, *Genjin hakkutsu*, *126–38*. Other important fossils were lost during World War II, including, most famously, the bones of the original Peking Man. See Schmalzer, *The People's Peking Man*, 46.

95. Endō, *Genjin hakkutsu*, 147–65. On the continuation of Japanese science and technology in postwar East Asia, see Ward, "Delaying Repatriation"; King,"Reconstructing China;" Mizuno, Moore, and DiMoia, *Engineering Asia*.

96. Seth, "Putting Knowledge in Its Place," 374.

97. This situation contrasts starkly with that of India. By 1920, India's science services were 90 percent European, and the few Indians employed in science suffered considerable discrimination. The Geological Survey of India employed no Indians at all. Arnold, *Science, Technology and Medicine in Colonial India*, 138–40.

98. Stilwell, "Trilobites and Linnaeus," 101.

99. Smil, *Energy Transitions*; Miller and Warde, "Energy Transitions as Environmental Events."

100. Zizzamia, "Restoring the Paleo-West," 132.

101. Zizzamia, 142.

102. Seow, "Forum," 505.

CHAPTER 6

1. Based on a combination of essays and poems included in Yamagata's volume about birdwatching in Manchuria. Yamagata Miyuki, *Manshū no yaseichō*. On the *koma hohojiro*, 14–19; on pet birds at home, 135–36.

2. Yamagata, 16.

3. Yamagata, 140–41.

4. Minami Manshū tetsudō kabushiki kaisha eiseika, *Kōtoku ni-nendo pesuto bōeki gaikyō*.

5. On the cultural activities of Manchuria's Japanese tourists and settler bourgeoisie, see McDonald, *Placing Empire*; O'Dwyer, *Significant Soil*; Fogel introduction to Yosano, *Travels in Manchuria and Mongolia*, 1–8.

6. On Manchukuo as the "jewel in the crown" of the Japanese empire, see Young, *Japan's Total Empire*, 21.

7. Nash, *Inescapable Ecologies*; Cumminsky, "An Ecological Experiment"; Anderson, *Colonial Pathologies*.

8. Tilley, *Africa as a Living Laboratory*; Anderson, *Colonial Pathologies*.

9. Tilley, "Ecologies of Complexity," 22–23.

10. Thomas Spencer Cobbold, *Parasites: A Treatise on the Entozoa of Man and Animals*, quoted in Li, "Natural History of Parasitic Disease," 202.

11. Wu's own account of his work best exemplifies this laudatory writing. Wu, *Plague Fighter*.

12. Representative works on Unit 731 in English include Harris, *Factories of Death*; Williams and Wallace, *Unit 731*; Barenblatt, *A Plague upon Humanity*; Gold, *Japan's Infamous Unit 731*.

13. Chang, *The Rape of Nanking*.

14. Koch, *Cartographies of Disease*; Koch, *Disease Maps*.

15. Hanson, "Visualizing the Geography of the Diseases of China," 222.

16. Kang et al., "Grassland Ecosystems in China"; World Wildlife Federation, "Mongolian-Manchurian Grassland."

17. For insights into the landscape in the first half of the twentieth century (in English), see Allen, *The Mammals of China and Mongolia*, 9–13.

18. Smith, *Guide to the Mammals of China*, 172–274.

19. Li and Dai, "Different Region Analysis for Genotyping *Yersinia pestis* Isolates from China"; *Atlas of Plague and Its Environment in the People's Republic of China*.

20. Ben-Ari et al., "Identification of Chinese Plague Foci from Long-Term Epidemiological Data."

21. Li and Dai, "Different Region Analysis for Genotyping *Yersinia pestis* Isolates from China."

22. See "Distribution Maps of the Reservoirs of Plague in China," in *Atlas of Plague and Its Environment in the People's Republic of China*.

23. The similarities between *Yersinia pesti*s and these other more common bacteria mean that it can sometimes be misidentified by automated identification tests. "Snapshots of Salmonella Serotypes | Salmonella Atlas | Reports and Publications | Salmonella | CDC."

24. The following discussion of *Yersinia pestis*, its hosts, and its vectors, primarily draws from Gage and Kosoy, "Natural History of the Plague."

25. Gage, "Factors Affecting the Spread and Maintenance of Plague."

26. Hinnebusch, "The Evolution of Flea-Borne Transmission in *Yersinia pestis*."

27. Pechous et al., "Pneumonic Plague."

28. Elleman and Kotkin, *Manchurian Railways*.

29. Summers, *The Great Manchurian Plague of 1910–1911*, 116–24.

30. For descriptions of the tarbagan and its natural history, see Shen, "Pneumonic Plagues, Environmental Changes, and the International Fur Trade."

31. Wu Lien-the and Dongbei fang yi chu, *North Manchurian Plague Prevention Service Reports (1911–1913)*, 23.

32. Summers, *The Great Manchurian Plague of 1910–1911*; Nathan, *Plague Prevention and Politics in Manchuria, 1910–1931*; Gamsa, "The Epidemic of Pneumonic Plague in Manchuria 1910–1911."

33. On the unfolding of plague in Harbin, see Gamsa, "The Epidemic of Pneumonic Plague in Manchuria 1910–1911."

34. Andrews, *The Making of Modern Chinese Medicine, 1850–1960*, 96–105.

35. Wu, *Plague Fighter*, 131–219.

36. Lynteris, "Plague Masks."

37. Lei, "Sovereignty and the Microscope."

38. Yang Ruisong, *Bing fu, huang huo yu shui shi*; Amelung, *Revisiting the Sick Man of Asia*.

39. Wu Lien-teh and Dongbei fang yi chu, *North Manchurian Plague Prevention Service Reports (1911–1913)*, 12.

40. On similar improvisation during the 1894 Hong Kong plague, see Peckham, "Matshed Laboratory."

41. Wu Lien-teh and Dongbei fang yi chu, *North Manchurian Plague Prevention Service Reports (1911–1913)*, 22.

42. Wu Lien-teh and Dongbei fang yi chu, 32–36.

43. Wu Lien-teh and Dongbei fang yi chu, 49.

44. Wu Lien-teh and Dongbei fang yi chu, 50–51. For photograph of researchers taking the animal's temperature, see plate 11 opposite page 40.

45. Wu Lien-teh and Dongbei fang yi chu, 51.

46. Kohler, *Landscapes and Labscapes*, 92–94, 205–10.

47. Wu, "The Second Pneumonic Plague Epidemic in Manchuria, 1920–21"; Wu, Chun, and Pollitzer, "Clinical Observations upon the Manchurian Plague Epidemic, 1920–21"; Wu, Han, and Pollitzer, "Plague in Manchuria."

48. For photographs of the Harbin clinic, see Yu-lin Wu, *Memories of Dr. Wu Lien-teh, Plague Fighter*, frontpiece and 50–51; description of the experimental ward in Wu Lien-teh, *Manchurian Plague Prevention Service Memorial Volume, 1912–1932*, 199–202.

49. Wu, Han, and Pollitzer, "Plague in Manchuria," 308–19.

50. Wu, Han, and Pollitzer, 315.

51. Wu, Han, and Pollitzer, 333.

52. Wu, Han, and Pollitzer, 331–39.

53. Wu, Han, and Pollitzer, 319–21.

54. Wu, Pollitzer, and Lin, "Studies upon the Plague Situation in North China." This lengthy report was divided into three parts, representing epidemiological findings; laboratory, clinical, and pathology findings; and field research: "General Survey of the Outbreaks," "Clinical and Laboratory Observations," and "Report on an Expedition into the Plague Focus of Tungliao."

55. Wu, Pollitzer, and Lin, 321.

56. Wu, Pollitzer, and Lin, 283–84.

57. Wu, Pollitzer, and Lin, "Studies upon the Plague Situation in North China," "fresh" quote, 290; "ordeal," 335.

58. Wu, Pollitzer, and Lin, 335–37.

59. Wu, Pollitzer, and Lin, 321.

60. Wu, Pollitzer, and Lin, 340–50.

61. Wu, Pollitzer, and Lin, 359. Inhalation experiments, 361; "comrades" experiment, 365.

62. Wu, Pollitzer, and Lin, 362–63.

63. Swellings located in the groin, leg, armpit, neck, knee, and elbow. Wu, Pollitzer, and Lin, 313. In the past, observers frequently noted the location of buboes and sought clinical significance in the proportion of buboes in different places. Today, the location of the swellings is considered to have no particular significance other than suggesting the general area where the infecting bite may have occurred. On the question of bubo location, see Benedictow, *What Disease Was Plague?*, 312–381.

64. On the use of such photographs by Western physicians in China, see Heinrich, *The Afterlife of Images*, 73–112.

65. Wu Lien-teh, "The Second Pneumonic Plague Epidemic in Manchuria, 1920–21," 274.

66. Wu, Pollitzer, and Lin, "Studies upon the Plague Situation in North China," 308.

67. See "Notes on the Histology of Some of the Lesions Found in Pneumonic Plague," in Wu Lien-teh, Dongbei fang yi chu, and China National Quarantine Service, *Manchurian Plague Prevention Service Memorial Volume*, 21–50.

68. Lynteris, *Ethnographic Plague*, 121–41.

69. Lynteris, 274. On Chinese reactions to and rumors about invasive plague prevention procedures, see Gamsa, "The Epidemic of Pneumonic Plague"; and Rogaski, "Vampires in Plagueland."

70. Iijima Wataru, *Mararia to teikoku*; Iijima Wataru, *Pesuto to kindai Chūkoku*; Jannetta, *Epidemics and Mortality in Early Modern Japan*; Burns, "Constructing the National Body."

71. On Japan's early medical establishment and Kitasato's career, see Kim, *Doctors of Empire*; Bartholomew, *The Formation of Science in Japan*; Peckham, "Matshed Laboratory"; Liu, "The Ripples of Rivalry."

72. Liu, *Prescribing Colonization*.

73. Perrins, "Doctors, Disease and Development," 109.

74. Perrins, 109–10; Summers, 71–75.

75. On Gotō see Liu, *Prescribing Colonization*; Kim, *Doctors of Empire*.

76. Minami Manshū tetsudō kabushiki kaisha eiseika, *Manshū fūdo eisei kenkyū gaiyō*.

77. See, for example, the discussion of climate in Toyoda Tarō, *Manshū densenbyō*, 12–22.

78. Toyoda Tarō, *Manshū densenbyō*, 55.

79. On *densenbyō* and *chihōbyō*, see Uruno Katsuya, *Manshū no chihōbyō to densenbyō*.

80. Christmas, "Japanese Imperialism and Environmental Disease on a Soy Frontier, 1890–1940."

81. Toyoda Hidezō, *Toman to eisei*, 209–12, encapsulates this sensibility of colonial "hygienic uplift."

82. Toyoda Tarō, *Manshū densenbyō* is a prime example of this pessimistic approach.

83. Andō and Kurauichi, "A New Plague Endemic Area in the Northeastern Part of Inner Mongolia"; Andō Kōji, "Tōhoku Nai Mōko ni okeru pesuto no ryūkō ni tsuite."

84. Lynteris, *Ethnographic Plague*.

85. Andō et al., 36. The survey was so extensive that two new species of rodents were discovered.

86. Andō et al., 38.

87. Driscoll, *Absolute Erotic, Absolute Grotesque*, xliii. "Profoundly inhumane modernity" from Erik Esselstrom, review of Driscoll in the *American Historical Review* 116, no. 2 (2011): 432–33.

88. Xie Xueshi and Matsumura Takao, "Xinjing shuyi moulue—1940."

89. Young, *Japan's Total Empire*, 249. On Xinjing, see Sewell, *Constructing Empire*.

90. On the stark contrast between ideal and reality in Manchukuo, see Driscoll, *Absolute Erotic, Absolute Grotesque*; Culver, *Glorify the Empire*; Shepherdson-Scott, "Utopia/Dystopia."

91. Hiroki Masaji, "Shōwa hachinen Kitsurin-shō Nōan ni okeru pesuto ryūkō ni tsuite."

92. On Changchun's neighborhoods, see Liu, "Competing Visions of the Modern Urban Transformation and Social Change of Changchun, 1932–1957."

93. For the following narrative of the Hsinking outbreak, see Tsuneishi Keiichi, *Senjō no ekigaku*, 67–79; Xie Xueshi and Matsumura Takao, "Xinjing shuyi moulue—1940"; Matsumura Takao, "Shinkyō, No-an pesuto ryūkō (1940-nen) to 731 butai." The second half of Matsumura's article provides primary-source texts from the original epidemiological investigation, Takahashi Masahiko's "Shōwa 15 nen No-an to Shinkyō ni hassei seru pesuto ryūkō ni tsuite," *Rikugun Gun'i Gakkō bōeki kenkyū hōkoku 2*, 1943. On Takahashi's report and other Unit 731 publications from the Army Medical School, see *Sensō 731 to daigaku, ika daigaku*.

94. On plague control, see Tsuneishi Keiichi, *Senjō no ekigaku*, 121–35; Xie Xueshi and Matsumura Takao, "Xinjing shuyi moulue—1940."

95. Matsumura Takao, "Shinkyō, No-an pesuto ryūkō (1940-nen) to 731 butai," pt. 1, 650–51.

96. Tsuneishi Keiichi, *Senjō no ekigaku*, 79–86.

97. Writing on Unit 731 is vast. Williams and Wallace, *Unit 731*; Harris, *Factories of Death*; Nie et al., *Japan's Wartime Medical Atrocities*; Tsuneishi Keiichi, *Hyōteki, Ishii*; Matsumura Takao, *"Ronsō" 731 Butai*; Xie Xueshi and Matsumura Takao, *Zhanzheng yu eyi*.

98. For a concise overview, see Takashi Tsuchiya, "Imperial Japanese Experiments in China."

99. For this report, see Part II of Matsumura Takao, "Shinkyō, No-an pesuto ryūkō (1940-nen) to 731 butai."

100. On scholarly debates, see Tsuneishi Keiichi, *Senjō no ekigaku*, 157–84.

101. Takashi Masahiko, "Showa jūgo-nen," 2–515, quoted in Matsumura Takao, "Shinkyō, No-an pesuto ryūkō (1940-nen) to 731 butai," pt. 2, 406–7.

102. Takashi Masahiko, "Showa jūgo-nen," 2–515, quoted in Matsumura Takao, pt. 2, 408–10.

103. Matsumura Takao, pt. 1, 31.

104. Takahashi's report contained extensive calculations about the number of fleas (and amount of time) necessary to start a plague outbreak, calculations that map the travels of *Yersinia pestis* through a variety of spaces. Matsumura Takao, pt. 2, 409.

105. When queried by examiners at the Khabarovsk War Crimes trials: "What infectious diseases did the 1st Division employ most frequently in its experiments on human beings?" the former director of Unit 731, Kawashima Kiyoshi, answered: "Chiefly plague." Yamada Otozō (defendant), *Materials on the Trial of Former Servicemen of the Japanese Army*, 253.

106. According to testimony, Pingfang's eight massive, one-ton culture medium tanks could produce up to 660 pounds of *Yersinia pestis* per production cycle. Yamada Otozō (defendant), 255.

107. The Japanese researcher Tsuenishi Keiichi has enumerated over thirty wartime studies on flea cultivation produced by Unit 731–affiliated researchers, including dissertations at Japanese universities and articles in Japanese scientific publications. See Tsuneishi Keiichi, *Senjō no ekigaku*, 189–92.

108. On the flea-dispensing "Uji bomb" and crop-duster-like aerosols for dispersing *Y. pestis* through the air, see Harris, *Factories of Death*, 79–80.

109. Harris, 103–4.

110. N. H. Fell, "Brief Summary of New Information about Japanese B. W. Activities," 1947, RG 175, box 196, JWC #123, National Archives, College Park, MD.

111. Fell, "Brief Summary."

112. Fell, "Brief Summary."

113. Yamada Otozō (defendant), *Materials on the Trial of Former Servicemen of the Japanese Army*, 259.

114. Harris, *Factories of Death*, 88–89.

115. "The Report of Q," 1947, Document #56-FDTS-197, JWC 253, National Archives, College Park, MD. While the data contained in the Report of Q is clearly from Japanese scientists of Unit 731, the exact authorship of "The Report of Q" itself is not entirely clear. Japanese scholars including Tsuneichi Keiichi and Shōji Kondō state that the report was an English translation prepared by Unit 731 personnel at the behest of US Army investigators.

116. "The Report of Q" and other Unit 731 documents were transferred to the US government after World War II in exchange for clemency for Unit 731 researchers. After the documents were declassified in 1960, they were held for years at the US military's Dugway Proving Grounds in Utah, where researchers had to go through representatives of Fort Detrick, the US military's biological defense program, to access them. In the early 2000s, the documents were moved to the National Archives.

Ironically, Chinese researchers learned of "The Report of Q" through Western and Japanese monographs on Unit 731 and have only recently undertaken their own analysis. On the movement of the Unit 731 documents, see Yang Yajun, "Preliminary Interpretation on the Report of Plague by Unit 731."

117. Author's observation during visit to the "Exhibition Hall of Evidences of Crimes Committed by Unit 731 of the Japanese Imperial Army" outside of Harbin, August 2017 (http://www.731museum.org.cn/index.html).

118. The Japanese historian Tsuneishi Keiichi made the connection between the cases in the Report of Q and Takahashi's plague report. See Tsuneishi Keiichi, *Senjō no ekigaku*, 171–75.

119. Lederer, *Subjected to Science*; Hoppe, *Lords of the Fly*.

CHAPTER 7

1. Levine, *Anvil of Victory*.

2. The names of the villages are mentioned in *Report of the International Scientific Commission for the Investigation of the Facts Concerning Bacterial Warfare in Korea and China*, 217. The report does not mention that these "villages" are within the borders of the Chahayang State Farm, one of the first state farms in the region, established in 1948 as part of the Chinese Communist Party's Civil War–era grain-production effort. On the history of the Chahayang State Farm, see Chahayang nongchang zhi bianshen weiyuanhui, *Chahayang nongchang zhi*.

3. Rogaski, *Hygienic Modernity*, 285–99.

4. Endicott and Hagerman, *The United States and Biological Warfare*, 37–42.

5. On the World Peace Council, see Olšáková, "Pugwash in Eastern Europe."

6. Guillemin, *Biological Weapons*, 99–100.

7. For composition of the commission, see *Report*, 3. The Soviet member, Dr. Zhukov-Verezhnikov, had been the chief medical expert at the Soviet war crimes trial of captured Unit 731 members held at Khabarovsk in 1949.

8. On the role of germ warfare accusations in peace talks, see Foot, *A Substitute for Victory*, 91–92. For a Western study that supports the veracity of the 1952 allegations, see Endicott and Hagerman, *The United States and Biological Warfare*. A neutral stance can be found in Rogaski, "Nature, Annihilation, and Modernity."

9. Dai's team (which included Cornell and Cambridge graduates) also identified different pathogenic fungi from samples of peach tree leaves, oak leaves, and corn kernels. *Report*, appendix Ja, 181–88.

10. *Report*, appendix Jb, 191–93. Hu Xiansu was considered China's leading botanist—founder of the Botanical Society of China and the Fan Memorial Institute of Biology in Beijing. On Hu, see Haas, *China Voyager*.

11. Chen's name appears frequently in *Report*, appendix AA, 361–416.

12. On the flies of Shenyang, see *Report*, appendix E, 115–20.

13. See, for example, reports on plague investigations in appendix T, 307–16. Chinese media reported that some scientists had been mobilized to travel to points in the northeast to examine evidence in the field from winter 1951 to the spring of 1952.

For English-language discussion, see Hagerman and Endicott, *The United States and Biological Warfare*, 21–26.

14. Plague prevention measures detailed in *Report*, appendixes M and M-1, 217–35.

15. *Report*, appendix M-2, 236–41.

16. *Report*, appendix M-4, 243, which instructs readers to see appendix O, 271–76. Appendix O gives further detailed measurements and analysis in response to a query about the vole's identity from the International Commission.

17. *Report*, appendixes M-5 and M-6, 244–52.

18. Joseph Needham, "Notes on Kan-nan visit," Papers and Correspondence of Joseph Needham, Ch Frs, Material Relating to Chemical and Biological Warfare, folder 30, London: Imperial War Museum. Quoted in Rogaski, "Nature, Annihilation, and Modernity." The "very pretty female tractor driver" may have been the famous Liang Jun, a celebrity model worker of Gannan's Chahayang State Farm whose image was later featured on the PRC one-yuan note (see chapter 8).

19. For biographical information on the plague prevention personnel (and all other scientists involved in the investigations), see *Report*, appendix TT, 635–65.

20. *Report*, appendix K, 195–212.

21. *Report*, appendix L, 213.

22. On the continuation of certain aspects of Japanese science within its former colonies after the war, see Mizuno, Moore, and DiMoia, *Engineering Asia*.

23. Fang, *China and the Cholera Pandemic*; Gross, *Farewell to the God of Plague*.

24. Harris, *Factory of Death*, blames the 1947 and 1948 outbreaks on rats and germs left over from Unit 731, but others see the chaos of war as the main culprit. The Communist siege of Changchun in 1948 was a particularly brutal episode that resulted in an estimated 150,000 civilian deaths from starvation and disease, including plague. See Tanner, *Where Chiang Kai-Shek Lost China*; Li Honghe, "Dongbei jiefangqu de shuyi liuxing ji jiuzhi."

25. Dongbei renmin zhengfu weisheng bu, *Shuyi yufang gongzuo shouce*. The 1951 epidemic was centered on Tongliao, the town that had been the center of the 1927–1928 plague outbreak studied by both Wu Lien-teh and Japanese scientists.

CHAPTER 8

1. Zhang Yuliang, *Guoying youyi nongchang de zhiwu diaocha*, 2.

2. Mullaney, *Coming to Terms with the Nation*; Minzu wenti wuzhong congshu Heilongjiang sheng bianji zu, *Hezhezu shehui lishi diaocha*.

3. Beidahuang literally means "Northern Large Wasteland." Beidahuang is often translated as "Great Northern Wilderness," but instead of "untouched nature," *huang* more closely connotes unproductive, uncultivated land. On cultural valences of *huang*, see Muscalino, "Refugees, Land Reclamation, and Militarized Landscapes in Wartime China," 454.

4. Zhang Yuliang, 2–13.

5. Blackbourn, *The Conquest of Nature*, 15.

6. A sampling of the massive Beidahuang-related cultural output includes the literary journal *Beidahuang wenxue* (1985–); films such as *Beidahuang ren* [People of the Great Northern Wasteland], directed by Cui Wei (1961), and *Shenqi de tudi* [Miraculous land], directed by Gao Tianhong (1985); and novels such as Liang Xiaosheng's *Nian lun* [Growth rings] (Beijing: Kaiming chubanshe, 2010) (also serialized as a television drama).

7. Gao Yuehui and Huang Hong, *Beidahuang jingshen*.

8. Williams, *The Country and the City*, 105.

9. Cronon, *Changes in the Land*.

10. Andrews, *Killing for Coal*, 115.

11. White, "Are You an Environmentalist, or Do You Work for a Living?," 172.

12. White, *The Organic Machine*; Morse, *The Nature of Gold*; Rogers, *Deepest Wounds*; Klubock, *La Frontera*; Nash, *Inescapable Ecologies*; Mercier, *Anaconda*; Brown, *Plutopia*; Andrews, *Killing for Coal*.

13. Rogers, *Deepest Wounds*.

14. Brown and Klubock, "Environment and Labor."

15. Andrews, *Killing for Coal*, 17.

16. For examples of mass labor mobilizations, see Shapiro, *Mao's War against Nature*.

17. Schmalzer, *Red Revolution, Green Revolution*.

18. Pritchard, *Confluence*, 1–4.

19. For basic stats on Sanjiang, see Liu Jiping, "Landscape Pattern Dynamics and Driving Forces Analysis in the Sanjiang Plain from 1954 to 2010."

20. Allaby, "Chernozem."

21. Legros, *Major Soil Groups of the World*, 448.

22. Fedotova, "The Origins of the Russian Chernozem Soil (Black Earth)," 272.

23. Lavelle, *The Profits of Nature*, 170.

24. Isett, *State, Peasant, and Merchant*, 75–144.

25. Song, *Making Borders in Modern East Asia*, 12–126.

26. Shan, *Taming China's Wilderness*, chapter 2; Reardon-Anderson, *Reluctant Pioneers*, 78–83.

27. Ma, "Transforming of Forest Resources from Qing Dynasty to the Republic of China in the Sanjiang Plain."

28. Shan, *Taming China's Wilderness*, 52–55.

29. Xu Shuming, "Qingdai dongbei diqu tudi kaiken."

30. Reardon-Anderson, *Reluctant Pioneers*, 78–84.

31. *Huachuan xian zhi*, wheat, 208; carts, 136.

32. *Huachuan xian zhi*, 199.

33. *Huachuan xian zhi*, 199.

34. On the "primitive" techniques of north Manchurian farmers, see Readeron-Anderson, *Reluctant Pioneers*, 123–24. For detailed descriptions of the farmers' tools, see Chinese Eastern Railway et al., *North Manchuria and the Chinese Eastern Railway*.

35. Lavelle, "Tools for Overcoming Crisis."

36. Shan, *Taming China's Wilderness*, 32–33; Sun Zhanwen, *Heilongjiang sheng shi tansuo*, 288–90.

37. Moore, *Defining and Defending the Open Door Policy*, 53–86.

38. Rauchway, "Willard Straight and the Paradox of Liberal Imperialism"; Roberts, "Willard Straight, the First World War, and 'Internationalism of All Sorts.'"

39. Harvard University, Class of 1910, *Report*.

40. Palmer, "Colonising in Manchuria," 271.

41. Palmer, 272.

42. Young, *Japan's Total Empire*, 307–98; Culver, "Constructing a Rural Utopia: Propaganda Images of Japanese Settlers in Northern Manchuria, 1936–43"; Shepherdson-Scott, "Utopia/Dystopia: Japan's Image of the Manchurian Ideal"; "Japanese Expatriate Farmers Plow and Plant on a Manchukuo Field."

43. Iyasaka itself was founded directly on the site of the "Hunan Camp" settlement. Scherer, "Japanese Emigration to Manchuria." For an extensive overview of the village history, see Iyasaka-son shi kankō iinkai, *Iyasaka-son shi*.

44. Young, *Japan's Total Empire*, 402.

45. On various forms of Japanese-enforced "collective living" in Manchuria, see Park, *Two Dreams in One Bed*, 145–80. On "concentration villages" for Mongols, see Christmas, "The Cartographic Steppe," 105–25.

46. Optimistic plans called for the establishment of fifty-five "mechanical farms" to be opened, with each farm receiving eight tractors. Wartime exigencies after 1941 limited the distribution of tractors. "Mechanical Farming."

47. On the difficulties faced by Japanese settlers, see Tamanoi, *Memory Maps*.

48. Levine, *Anvil of Victory*.

49. Zheng, *Zhongguo dongbeijiao*, chap. 2.

50. Rohlf, "The Soviet Model and China's State Farms"; Brown, *City versus Countryside in Mao's China*, 169–99.

51. Heilongjiang sheng difangzhi bianzhuan weiyuanhui, *Heilongjiang sheng zhi: guoying nongchang zhi* (GNZ), 16.

52. Zheng, *Zhongguo dongbeijiao*, 3.

53. On the use of Japanese experts after 1949, see King, "Reconstructing China"; Seow, *Carbon Technocracy*.

54. *Guoying Youyi nongchang zai jianshe zhong*.

55. Smith, *Thought Reform and China's Dangerous Classes*.

56. GNZ, 75.

57. GNZ, 87–95; Sun, "'War against the Earth.'"

58. Bawusan nongchang zhi bianshen weiyuanhui, *Bawusan nongchang zhi, 1956–1985*, 377–81.

59. This well-known image, of uncertain provenance but reproduced in many histories of Beidahuang, has even been replicated as a half-scale statue in the Beidahuang museum in Harbin. Author's photograph, August 2017.

60. *Beidahuang ren* [People of the Great Northern Wasteland], directed by Cui Wei (1961), https://www.bilibili.com/video/av4700532/.

61. GNZ, 42.

62. *GNZ*, 16, 40.

63. Zhou, *Heitushang de Zhongguo*, 53–73.

64. Zheng, *Zhongguo dongbei jiao*, 57–58.

65. For media images of Soviet equipment, see front matter in *Guoying Youyi nongchang zai jianshe zhong*.

66. On the Soviet factory that produced the famous Stalinets tractors (and tanks), see Samuelson, *Tankograd*.

67. Details on farming technology from Yu and Han, *Guoying Youyi nongchang fangwen ji*, 11–17.

68. Yu and Han, 11–17.

69. On Liu Shene, see *Zhongguo kexuejia cidian, xiandai ce*, 4:90.

70. Zhang Yuliang, *Guoying Youyi nongchang de zhiwu diaocha*, 2–15.

71. Zheng Jiazhen, *Zhongguo dongbei jiao*, 61.

72. Bawusan nongchang zhi bianshen weiyuanhui, *Bawusan nongchang zhi, 1956–1985*, 382–83. The Yanwodao tractor rescue is lauded through photographs and diaoramas in museums throughout the northeast.

73. Donnithorne, *China's Economic System*, 112.

74. While *zhiqing* volunteers are usually thought of as a Cultural Revolution phenomenon, youth "volunteers" to Beidahuang began as early as 1954. For a fictionalized account of early *zhiqing* work in the 1950s, see Zheng Jiazhen, *Zhan dou zai Beidahuang*.

75. This section is drawn from a reading of approximately two hundred interviews and essays primarily from two published collections: Jia Hongtu, *Women de gushi*; and Zhongguo renmin zhengzhi xieshang huiyi, *Zhishi qingnian zai Heilongjiang (ZQHL)*.

76. For representative photos, see Jia Hongtu, *Women de gushi*; Heilongjiang sheng shengchan jianshe dui, zhengzhi bu, *Zhishiqingnian zai Beidahuang*; Huang Chengjiang, *Beidahuang zhiqing*.

77. For specific information on which farms hosted which groups of *zhiqing*, see Jin Guanghui, *Zhongguo xin fangzhi zhishi qingnian shangshan xiaxiang shiliao jilu*.

78. Guo Jinxue, "Ganxiao shenghuo huiyi" [Recollections from life in a cadre school], *ZQHL*, 790–98.

79. On Liang Jun and early PRC female tractor drivers, see Li Tailing, "Zou jin wang shi" [Out into the world], *ZQHL*, 639; Finnane, *Changing Clothes in China*, 205.

80. Xie Xiaoli, "Zihaode di yidai Beidahuang ren" [A proud first generation Beidahuanger], *ZQHL*, 1–3.

81. Guo Jinxue, *ZQHL*, 798. On clearing land with explosives and tractors, camping out, and being pursued by wolves, see Zhang Baoguo, "Xiao Gou dadui huiyi" [Recollection of Little Ditch Platoon], *ZQHL*, 744; for *flotant* encounter, see Zhou Hongxin, "Piaofa dianzi," *ZQHL*, 651.

82. On operating seeding machines, see Lu Yongzhen, "Zai guangkuo tiandili" [On the vast fields], *ZQHL*, 292–311.

83. On hoeing as women's work, see Hershatter, *The Gender of Memory*, 142–48.

84. Although these dirt piles and ditches were far removed from any visions of

imperial grandeur, their names were formed by included the character for dragon (*long*) on top of the character for soil (*tu*).

85. From an essay aptly entitled "Duiyu hei tudi de guangda he liaorun zhuyao shi tongguo chandi lai renshide" [An understanding of the vastness of the "black earth" was mostly gained through hoeing]. See Zhang Kangkang, "Ku zhong zuo le" [Finding happiness in suffering], *ZQHL*, 515–16. Zhang's observations were echoed by Li Chang, "Ku le shiyi nian" [Bittersweet ten years], who in somewhat more direct language called the repetitive, machinelike work *fansi* (deathly annoying). Li Chang, 172.

86. Yang Yong, "Mai shou" [The harvest], *ZQHL*, 555; Ma Shuyi, "Xiao liandao de baochou" [Revenge of the sickle], *ZQHL*, 586.

87. Guo Jinxue, *ZQHL*, 799.

88. Zhou Xiangyang, "Si nian zhiqing liandui shenghuo" [My four years as a sent-down youth], *ZQHL*.

89. Youyi nongchang, *Beifang hanliang chanqu nongye xiandaihua zonghe kexue shiyan jidi chengguo qiandinghui cailiao huibian* [Committee to evaluate the results of the scientific experimental station for the agricultural modernization of the northern dry-grain production area; hereafter *BFHL*], report 4, "Peifei dili shiyan yanjiu congjie," unpublished manuscript in possession of author, August 1985.

90. *BFHL*, report 1, "Youyi wufenchang butong jingfa xiaoguo fenxi ji heli shaojing tixi de jianyi."

91. *BFHL*, report 10, "Chun xiaomai damianji pengguan yingyong jishu yanjiu."

92. *BFHL*, report 1 "Youyi wufenchang butong jingfa xiaoguo fenxi ji heli shaojing tixi de jianyi."

93. William Hinton is best known for his observations of the north China village of Long Bow in the 1940s (and again in the 1970s), but in the late 1970s he lived at the Youyi Farm advising the Fifth Farm Second Team Experimental Station. Many older Youyi residents have vivid memories of "Han Ding."

94. "Nongye jixie shebei shengchan shiyan baogao," 3.

95. "Nongye jixie shebei shengchan shiyan baogao," 4.

96. "Heilongjiangsheng Youyi nongchang wufenchang erdui yinjin chengtao nongji shebei shengchang shiyan baogao" (hereafter "Erdui baogao") 10, chart 1.

97. "Erdui baogao," 11, chart 2.

98. This sudden "great leap" in the mechanization of farm work was a tremendous shock that some, particularly the farmworkers who lost their jobs due to increased productivity, approached with great skepticism. Some workers criticized the mechanization as "fine for the United States, but not suitable for China's circumstances." Some excess workers were reassigned to a newly created production team tasked with opening up new land—using the decommissioned Chinese-made tractors. The largest number of displaced workers were organized into a *shuili dui* (a water control team) in other words, they were assigned to ditch digging. While a handful of workers operated high-end American equipment, the majority of workers were tasked with re-creating the primitive process of opening wastelands from the 1950s and 1960s. The introduction of American technologies in some ways replicated

the hierarchy of labor that was present from the inception of different types of state farms on the Sanjiang plain at the beginning of the PRC. "Erdui baogao."

99. "Hongxinglong nongkenqu nongye nongji zhuangbei yilan biao," displayed at the Hongxinglong museum, photograph by author, August 2017.

100. Liu Jiping et al., "A Dynamic Change Map of Marshes in the Small Sanjiang Plain, Heilongjiang, China, from 1955 to 2005"; Liu, "Characterizing the Spatial Pattern of Marshlands in the Sanjiang Plain, Northeast China"; Liu Jiping, "1954–2010 nian Sanjiang pingyuan tudi liyong jingguan geju dongtai bianhua ji qudong li."

101. Lu, "Rice Cultivation Changes and Its Relationships with Geographical Factors in Heilongjiang Province, China."

102. Guan Yanbo, "Shi lun Dongbei Chaoxianzu de dao zuo."

103. Shapiro, *Mao's War against Nature*.

104. Worster, *Dust Bowl*.

105. Prince, *Wetlands of the American Midwest*, 206.

106. Leopold, *A Sand County Almanac and Sketches Here and There*.

107. On the disparate memories of sent-down youth, see Xu Bin, *Chairman Mao's Children*.

108. Schmalzer, "Layer upon Layer," 422.

CONCLUSION

1. Rieppel, *Assembling the Dinosaur*, 220–52.

2. Gao et al., "Exceptional Fossil Material of a Semi-Aquatic Reptile from China."

3. For memories of earlier PRC plague control activities in the grasslands, see Liu Jiyou, *Xiaomie shuhai*. News of the plague cases of 2019 was rapidly overshadowed by the specter of the SARS-CoV-2 virus, which emerged in late 2019 from China's southwest borderlands. On historical similarities between plague and COVID, see Rogaski, "The Manchurian Plague and COVID-19."

4. Author's observations, August 2017. The museum's exhibits discuss the US military's deal with Unit 731 after World War II but do not press the case that the United States used Japanese techniques of germ warfare during the Korean War. Another PRC museum in Dandong on the border with North Korea still maintains that the allegations were true. See Jung, "China's Memory and Commemoration of the Korean War."

5. Xu et al., "Design of Nature Reserve System for Red-Crowned Crane in China."

6. Haruyama and Shiraiwa, *Environmental Change and the Social Response in the Amur River Basin*; Higgins, "Russia's Acres, If Not Its Locals, Beckon Chinese Farmers."

BIBLIOGRAPHY

15-nen sensō to Nihon no igaku iryō kenkyūkai. *Sensō 731 to daigaku ika daigaku* [The War, Unit 731 and universities, medical universities]. Tokyo: Bunrikaku, 2016.

Adams, Jad. *Hideous Absinthe: A History of the Devil in a Bottle*. London: I. B.Tauris, 2004.

Adas, Michael. *Machines as the Measure of Men: Science, Technology, and Ideologies of Western Dominance*. Ithaca, NY: Cornell University Press, 1989.

Ahn, Yonson. "China and Korea Clash over Mount Paekdu/Changbai: Memory Wars Threaten Regional Accommodation." *Japan Focus*, July 27, 2007. http://japanfocus.org/-Yonson-Ahn/2483.

Ahnert, E. E. "The Geological Society and Science in Asia." *Bulletin of the Geological Society of China* 1, nos. 1–4 (1922): 10.

Ahnert, E. E., and A. I. Lauroushin. "Subdivisions of the Jurassic, Cretaceous and Tertiary Coal-Bearing Strata of Russian Maritime and Amur Provinces, and of Sakhalin Island." *Bulletin of the Geological Society of China* 3, nos. 3–4 (1924): 195–206. https://doi.org/10.1111/j.1755-6724.1924.mp33-4002.x.

Alexander, Zeynep Çelik. *Kinaesthetic Knowing: Aesthetics, Epistemology, Modern Design*. Chicago: University of Chicago Press, 2017.

Allaby, Michael. "Chernozem." In *A Dictionary of Ecology*. New York: Oxford University Press, 2010.

Allen, Glover M. *The Mammals of China and Mongolia*. New York: American Museum of Natural History, 1938.

Amelung, Iwo, ed. *Revisiting the "Sick Man of Asia": Discourses of Weakness in Late 19th and Early 20th Century China*. Frankfurt: Campus Verlag, 2020.

An Shuangcheng, ed. *Qingchu xiyang chuanjiaoshi manwen dangan yiben* [Translations of Manchu archives regarding Western missionaries in the early Qing]. Zhengzhou: Daxiang chubanshe, 2015.

Anderson, Warwick. *Colonial Pathologies: American Tropical Medicine, Race, and Hygiene in the Philippines*. Durham: Duke University Press, 2006.

Andō Kōji. "Tōhoku Nai Mōko ni okeru pesuto no ryūkō ni tsuite" [On the spread of plague in the northeastern part of Inner Mongolia]. *Kansenshogaku zasshi* 4, no. 5 (1930): 411–26.

Andō, K., and K. Kurauichi. "A New Plague Endemic Area in the Northeastern Part of Inner Mongolia: Plague Studies I." *Kitasato Archives of Experiment Medicine* 8 (1931): 24–38.

Andrade, Tonio. *The Gunpowder Age: China, Military Innovation, and the Rise of the West in World History.* Princeton, NJ: Princeton University Press, 2016.

Andrews, Bridie. *The Making of Modern Chinese Medicine, 1850–1960.* Vancouver: University of British Columbia Press, 2014.

Andrews, Thomas G. *Killing for Coal: America's Deadliest Labor War.* Cambridge, MA: Harvard University Press, 2008.

Arnold, David. *Science, Technology and Medicine in Colonial India.* Cambridge: Cambridge University Press, 2000.

———. "Touching the Body: Perspectives on Indian Plague, 1896–1900." In *Selected Subaltern Studies,* edited by Ranajit Guha, 391–426. Columbus: Ohio University Press, 1988.

Arsen'ev, Vladimir Klavdievich. *Dersu the Trapper.* Kingston, NY: McPherson, 1996.

Arsenyev, Vladimir K. *Across the Ussuri Kray: Travels in the Sikhote-Alin Mountains.* Bloomington: Indiana University Press, 2016.

Atlas of Plague and Its Environment in the People's Republic of China. Beijing: Science Press, 2000.

Baranov, A. "Na reke Amure 1854–1855" [On the Amur River]. *Russkaya Starina* 71 (March 1891): 327–54. Translated by Mark Conrad. http://www .marksrussianmilitaryhistory.info/AmurBaranov.html.

Barenblatt, Daniel. *A Plague upon Humanity: The Secret Genocide of Axis Japan's Germ Warfare Operation.* New York: HarperCollins, 2004.

Barrera-Osorio, Antonio. *Experiencing Nature: The Spanish American Empire and the Early Scientific Revolution.* Austin: University of Texas Press, 2006.

Bartholomew, James R. *The Formation of Science in Japan: Building a Research Tradition.* New Haven, CT: Yale University Press, 1989.

Bassin, Mark. *Imperial Visions: Nationalist Imagination and Geographical Expansion in the Russian Far East, 1840–1865.* Cambridge: Cambridge University Press, 1999.

———. "The Russian Geographical Society, the 'Amur Epoch,' and the Great Siberian Expedition 1855–1863." *Annals of the Association of American Geographers* 73, no. 2 (1983): 240–56.

Bawuer nongchang zhi bianshen weiyuanhui. *Bawuer nongchang zhi* [Gazetteer of the 852 farm]. Heilongjiang: Bawuer nongchang, 1998.

Bawusan nongchang zhi bianshen weiyuanhui. *Bawusan nongchang zhi, 1956–1985* [Gazetteer of the 853 Farm, 1956–1985]. Baoqing: Bawusan nongchang zhi biansheng weiyuan hui, 1986.

Bello, David A. *Across Forest, Steppe, and Mountain: Environment, Identity, and Empire in Qing China's Borderlands.* Cambridge: Cambridge University Press, 2015.

———. "The Cultured Nature of Imperial Foraging in Manchuria." *Late Imperial China* 31, no. 2 (2010): 1–33.

Ben-Ari, Tamara, Simon Neerinckx, Lydiane Agier, Bernard Cazelles, Lei Xu, Zhibin Zhang, Xiye Fang, Shuchun Wang, Qiyong Liu, and Nils C. Stenseth. "Identification of Chinese Plague Foci from Long-Term Epidemiological Data." *PNAS* 109, no. 21 (May 2012): 8196–8201. https://doi.org/10.1073/pnas.1110585109.

Bennett, B., and J. Hodge, eds. *Science and Empire: Knowledge and Networks of Science across the British Empire, 1800–1970*. New York: Palgrave Macmillan, 2011.

Berger, Patricia Ann. *Empire of Emptiness: Buddhist Art and Political Authority in Qing China*. Honolulu: University of Hawai'i Press, 2003.

Bernstein, Andrew. "Whose Fuji? Religion, Region, and State in the Fight for a National Symbol." *Monumenta Nipponica* 63, no. 1 (2008): 51–99.

Bertoni, Filippo. "Charming Worms: Crawling between Natures." *Cambridge Journal of Anthropology* 30, no. 2 (2012): 65–81.

Bian, He. "Governmental Science before Missionary Science: Another Look at the Reinvention of Botany in 19th Century China." Presentation at the History of Science Annual Workshop, "Trading Objecthood: Global Business and the Language of Natural History in the Long Nineteenth Century," Princeton University, February 2019.

———. *Know Your Remedies: Pharmacy and Culture in Early Modern China*. Princeton, NJ: Princeton University Press, 2020.

Blackbourn, David. *The Conquest of Nature: Water, Landscape, and the Making of Modern Germany*. New York: Norton, 2006.

Blaesing, Uwe. "The Tungusic Plant Names in Primitiae Florae Amurensis, Versuch Einer Flora Des Amurlandes by Carl Joh. Maximowicz." In *Man and Nature in the Altaic World, Proceedings of the 49th Permanent International Altaistic Conference*, 37–47. Berlin: Klaus Schwarz Verlag, 2012.

Bleichmar, Daniela. *Visible Empire: Botanical Expeditions and Visual Culture in the Hispanic Enlightenment*. Chicago: University of Chicago Press, 2012.

Braun, Bruce. "Producing Vertical Territory: Geology and Governmentality in Late Victorian Canada." *Ecumene* 7, no. 1 (January 2000). https://doi.org/10.1177/096746080000700102.

Bray, Francesca. "Millet Cultivation in China: A Historical Survey." *Journal d'agriculture traditionnelle et de botanique appliquée* 28, no. 3 (1981): 291–307. https://doi.org/10.3406/jatba.1981.3848.

Bretschneider, E. *History of European Botanical Discoveries in China*. London: Sampson Low, Marston and Co., 1898.

Breuker, Remco E. *Establishing a Pluralist Society in Medieval Korea, 918–1170: History, Ideology and Identity in the Koryŏ Dynasty*. Leiden: Brill, 2010.

Brook, Timothy. *The Confusions of Pleasure: Commerce and Culture in Ming China*. Berkeley: University of California Press, 1999.

———. *The Troubled Empire*. Cambridge, MA: Harvard University Press, 2010.

Brown, Jeremy. *City versus Countryside in Mao's China: Negotiating the Divide*. Cambridge: Cambridge University Press, 2012.

Brown, Kate. *A Biography of No Place: From Ethnic Borderland to Soviet Heartland.* Cambridge, MA: Harvard University Press, 2004.

———. *Plutopia: Nuclear Families, Atomic Cities, and the Great Soviet and American Plutonium Disasters.* Oxford: Oxford University Press, 2013.

Brown, Kate, and Thomas Klubock. "Environment and Labor: Introduction." *International Labor and Working-Class History* 85 (March 2014): 4–9. https://doi.org/10.1017/S0147547913000513.

Browne, Janet. "Biogeography and Empire." In *Cultures of Natural History*, edited by Nicolas Jardine, James Secord, and E. C. Spary, 305–21. Cambridge: Cambridge University Press, 1996.

———. *The Secular Ark: Studies in the History of Biogeography.* New Haven, CT: Yale University Press, 1983.

Bruun, Ole. *Fengshui in China: Geomantic Divination between State Orthodoxy and Popular Religion.* Copenhagen: NIAS Press, 2003.

Burnett, D. Graham. *Masters of All They Surveyed: Exploration, Geography, and a British El Dorado.* Chicago: University of Chicago Press, 2000.

Burns, Susan. "Constructing the National Body: Public Health and the Nation in Nineteenth-Century Japan." In *Nation Work: Asian Elites and National Identities*, edited by Timothy Brook and André Schmid, 17–50. Ann Arbor: University of Michigan Press, 2000.

Byington, Mark E., ed. *The History and Archaeology of the Koguryŏ Kingdom.* Cambridge, MA: Korea Institute, Harvard University.

Campany, Robert Ford. *Making Transcendents: Ascetics and Social Memory in Early Medieval China.* Honolulu: University of Hawaiʻi Press, 2009.

Cams, Mario. *Companions in Geography: East-West Collaboration in the Mapping of Qing China (c. 1685–1735).* Leiden: Brill, 2017.

Candolle, Augustin Pyramus de. *Prodromus systematis naturalis regni vegetabilis . . .* Paris: Treuttel et Würtz, 1849.

Carroll, Peter J. *Between Heaven and Modernity: Reconstructing Suzhou, 1895–1937.* Palo Alto, CA: Stanford University Press, 2006.

Carter, Paul. *The Road to Botany Bay: An Essay in Spatial History.* London: Faber, 1987.

Cha Zhilong and Zhang Jinyan. *Dai shi: 18 juan* [History of Taishan in eighteen volumes]. National Library of China—Harvard-Yenching Library Chinese rare book digitization project. http://nrs.harvard.edu/urn-3:FHCL:5110523.

Chahayang nongchang zhi bianshen weiyuanhui. *Chahayang nongchang zhi* [Gazetteer of Chahayang farm]. Heilongjiang sheng: Chahayang nongchang, 2004.

Chan, Timothy Wai Keung. "Beyond Border and Boudoir: The Frontier in the Poetry of Four Elites of the Early Tang." In *Reading Medieval Chinese Poetry: Text, Context, and Culture*, edited by Paul Kroll, 130–68. Leiden: Brill, 2014.

Chang, Michael G. *A Court on Horseback: Imperial Touring & the Construction of Qing Rule, 1680–1785.* Cambridge, MA: Harvard University Asia Center, 2007.

Chaves, Jonathan. *Every Rock a Universe: The Yellow Mountains and Chinese Travel Writing.* Warren, CT: Floating World Editions, 2013.

———. "The Yellow Mountain Poems of Ch'ien Ch'ien-i (1582–1664): Poetry as Yu-Chi." *Harvard Journal of Asiatic Studies* 48, no. 2 (December 1988): 465–92.

Chen Bochao and Piao Yushun. *Shengjing gongdian jianzhu* [The palace architecture of Shengjing]. Beijing: Zhongguo jianzhu gongye chubanshe, 2007.

Chen Hongzhou. "Volcanic Eruptive Processes and Characteristics of the Current Volcanoes in the Wudalianchi Volcano Clusters Known from Manchurian-Language Historical Archives." *Geological Review* 45, no. 1 (1999): 409–14.

Chen Hui. "Changbaishan chongbai kao" [Investigation into the worship of Mt. Changbai]. *Shehui kexue zhanxian*, no. 3 (2011): 104–10.

Chen, Pei-ji. *The Jehol Fossils: The Emergence of Feathered Dinosaurs, Beaked Birds and Flowering Plants*. Cambridge, MA: Academic Press, 2011.

Chen, Peiji, Zhi-ming Dong, and Shuo-nan Zhen. "An Exceptionally Well-Preserved Theropod Dinosaur from the Yixian Formation of China." *Nature* 391, no. 6663 (January 8, 1998): 147–52. https://doi.org/10.1038/34356.

Chengzhi (Kicengge). *Daichin Gurun to sono jidai: Teikoku no keisei to hakki shakai* [The Qing dynasty and its era: Eight banner society in the formation of empire]. Nagoya: Nagoya Daigaku Shuppankai, 2009.

———. "'Huangyu quanlan tu' Dongbei dadi cehui kao: Yi Manwen dangan wei zhongxin" [An investigation into the survey of the Northeast for the "Complete View of the Imperial Realm" map: Centered on Manchu archives]. In *Xi yu lishi yuyan yanjiu jikan*, 10:479–519. Beijing: Kexue chubanshe, 2018.

———. "The Illusion of the Nerchinsk Treaty Boundary-stone: The Map of the Amur Region in Manchu." *Gugong xueshu jikan* 29, no. 1 (2011): 147–236.

———. "Manwen 'Wula deng chu difang tu' kao" [An investigation on the Manchu-language map of Ula and environs]. *Gugong xueshu jikan* 26, no. 4 (2009): 1–74.

Chinese Eastern Railway, I. A. Mikhoĭlov, T. L. Lilliestrom, and A. G. Skerst. *North Manchuria and the Chinese Eastern Railway*. Harbin, China: C. E. R. Print Office, 1924.

Cho, Hyun-soul. "The Significance in Perception of Baekdusan in Baekdu-Related Myths." *Review of Korean Studies* 13, no. 4 (December 2010): 33–52.

Chongmyo ŭigwe [Protocol of the temple for the royal ancestors]. Manuscript version (c. 1900). https://iiif.lib.harvard.edu/manifests/view/drs:7551429$1i.

Chosŏn wangjo sillok [Veritable records of the Chosŏn dynasty]. Kwachŏn: Kuksa p'yŏnch'an wiwŏnhoe, 2016.

Christmas, Sakura. "The Cartographic Steppe: Mapping Environment and Ethnicity in Japan's Imperial Boderlands." PhD diss., Harvard University, 2016.

———. "Japanese Imperialism and Environmental Disease on a Soy Frontier, 1890–1940." *Journal of Asian Studies* 78, no. 4 (November 2019): 809–36. https://doi.org/10.1017/S0021911819000597.

Chu, Pingyi. "Against Prognostication: Ferdinand Verbiest's Criticisms of the Chinese Mantic Arts." In *Coping with the Future: Theories and Practices of Divination in East Asia*, edited by Michael Lackner, 433–50. Leiden: Brill, 2018.

———. "Scientific Dispute in the Imperial Court: The 1664 Calendar Case." *Chinese Science*, no. 14 (1997): 7–34.

Clunas, Craig. *Screen of Kings: Royal Art and Power in Ming China*. London: Reaktion Books, 2013.

———. *Superfluous Things: Material Culture and Social Status in Early Modern China*. Honolulu: University of Hawai'i Press, 2004.

Coble, Parks M. *Facing Japan: Chinese Politics and Japanese Imperialism, 1931–1937*. Cambridge, MA: Council on East Asian Studies, Harvard University, 1991.

Cocco, Sean. *Watching Vesuvius: A History of Science and Culture in Early Modern Italy*. Chicago: University of Chicago Press, 2012.

Coen, Deborah R. *Climate in Motion: Science, Empire, and the Problem of Scale*. Chicago: University of Chicago Press, 2018.

Coggins, Chris. *The Tiger and the Pangolin: Nature, Culture, and Conservation in China*. Honolulu: University of Hawai'i Press, 2003.

Cook, Harold John. *Matters of Exchange: Commerce, Medicine, and Science in the Dutch Golden Age*. New Haven, CT: Yale University Press, 2007.

Corrado, Sharyl. "The 'End of the Earth': Sakhalin Island in the Russian Imperial Imagination, 1849–1906." PhD diss., University of Illinois at Urbana–Champaign, 2010. Proquest (AAT 3430854).

———. "A Land Divided: Sakhalin and the Amur Expedition of G. I. Nevel'skoi, 1848–1855." *Journal of Historical Geography* 45 (July 2014): 70–81. https://doi.org/10.1016/j.jhg.2014.05.030.

Cosgrove, Denis, and Veronica della Dora, eds. *High Places: Cultural Geographies of Mountains, Ice and Science*. London: I. B. Tauris, 2009.

Cronon, William. *Changes in the Land: Indians, Colonists, and the Ecology of New England*. New York: Hill and Wang, 1983.

Crossley, Pamela Kyle. "An Introduction to the Qing Foundation Myth." *Late Imperial China* 6, no. 2 (1985): 13–24.

———. "Manzhou Yuanliu Kao and the Formalization of the Manchu Heritage." *Journal of Asian Studies* 46, no. 4 (1987): 761–90. https://doi.org/10.2307/2057101.

———. "The Rulerships of China." *American Historical Review* 97, no. 5 (1992): 1468–83. https://doi.org/10.2307/2165948.

———. *A Translucent Mirror: History and Identity in Qing Imperial Ideology*. Berkeley: University of California Press, 2000.

Cua, Antonio S., ed. *Encyclopedia of Chinese Philosophy*. Abington, UK: Routledge, 2013.

Culver, Annika A. "Constructing a Rural Utopia: Propaganda Images of Japanese Settlers in Northern Manchuria, 1936–43." In *Empire and Environment in the Making of Manchuria*, edited by Norman Smith, 152–78. Vancouver: University of British Columbia Press, 2017.

———. *Glorify the Empire: Japanese Avant-Garde Propaganda in Manchukuo*. Vancouver: University of British Columbia Press, 2013.

Cummiskey, Julia. "'An Ecological Experiment on the Grand Scale': Creating an Experimental Field in Bwamba, Uganda, 1942–1950." *Isis* 111, no. 1 (March 2020): 3–21.

Cutler, Alan. *The Seashell on the Mountaintop: A Story of Science, Sainthood, and the*

Humble Genius Who Discovered a New History of the Earth. Cambridge: Cambridge University Press, 2003.

Daston, Lorraine, and Peter Galison. *Objectivity*. New York: Zone Books, 2007.

Daston, Lorraine, and Katharine Park. "The Age of the New." In *The Cambridge History of Science*, vol. 3, *Early Modern Science*, edited by Lorraine Daston and Katharine Park, 1–17. Cambridge: Cambridge University Press, 2003.

David, Armand. *Abbé David's Diary*. Edited by Helen M. Fox. Cambridge, MA: Harvard University Press, 1949.

Deane, Thatcher Elliott. "The Chinese Imperial Astronomical Bureau: Form and Function of the Ming Dynasty Qintianjian from 1365 to 1627." PhD diss., University of Washington, 1989.

Debarbieux, Bernard, Gilles Rudaz, and Martin F. Price. *The Mountain: A Political History from the Enlightenment to the Present*. Translated by Jane Marie Todd. Chicago: University of Chicago Press, 2015.

Delbourgo, James. "The Knowing World: A New Global History of Science." *History of Science* 57, no. 3 (September 2019): 373–99. https://doi.org/10.1177/0073275319831582.

Deloria, Philip. *Indians in Unexpected Places*. Lawrence: University Press of Kansas, 2004.

Dennis, Joseph. *Writing, Publishing, and Reading Local Gazetteers in Imperial China, 1100–1700*. Cambridge, MA: Harvard University Asia Center, 2015.

Ding Yizhuang, and Mark Elliott. "How to Write Chinese History in the Twenty-First Century: The Impact of the 'New Qing History' Studies and Chinese Responses." *Chinese Studies in History* 51, no. 1 (January 2018): 70–95. https://doi.org/10.1080/00094633.2018.1466565.

"Dinosaur Hunter." http://paleobiology.si.edu/history/gilmore.html.

Dongbei renmin zhengfu weisheng bu. *Shuyi yufang gongzuo shouce* [Handbook for plague prevention work]. Shenyang: Dongbei renmin zhengfu weisheng bu, 1951.

Donnithorne, Audrey. *China's Economic System*. Abington, UK: Routledge, 2013.

Dott, Brian Russell. *Identity Reflections: Pilgrimages to Mount Tai in Late Imperial China*. Cambridge, MA: Harvard University Asia Center, 2004.

Dott, Robert H. "An Introduction to the the Ups and Downs of Eustasy." In *Eustasy: The Historical Ups and Downs of a Major Geological Concept*, edited by Robert Dott, 1–17. Boulder, CO: Geological Society of America, 1999.

Driscoll, Mark. *Absolute Erotic, Absolute Grotesque: The Living, Dead, and Undead in Japan's Imperialism, 1895–1945*. Durham, NC: Duke University Press, 2010.

Drouin, Jean-Marc. "Principles and Uses of Taxonomy in the Works of Augustin-Pyramus de Candolle." *Studies in History and Philosophy of Science Part C* 32, no. 2 (2001): 255–75.

Du Erwei. *Zhongguo gudai zongjiao xitong* [The religious system of ancient China]. Taipei: Taiwan xuesheng shuju, 1983.

Duara, Prasenjit. *Sovereignty and Authenticity: Manchukuo and the East Asian Modern*. Lanham, MD: Rowman & Littlefield Publishers, 2004.

Dunnell, Ruth W., Mark C. Elliott, Philippe Foret, and James A. Millward, eds. *New Qing Imperial History: The Making of Inner Asian Empire at Qing Chengde.* Abingdon, UK: Taylor & Francis, 2004.

Edmonds, Richard L. *Northern Frontiers of Qing China and Tokugawa Japan: A Comparative Study of Frontier Policy.* Chicago: University of Chicago, Department of Geography, 1985.

Elleman, Bruce A., and Stephen Kotkin, eds. *Manchurian Railways and the Opening of China: An International History.* Armonk, NY: M. E. Sharpe, 2010.

Ellis, Steven J. R. "The Distribution of Bars at Pompeii: Archaeological, Spatial and Viewshed Analyses." *Journal of Roman Archaeology* 17 (2004): 371–84. https://doi.org/10.1017/S104775940000831X.

Elliott, Mark. "Frontier Stories: Periphery as Center in Qing History." *Frontiers of History in China* 9, no. 3 (2014): 336–60.

———. "The Limits of Tartary: Manchuria in Imperial and National Geographies." *Journal of Asian Studies* 59, no. 3 (August 2000): 603–46.

———. *The Manchu Way: The Eight Banners and Ethnic Identity in Late Imperial China.* Palo Alto, CA: Stanford University Press, 2001.

Elman, Benjamin. *A Cultural History of Civil Examinations in Late Imperial China.* Berkeley: University of California Press, 2000.

———. *Civil Examinations and Meritocracy in Late Imperial China.* Cambridge, MA: Harvard University Press, 2013.

———. *From Philosophy to Philology: Intellectual and Social Aspects of Change in Late Imperial China.* Cambridge, MA: Council on East Asian Studies, 1984.

———. *On Their Own Terms.* Cambridge, MA: Harvard University Press, 2005.

Elvin, Mark. *Another History: Essays on China from a European Perspective.* Broadway, NSW: Wild Peony, 1996.

———. "The Man Who Saw Dragons: Science and Styles of Thinking in Xie Zhaozhe's Fivefold Miscellany." *Journal of the Oriental Society of Australia* 25–26 (1993–1994): 1–41.

Endicott, Stephen Lyon, and Edward Hagerman. *The United States and Biological Warfare: Secrets from the Early Cold War and Korea.* Bloomington: Indiana University Press, 1998.

Endō, Ryūji. *The Canadian and Ordovician Formations and Fossils of South Manchuria.* Washington, DC: US Government Printing Office, 1932.

———. *Genjin hakkutsu: Ichi koseibutsu gakusha no Manshū 25-nen* [Excavating primitive man: A paleontologist's 25 years in Manchuria]. Tokyo: Shunjūsha, 1965.

———. "A New Genus of Thecodontia from the Lycoptera Beds of Manchoukuo." *Bulletin of the Central National Museum of Manchoukuo* 2 (1940): 1–14.

Eyferth, Jan Jacob Karl. *Eating Rice from Bamboo Roots: The Social History of a Community of Handicraft Papermakers in Rural Sichuan, 1920–2000.* Cambridge, MA: Harvard University Asia Center, 2009.

Fan, Fa-ti. *British Naturalists in Qing China: Science, Empire, and Cultural Encounter.* Cambridge, MA: Harvard University Press, 2009.

———. "Science in Cultural Borderlands: Methodological Reflections on the Study of Science, European Imperialism, and Cultural Encounter." *East Asian Science, Technology and Society: An International Journal* 1, no. 2 (December 2007): 213–31. https://doi.org/10.1007/s12280-007-9022-5.

Fang Gongqian. *Helouju ji* [Collected writings from the "what uncouthness?" abode]. Ha'erbin: Heilongjiang daxue chubanshe, 2010.

———. *Ningguta zhi* [Ningguta gazetteer]. Taibei: Guang wen shu ju, 1968.

Fang, Xiaoping. *China and the Cholera Pandemic: Restructuring Society under Mao.* Pittsburgh: University of Pittsburgh Press, 2021.

Fedotova, Anastasia. "The Origins of the Russian Chernozem Soil (Black Earth): Franz Joseph Ruprecht's 'Geo-Botanical Researches into the Chernozem' of 1866." *Environment and History* 16, no. 3 (2010): 271–93.

Field, James. "In Search of Dragons: The Folk Ecology of Fengshui." In *Daoism and Ecology: Ways within a Cosmic Landscape*, edited by N. J. Girardot and James Miller, 285–300. Cambridge, MA: Harvard Divinity School, 2001.

Findlen, Paula. *Possessing Nature: Museums, Collecting, and Scientific Culture in Early Modern Italy.* Berkeley: University of California Press, 1996.

Finnane Antonia. *Changing Clothes in China: Fashion, History, Nation.* New York: Columbia University Press, 2008.

Foot, Rosemary. *A Substitute for Victory: The Politics of Peacemaking at the Korean Armistice Talks.* Ithaca, NY: Cornell University Press, 2018.

Frost, Orcutt. *Bering: The Russian Discovery of America.* New Haven, CT: Yale University Press, 2003.

Fuchs, Walter. *Der Jesuiten-Atlas der Kanghsi-Zeit: Seine Entstehungsgeschichte nebst Namensindices für die Karten der Mandjurei, Mongolei, Ostturkestan und Tibet, mit Wiedergabe der Jesuiten-Karten in Original Grösse.* Peking: Furen Universität, 1943.

Fujiyama Kazuo. *Shin hakubutsukan taisei* [The state of the modern museum]. Shinkyō: Man-Nichi bunka kyōkai, 1940.

Fukuoka, Maki. *The Premise of Fidelity: Science, Visuality, and Representing the Real in Nineteenth-Century Japan.* Stanford, CA: Stanford University Press, 2012.

Funing xian zhi [Gazetteer of Funing county]. Airusheng Zhongguo fangzhi ku, 1682.

Fushun shi renmin zhengfu difangzhi bangongshi. *Qing Yongling zhi* [Gazetteer of the Qing Yongling tomb]. Shenyang: Liaoning minzu chubanshe, 2008.

Gage, Kenneth L. "Factors Affecting the Spread and Maintenance of Plague." *Advances in Experimental Medicine and Biology* 954 (2012): 79–94. https://doi.org/10.1007/978-1-4614-3561-7_11.

Gage, Kenneth L., and Michael Y. Kosoy. "Natural History of the Plague: Perspectives from More Than a Century of Research." *Annual Review of Entomology* 50, no. 1 (January 2005): 505–28. https://doi.org/10.1146/annurev.ento.50.071803.130337.

Gallagher, Sean. "Zhalong Wetland, Heilongjiang." Asia Society: The Wetland Series, Sept. 8, 2011, http://sites.asiasociety.org/chinagreen/threatened-waters/.

Gamsa, Mark. "The Epidemic of Pneumonic Plague in Manchuria 1910–1911." *Past & Present*, no. 190 (2006): 147–83.

Gao Keqin, Susan Evans, Ji Qiang, Mark Norell, and Ji Shu'An. "Exceptional Fossil Material of a Semi-Aquatic Reptile from China: The Resolution of an Enigma." *Journal of Vertebrate Paleontology* 20, no. 3 (September 2000): 417–21. https://doi .org/10.1671/0272–4634(2000)0200417:EFMOAS2.0.CO;2.

Gao Shiqi. *Hucong dongxun rilu* [Daily record of serving in the imperial eastern tour]. Edited by Chen Jianwei. Changchun: Jilin wenshi chuban she, 1986.

———. *Saibei xiaochao* [Miscellaneous notes on the land north of the Great Wall]. Taipei: Guangwen shuju, 1968.

Gao Shuqiao. *Baishan heishui de zunyan* [The majesty of the white mountains and black waters]. Shenyang: Liaoning renmin chuban she, 1995.

Gao Yuehui and Huang Hong. *Beidahuang jingshen* [The spirit of the Great Northern Wasteland]. Beijing: Renmin chubanshe, 2012.

Geniusz, Wendy Makoons. *Our Knowledge Is Not Primitive: Decolonizing Botanical Anishinaabe Teachings.* Syracuse, NY: Syracuse University Press, 2009.

Getty Images. "Japanese Expatriate Farmers Plow and Plant on a Manchukuo Field." https://www.gettyimages.com/detail/video/japanese-expatriate-farmers-plow -and-plant-on-a-manchukuo-news-footage/505935243.

Geurts, Kathryn Linn. *Culture and the Senses: Bodily Ways of Knowing in an African Community.* Berkeley: University of California Press, 2003.

Giersch, C. Patterson. *Asian Borderlands: The Transformation of Qing China's Yunnan Frontier.* Cambridge, MA: Harvard University Press, 2006.

Gold, Hal. *Japan's Infamous Unit 731: Firsthand Accounts of Japan's Wartime Human Experimentation Program.* New York: Tuttle Publishing, 2011.

Golvers, Noël, ed. *Letters of a Peking Jesuit: The Correspondence of Ferdinand Verbiest SJ (1623–1688).* Leuven, Belgium: Ferdinand Verbiest Institute, 2017.

Gomez, Pablo. *The Experiential Caribbean.* Chapel Hill: University of North Carolina Press, 2017.

Goodman, Michael K., Jo Littler, Dan Brockington, and Maxwell Boykoff. "Spectacular Environmentalisms: Media, Knowledge and the Framing of Ecological Politics." *Environmental Communication* 10, no. 6 (November 2016): 677–88. https://doi.org/10.1080/17524032.2016.1219489.

Gordon, Michael. *Scientific Babel: How Science Was Done before and after Global English.* Chicago: University of Chicago Press, 2015.

Gottschang, Thomas R., and Diana Lary. *Swallows and Settlers: The Great Migration from North China to Manchuria.* Ann Arbor, MI: Center for Chinese Studies, University of Michigan, 2000.

Goudkoff, P. P. "Economic Geography of the Coal Resources of Asiatic Russia." *Geographical Review* 13, no. 2 (1923): 283–93. https://doi.org/10.2307/208453.

Gould, Stephen Jay. *Leonardo's Mountain of Clams and the Diet of Worms.* Cambridge, MA: Harvard University Press, 2011.

Grabau, Amadeus William. "Mollusca of North China." *Bulletin of the Geological Survey of China.* 6, no. 2 (1923): 183–97.

———. *Principles of Stratigraphy.* New York: A. G. Seiler, 1913. http://archive.org/ details/principlesofstra00grab.

Grafton, Anthony. *New Worlds, Ancient Texts: The Power of Tradition and the Shock of Discovery*. Cambridge, MA: Harvard University Press, 1992.

Grainger, John D. *The First Pacific War: Britain and Russia, 1854–1856*. Woodbridge, UK: Boydell Press, 2008.

Grayson, James H. *Myths and Legends from Korea: An Annotated Compendium of Ancient and Modern Materials*. Abingdon, UK: Routledge, 2012.

Greenblatt, Stephen. *Marvelous Possessions: The Wonder of the New World*. Chicago: University of Chicago Press, 1992.

Gross, Miriam. *Farewell to the God of Plague: Chairman Mao's Campaign to Deworm China*. Berkeley: University of California Press, 2016.

Guan Yanbo. "Shi lun Dongbei Chaoxianzu de dao zuo" [On the riziculture of the Korean people of northeast China]. *Heilongjiang minzu congkan* 85, no. 2 (2005): 97–102.

Guillemin, Jeanne. *Biological Weapons: From the Invention of State-Sponsored Programs to Contemporary Bioterrorism*. New York: Columbia University Press, 2004.

Guo, Qinghua. "Shenyang: The Manchurian Ideal Capital City and Imperial Palace, 1625–43." *Urban History* 27, no. 3 (2000): 344–59.

Guoying Youyi nongchang zai jianshe zhong [The emergence of the Friendship farm]. Shanghai: Xin zhishi chubanshe, 1955.

"H M Becher | British Museum." https://www.britishmuseum.org/collection/term/ BIOG121089.

Haas, Willliam J. *China Voyager: Gist Gee's Life in Science*. Armonk, NY: M. E. Sharpe, 1996.

Haeussler, Sonja. "Descriptions of the Baekdusan and the Surrounding Area in Russian and German Travel Accounts." *Review of Korean Studies* 13, no. 4 (December 2010): 151–86.

Hafner, Franz, and Klaus Feichtenberger. *Amur—Asia's Amazon*. Terra Factual Studios, 2015. https://www.terramater.at/productions/the-forbidden-river-asias-amazon/.

Han, Xuemei. "The Forest Stand Structures in Northeastern China; Their Potential Effects on the Suitability of Forests for Animals, Plants, and Other Values; and the Possible Relationships to Amur Tiger (*Panthera tigris altaica*) Conservation." PhD diss., Yale University, 2011.

Han, Yong-U., Hwi-Jun An, U.-Song Pae, and Byonghyon Choi. *The Artistry of Early Korean Cartography*. Larkspur, CA: Tamal Vista Publications, 2008.

Hansen, Valerie. *Changing Gods in Medieval China, 1127–1276*. Princeton, NJ: Princeton University Press, 2014.

Hanson, Marta. "Is the 2015 Nobel Prize a Turning Point for Traditional Chinese Medicine?" *The Conversation*. http://theconversation.com/is-the-2015-nobel-prize-a-turning-point-for-traditional-chinese-medicine-48643.

———. "Visualizing the Geography of the Diseases of China: Western Disease Maps from Analytical Tools to Tools of Empire, Sovereignty, and Public Health Propaganda, 1878–1929." *Science in Context* 30, no. 3 (September 2017): 219–80. https://doi.org/10.1017/S0269889717000205.

Harbsmeier, Christoph. "Towards a Conceptual History of Some Concepts of Nature in Classical Chinese: Zi Ran 自然 and Zi Ran Zhi Li 自然之理." In *Concepts of Nature: A Chinese-European Cross-Cultural Perspective*, edited by Hans Ulrich Vogel, 220–54. Leiden: Brill, 2010.

Harris, Sheldon H. *Factories of Death: Japanese Biological Warfare, 1932–1945, and the American Cover-Up*. Revised ed. New York: Routledge, 2002.

Haruyama, Shigeko, and Takayuki Shiraiwa. *Environmental Change and the Social Response in the Amur River Basin*. New York: Springer, 2014.

Harvard University, Class of 1910. *Report*. Cambridge, MA: Harvard University, 1911. http://archive.org/details/4threportclass1910harvuoft.

Heilongjiang sheng difangzhi bianzhuan weiyuanhui. *Heilongjiang sheng zhi: Guoying nongchang zhi* [Heilongjiang province gazetteer: State farm gazetteer]. Ha'erbin Shi: Heilongjiang renmin chubanshe, 1992.

Heilongjiang sheng Ning'an nongchang shi bianshen weiyuanhui. *Ning'an nongchang shi, 1947–1984* [History of Ning'an farm, 1947–1984]. Heilongjiang sheng Ning'an nongchang shi bianshen weiyuanhui, 1985.

Heilongjiang sheng shengchan jianshe dui, zhengzhi bu. *Zhishi qingnian zai Beidahuang* [Educated youth in the Great Northern Wasteland]. Beijing: Renmin meishu chubanshe, 1973.

Heinrich, Ari Larissa. *The Afterlife of Images: Translating the Pathological Body between China and the West*. Durham, NC: Duke University Press, 2008.

Hersey, Mark D., and Jeremy Vetter. "Shared Ground: Between Environmental History and the History of Science." *History of Science* 57, no. 4 (December 1, 2019): 403–40. https://doi.org/10.1177/0073275319851013.

Hershatter, Gail. *The Gender of Memory: Rural Women and China's Collective Past*. Berkeley: University of California Press, 2011.

Heshen. *Qin ding Rehe zhi* [Imperially ordered Rehe gazetteer]. Beijing: Wu ying dian, 1781. Airusheng Zhongguo fangzhi ku.

Hibbett, David, Robert Blanchette, Paul Kenrick, and Benjamin Mills. "Climate, Decay, and the Death of the Coal Forests." *Current Biology* 26, no. 13 (July 2016): R563–67. https://doi.org/10.1016/j.cub.2016.01.014.

Higgins, Andrew. "Russia's Acres, If Not Its Locals, Beckon Chinese Farmers." *New York Times*, July 31, 2016, sec. World. https://www.nytimes.com/2016/08/01/world/asia/russia-china-farmers.html.

Hillier, Amy, and Anne Kelly Knowles, eds. *Placing History: How Maps, Spatial Data, and GIS Are Changing Historical Scholarship*. Redlands, CA: Esri Press, 2008.

Hinchliffe, Steve. "Inhabiting—Landscapes and Natures." In *Handbook of Cultural Geography*, edited by Kay Anderson, 207–25. London: Sage Publications, 2003. https://doi.org/10.4135/9781848608252.

Hinnebusch, B. Joseph. "The Evolution of Flea-Borne Transmission in Yersinia Pestis." *Current Issues in Molecular Biology* 7, no. 2 (July 2005): 197–212.

Hiroki Masaji. "Shōwa hachinen Kitsurin-shō Nōan ni okeru pesuto ryūkō ni tsuite" [The plague epidemic in Nong'an, Jilin province, in the eighth year of the Showa reign]. *Manshū ikagakku zasshi* 21 (1934).

Hong Se-t'ae. "Paektusan gi" [Journey to Mount Paektu], In *Yuha chip*, 9:14b–22b. Database of Korean Classics, 1730. http://db.itkc.or.kr/inLink?DCI=ITKC_MO _0440A_0100_030_0020_2004_A167_XML.

Hoppe, Kirk A. *Lords of the Fly: Sleeping Sickness Control in British East Africa, 1900–1960*. Westport, CT: Greenwood, 2003.

Hoquet, Thierry. "Botanical Authority: Benjamin Delessert's Collections between Travelers and Candolle's Natural Method (1803–1847)." *Isis* 105, no. 3 (2014): 508–39. https://doi.org/10.1086/678169.

Horn, S. "Volatile Emissions during the Eruption of Baitoushan Volcano (China/ North Korea) c. 969 AD [*sic*]." *Journal of Vulcanology* 61, no. 8 (February 2000): 537–55.

Hsu, Elisabeth. "The History of Qing Hao in the Chinese Materia Medica." *Transactions of the Royal Society of Tropical Medicine and Hygiene* 100 (2006): 505–8.

Huachuan xian zhi [Huachuan county gazetteer]. 1928. Airusheng Zhongguo fangzhi ku.

Huang Chengjiang. *Beidahuang zhiqing* [Educated youth of the Great Northern Wasteland]. Beijing: Zhongguo sheying chubanshe, 1998.

Huang, Hongyu. "History, Romance, and Identity: Wu Weiye (1609—1672) and His Literary Legacy." PhD diss., Yale University, 2007.

Huang Jiqing. *Zhongguo xiandai dizhixuejia zhuan* [Biographies of China's modern geologists]. Changsha: Hunan kexue jishu chubanshe, 1990.

Hummel, Arthur W. *Eminent Chinese of the Ch'ing Period (1644–1912)*. Taipei: Ch'eng Wen, 1970.

Hung, Kuang-chi. "Finding Patterns in Nature: Asa Gray's Plant Geography and Collecting Networks (1830s–1860s)." PhD diss., Harvard University, 2013.

———. "'Plants That Remind Me of Home': Collecting, Plant Geography, and a Forgotten Expedition in the Darwinian Revolution." *Journal of the History of Biology* 50, no. 1 (February 2017): 71–132. https://doi.org/10.1007/ s10739-015-9431-6.

Iannaccone, Isaia. "Lo zoo dei gesuiti: La trasmissione scientifica del bestiario rinascimentale europeo alla Cina dei Qing in Kunyu tushuo di Ferdinand Verbiest (1674)." In *Studi in onore di Lionello Lanciotti*, edited by S. Carletti, 2:739–64. Naples: Istituto Universitario Orientale, 1996.

Iijima Wataru. *Mararia to teikoku: Shokuminchi igaku to Higashi Ajia no kōiki chitsujo* [Malaria and empire: Colonial medicine and the regional order of East Asia]. Tokyo: Tōkyō Daigaku shuppankai, 2005.

———. *Pesuto to kindai Chūgoku* [Plague and modern China]. Tokyo: Kenbun Shuppan, 2000.

Imperial Geological Survey of Japan. *A Catalogue of Articles and Analytical Results of the Specimens of Soils Exhibited at the Louisiana Purchase Exposition Held at St. Louis, Missouri, United States of America in 1904*. Tokyo: Imperial Geological Survey of Japan, Department of Agriculture and Commerce, 1904.

Ingold, Tim. *The Perception of the Environment: Essays on Livelihood, Dwelling and Skill*. London: Routledge, 2000.

Ingold, Tim, and Jo Lee. "Fieldwork on Foot: Perceiving, Routing, Socializing." In *Locating the Field: Space, Place and Context in Anthropology*, edited by Simon Coleman and Peter Collins, 67–85. Oxford, UK: Berg, 2006.

Injae, Lee, Owen Miller, Park Jinhoon, and Yi Hyun-Hae. *Korean History in Maps*. Cambridge: Cambridge University Press, 2014.

International Geological Congress. *Report of the XVI Session, United States of America, 1933*. Washington, DC: G. Banta Pub., 1936.

International Scientific Commission for the Investigation of the Facts Concerning Bacterial Warfare in Korea and China. *Report of the International Scientific Commission for the Investigation of the Facts Concerning Bacterial Warfare in Korea and China: With Appendices*. Peking, 1952.

Iriye, Akira. *The Origins of the Second World War in Asia and the Pacific*. London: Routledge, 1987.

Isett, Christopher Mills. *State, Peasant, and Merchant in Qing Manchuria, 1644–1862*. Stanford, CA: Stanford University Press, 2007.

Itō Seizō. *Hōten kyūden kenchiku no kenkyū* [Research on the architecture of the palaces at Fengtian]. N.p., 1920.

Itō Takeo. *Life along the South Manchurian Railway: The Memoirs of Itō Takeo*. Translated and with introduction by Joshua Fogel. Armonk, NY: ME Sharpe, 1988.

Iyasaka-son shi kankō iinkai. *Iyasaka-son shi: Manshū daiichiji kaitakudan no kiroku* [History of Iyasaka village: A record of Manchuria's first settler pioneers]. Tokyo: Seisaku Ātorando, 1986.

Jami, Catherine. *The Emperor's New Mathematics: Western Learning and Imperial Authority during the Kangxi Reign (1662–1722)*. Oxford: Oxford University Press, 2012.

———. "Imperial Control and Western Learning: The Kangxi Emperor's Performance." *Late Imperial China* 23, no. 1 (2002): 28–49.

Janhunen, Juha. *Manchuria: An Ethnic History*. Helsinki: Suomalais-ugrilainen Seura, 1996.

Jannetta, Ann Bowman. *Epidemics and Mortality in Early Modern Japan*. Princeton, NJ: Princeton University Press, 1987.

Ji Peng and Wu Zhiling. "Ding Wenjiang jingying Beipiao meikuang" [Ding Wenjiang's management of the Beipiao mines]. *Window of the Northeast Year*, no. 6 (2013): 32–33.

Jia Hongtu. *Women de gushi: Yibai ge Beidahuang lao zhiqing de rensheng xingtai* [Our stories: The lives of one hundred old educated youth in the Great Northern Wasteland]. Beijing: Zuojia chubanshe, 2008.

Jiang, Yonglin. *The Mandate of Heaven and the Great Ming Code*. Seattle: University of Washington Press, 2011.

Jilin sheng difangzhi bianxiu weiyuanhui. *Jilin sheng zhi: Ziran dili zhi* [Jilin provincial gazetteer: Geography section]. Changchun: Jilin renmin chuban she, 1992.

Jin Dongchun and Cui Zhongxie. "Changbaishan Tianchi huoshan pengfa lishi wenxian jizai de kaojiu" [A study of eruptions of the Changbai Tianchi volcano recorded in historical documents]. *Dili pinglun* [Geological Review] 45, no. 1 (1999): 304–7.

Jin Guanghui. *Zhongguo xin fangzhi zhishi qingnian shangshan xiaxiang shiliao jilu* [Historical resources about sent-down youth in China's new local gazetteers]. Shanghai: Shanghai renmin chubanshe, 2014.

Jin, Jong-Heon. "Paektudaegan: Science and Colonialism, Memory and Mapping in Korean High Places." In *High Places: Cultural Geographies of Mountains, Ice and Science*, edited by Denis Cosgrove and Veronica della Dora, 196–215. London: I. B. Tauris, 2009.

Johnson, Markes. "A. W. Grabau's Embryonic Sequence Stratigraphy and Eustatic Curve." In *Eustasy: The Historical Ups and Downs of a Major Geological Concept*, 43–54. Boulder, CO: Geological Society of America, 1992.

Jöns, Heike. "'Centre of Calculation.'" In *The SAGE Handbook of Geographical Knowledge*, edited by John Agnew and David N. Livingstone, 158–70. London: SAGE, 2011.

Jordheim, Helge, and David Gary Shaw. "Opening Doors: A Turn to Knowledge." *History and Theory* 59, no. 4 (2020): 3–18. https://doi.org/10.1111/hith.12179.

Jørgensen, Dolly, Finn Arne Jørgensen, and Sara B. Pritchard, eds. *New Natures: Joining Environmental History with Science and Technology Studies*. Pittsburgh, PA: University of Pittsburgh Press, 2013.

Jung, Chi-yong. "The People of Joseon's Perception of Baekdusan Viewed through Geographical Materials." *Review of Korean Studies* 13, no. 4 (December 2010): 105–32.

Jung, Dong-min. "Liaoxi: The Contact Zone between the Chinese Dynasty, the Nomadic Tribes, and Goguryeo in Sixth and Seventh Century East Asia." In *Borderlands of China and Korea: Historical Changes in the Contact Zones*, edited by Yong-ku Cha, 49–80. Lanham, MD: Lexington Books, 2020.

Jung, Keun-sik. "China's Memory and Commemoration of the Korean War in the Memorial to Resist America and Aid (North) Korea." *Cross-Currents: East Asian History and Culture Review* 14 (March 2015). https://cross-currents.berkeley.edu/e-journal/issue-14/jung.

Kang, Le, et al. "Grassland Ecosystems in China: Review of Current Knowledge and Research Advancement." *Philosophical Transactions of the Royal Society B: Biological Sciences* 362, no. 1482 (June 2007): 997–1008. https://doi.org/10.1098/rstb.2007.2029.

Kantō tōtokufu minseibu. *Manshū sangyō chōsa shiryō: Kōsan* [Survey of Manchuria's resources: Section on mining]. Dairen: Kantō tōtokufu, 1906.

Kilpatrick, Jane. *Fathers of Botany: The Discovery of Chinese Plants by European Missionaries*. Chicago: University of Chicago Press, 2014.

Kim, Hoi-eun. *Doctors of Empire: Medical and Cultural Encounters between Imperial Germany and Meiji Japan*. Toronto: University of Toronto Press, 2014.

Kim, Loretta Eumie. *Ethnic Chrysalis: China's Orochen People and the Legacy of Qing Borderland Administration*. Leiden: Brill, 2019.

———. "Marginal Constituencies: Qing Borderland Policies and Vernacular Histories of Five Tribes on the Sino-Russian Frontier." PhD diss., Harvard University, 2009.

——. "Saints for Shamans? Culture, Religion and Borderland Politics in Amuria from the Seventeenth to Nineteenth Centuries." *Central Asiatic Journal* 56, nos. 1–2 (2012–2013): 169–201.

Kim, Seonmin. *Ginseng and Borderland: Territorial Boundaries and Political Relations Between Qing China and Chosŏn Korea, 1636–1912.* Berkeley: University of California Press, 2017.

Kim, Yung Sik. *Questioning Science in East Asian Contexts: Essays on Science, Confucianism, and the Comparative History of Science.* Leiden: Brill, 2014.

King, Amy. "Reconstructing China: Japanese Technicians and Industrialization in the Early Years of the People's Republic of China." *Modern Asian Studies* 50, no. 1 (January 2016): 141–74. https://doi.org/10.1017/S0026749X15000074.

King, Ursula. *Spirit of Fire: The Life and Vision of Pierre Teilhard de Chardin.* Maryknoll, NY: Orbis Books, 2015.

Klubock, Thomas Miller. *La Frontera: Forests and Ecological Conflict in Chile's Frontier Territory.* Durham, NC: Duke University Press, 2014.

Knechtges, David. *The Han Rhapsody: A Study of the Fu of Yang Hsiung.* Cambridge: Cambridge University Press, 1976.

Kobayashi, Teiichi. *Geology of Japan.* Tokyo: University of Tokyo Press, 1963.

Koch, Tom. *Cartographies of Disease: Maps, Mapping, and Medicine.* New York: Esri Press, 2005.

——. *Disease Maps: Epidemics on the Ground.* Chicago: University of Chicago Press, 2011.

Köhle, Natalie. "Why Did the Kangxi Emperor Go to Wutai Shan? Patronage, Pilgrimage, and the Place of Tibetan Buddhism at the Early Qing Court," *Late Imperial China* 29, no. 1 (June 2008): 73–119.

Kohler, Robert E. *All Creatures: Naturalists, Collectors, and Biodiversity, 1850–1950.* Princeton, NJ: Princeton University Press, 2006.

——. *Landscapes and Labscapes: Exploring the Lab-Field Border in Biology.* Chicago: University of Chicago Press, 2002.

Komatsu Naomoto. "Manshū ni okeru Nihon no sekiyu tankō" [Japan's oil exploration in Manchuria]. *Sekiyu gijutsu kyōkai shi* 70, no. 3 (2005): 250–58.

Kong Jingwei. *Qing dai Dongbei diqu jingji shi* [Economic history of Northeast China during the Qing]. Haerbin Shi: Heilongjiang renmin chuban she, 1990.

Krykhtin, M. L., and V. G. Svirskii. "Endemic Sturgeons of the Amur River: Kaluga (*Huso dauricus*) and Amur Sturgeon (*Acipenser schrenckii*)." In *Sturgeon Biodiversity and Conservation*, edited by Vadim J. Birstein, 231–39. New York: Springer, 2006.

Kukťo chiri chŏngbowŏn. *The National Atlas of Korea.* Suwon-si, Korea: National Geographic Information Institute, 2009.

Kwon, N. "Disputes about the Holding of Rituals at Baekdusan During King Youngjo's Reign of Joseon Dynasty—Mostly Referring to Opinions of the Participants." *Han'guk Inmulsa yon'gu*, no. 15 (March 2011): 273–301.

Kwon, Nae-hyun. "Changes in Perception of Baekdusan in the Late Joseon Period." *Review of Korean Studies* 13, no. 4 (December 2010): 73–103.

Kwong, Chi Man. *War and Geopolitics in Interwar Manchuria: Zhang Zuolin and the Fengtian Clique during the Northern Expedition*. Leiden: Brill, 2017.

Latour, Bruno. *Reassembling the Social: An Introduction to Actor-Network-Theory*. Oxford: Oxford University Press, 2007.

———. "Visualization and Cognition: Thinking with Eyes and Hands." Edited by H. Kuklick. *Knowledge and Society: Studies in the Sociology of Culture Past and Present* 6 (1986): 1–40.

Lattimore, Owen. *Manchuria: Cradle of Conflict*. New York: Macmillan, 1932.

———. "The Unknown Frontier of Manchuria." *Foreign Affairs* 11, no. 2 (January 1933): 315–30.

Lavelle, Peter B. *The Profits of Nature: Colonial Development and the Quest for Resources in Nineteenth-Century China*. New York: Columbia University Press, 2020.

———. "Tools for Overcoming Crisis: Agriculture, Scarcity, and Ideas of Rural Mechanization in Late Qing China." *Agricultural History* 94, no. 3 (2020): 386–412. https://doi.org/10.3098/ah.2020.094.3.386.

Le Lievre, Audrey. "Carl Johann Maximowicz, 1827–91 Explorer and Plant Collector." *New Plantsman* 4 (1997): 131–42.

Lederer, Susan E. *Subjected to Science: Human Experimentation in America before the Second World War*. Baltimore: Johns Hopkins University Press, 1997.

Ledyard, Gari. "Cartography in Korea." In *Cartography in the Traditional East and Southeast Asian Societies*, edited by J. B. Harley and David Woodward, 2:235–345. Chicago: University of Chicago Press, 1994.

Lee Hun. "Chŏng chʼogi Changpaeksan tʼamsawa hwangjegwon" [The expedition to Mt. Changbai and imperial authority in the early Qing era]. *Tongyang sahak yŏnʼgu* 126 (March 2014): 236–75.

Lee, Hyungdae. "Korean Intellectuals' Perceptions of Baekdusan and the Historical Significance Thereof: Focusing on Travelogues Produced during the 1920s and 1930s." *Review of Korean Studies* 13, no. 4 (December 2010): 53–71.

Lee, Kyungsoon, and Se-Woong Koo. "The Confucian Transformation of Mountain Space: Travels by Late-Chosŏn Confucian Scholars and the Attempted Confucianization of Mountains." *Journal of Korean Religions* 5, no. 2 (October 2014): 119–43.

Lee, Robert H. G. *The Manchurian Frontier in Chʼing History*. Cambridge, MA: Harvard University Press, 1970.

Legge, James. *The Chinese Classics*. Vol. 1. Philadelphia: J. B. Lippincott & Co., 1867.

Legros, Jean-Paul. *Major Soil Groups of the World: Ecology, Genesis, Properties and Classification*. Boca Raton, FL: CRC Press, 2012.

Lei, Sean Hsiang-lin. "Sovereignty and the Microscope: Constituting Notifiable Infectious Disease and Containing the Manchurian Plague." In *Health and Hygiene in Chinese East Asia: Policies and Publics in the Long Twentieth Century*, edited by Angela Ki Che Liang and Charlotte Furth, 73–106. Durham NC: Duke University Press, 2010.

Lensen, George Alexander. *Russia's Japan Expedition of 1852 to 1855*. Gainesville: University of Florida Press, 1955.

Leopold, Aldo. *A Sand County Almanac and Sketches Here and There*. 2nd ed. New York: Oxford University Press, 1968.

Levine, Neil. "Questioning the View: Seaside's Critique of the Gaze of Modern Architecture." In *Seaside: Making a Town in America*, edited by David Mohney and Keller Easterling, 240–59. New York: Princeton Architectural Press, 1996.

Levine, Stephen I. *Anvil of Victory: The Communist Revolution in Manchuria, 1945–1948*. New York: Columbia University Press, 1987.

Lewis, Mark Edward. "The Feng and Shan Sacrifices of Emperor Wu of the Han." In *State and Court Ritual in China*, edited by Michael Loewe and Edward Shaughnessy, 50–80. Cambridge: Cambridge University Press, 1999.

Li Di, ed. *Kangxi jixia gewu bianyi zhu* [The Kangxi emperor's "scientific essays written in leisure," translated and annotated]. Shanghai: Shanghai guji chubanshe: 1993.

Li Honghe. "Dongbei jiefangqu de shuyi liuxing ji jiuzhi" [The outbreak and control of plague in the Northeastern liberated areas]. *Zhonggong dangshi yanjiu*, no. 3 (2007): 111–18.

Li Huazi. "Ming-Qing shiqi Zhong-Chao dilizhi dui Changbaishan ji shuixi de jishu" [Descriptions of the riverine system of Mount Changbai in Chinese and Korean geographical writings from the Ming and Qing periods]. In *Zhongguo Chaoxianzu shi yanjiu, 2008*, edited by Huang Youfu. Beijing: Minzu chubanshe, 2009.

Li, Shang-Jen. "Natural History of Parasitic Disease: Patrick Manson's Philosophical Method." *Isis* 93, no. 2 (June 2002): 206–28. https://doi.org/10.1086/344961.

Li Shizhen. *Bencao gangmu* [Systematic materia medica]. Beijing: Zhongyi guji chubanshe, 1994.

Li, Wai-yee. "History and Memory in Wu Weiye's Poetry." In *Trauma and Transcendence in Early Qing Literature*, edited by W. L. Idema and Ellen Widmer, 99–148. Cambridge, MA: Harvard University Asia Center, 2006.

Li Xingsheng. *Dongbei liuren shi* [History of exiles to the Northeast]. Harbin: Heilongjiang renmin chubanshe, 1990.

———. *Jiangnan caizi saibei mingren Wu Zhaoqian zhuan* [Biography of Wu Zhaoqian, genius of the southlands, celebrity of the northlands]. Harbin: Heilongjiang renmin chubanshe, 2000.

———. *Wu Zhaoqian nianpu* [Chronological biography of Wu Zhaoqian]. Harbin: Heilongjiang daxue chubanshe, 2014.

Li, Yanjun, and Erhei Dai. "Different Region Analysis for Genotyping *Yersinia pestis* Isolates from China." *PLOS One* 3, no. 5 (May 2008): e2166. https://doi.org/10.1371/journal.pone.0002166.

Li, Yujia. "The Role of the Museum in the Rise of Modern Nationalism in East Asia: Competing Identity Politics in Shaping Historical Memory, 'Manchukuo' (1932–1945) as a Case Study." Master's thesis, Leiden University, 2019.

Liang Bo, and Feng Wei. "Mantie dizhi diaochasuo" [The SMRC geographical survey institute]. *Kexue xue yanjiu* 20, no. 3 (2002): 251–55.

Liang Xiaodong. "Kangxi, Qianlong huangdi dui huashi de yanjiu ji huashi wenhua" [Kangxi and Qianlong's research into fossils and fossil culture]. *Journal of Bohai University: Philosophy and Social Science Edition*, no. 5 (2012): 87–90.

Liaoning sheng dang'anguan, ed. *Qingdai Sanxing fu dutong yamen Man-Han wen dang'an xuanbian* [Selected Manchu and Chinese archival documents from the Sanxing lieutenant-general yamen]. Shenyang: Liaoning guji chubanshe, 1995.

———, ed. *Sanxing fu dutong yamen Manwen dang'an yibian* [Translated documents from the Sanxing lieutenant-general yamen's Manchu language archives]. Shenyang: Liaoning sheng Xinhua shudian, 1984.

Liu, Dianwei. "Land Use/Cover Changes and Environmental Consequences in Songnen Plain, Northeast China." *Chinese Geographical Science* 19, no. 4 (October 2009): 299. https://doi.org/10.1007/s11769-009-0299-2.

Liu Jiping. "1954–2010 nian Sanjiang pingyuan tudi liyong jingguan geju dongtai bianhua ji qudong li" [Land use, landscape pattern dynamics and driving force analysis on the Sanjiang plain from 1954 to 2010]. *Acta Ecologica Sinica (Shengtai xuebao)* 34, no. 12 (June 2014): 3234–44. https://doi.org/10.5846/stxb201306101639.

Liu, Jiping, et al. "A Dynamic Change Map of Marshes in the Small Sanjiang Plain, Heilongjiang, China, from 1955 to 2005." *Wetlands Ecological Management* 23 (2015): 419–37.

Liu Jiyou. *Xiaomie shuhai* [The eradication of the plague]. Hohot: Nei Menggu renmin chubanshe, 1983.

Liu, Michael Shiyung. *Prescribing Colonization: The Role of Medical Practices and Policies in Japan-Ruled Taiwan, 1895–1945*. Ann Arbor, MI: Association for Asian Studies, 2009.

Liu, Shiyung. "The Ripples of Rivalry: The Spread of Modern Medicine from Japan to Its Colonies." *East Asian Science, Technology and Society* 2, no. 1 (March 2008): 47–71. https://doi.org/10.1215/s12280-008-9030-0.

Liu, Xiaohui. "Characterizing the Spatial Pattern of Marshlands in the Sanjiang Plain, Northeast China." *Ecological Engineering* 53 (April 2013): 335–42. https://doi.org/10.1016/j.ecoleng.2012.12.071.

Liu, Yishi. "Competing Visions of the Modern Urban Transformation and Social Change of Changchun, 1932–1957." PhD diss., University of California, Berkeley, 2011. http://digitalassets.lib.berkeley.edu/etd/ucb/text/Liu_berkeley_0028E_12048.pdf.

Livingstone, David. *Putting Science in Its Place: Geographies of Scientific Knowledge*. Chicago: University of Chicago Press, 2003.

Loewe, Michael, and Edward L. Shaughnessy, eds. *The Cambridge History of Ancient China: From the Origins of Civilization to 221 BC*. Cambridge: Cambridge University Press, 1999.

Lomolino, Mark, Dov Sax, and James Brown, eds. *Foundations of Biogeography: Classic Papers with Commentaries*. Chicago: University of Chicago Press, 2004.

Low, Morris. *Japan on Display: Photography and the Emperor*. New York: Routledge, 2006.

Lü Xiulian and Zhao Kunyu. *Hezhezu nüxing lishi wenhua yanjiu* [Studies on the history and culture of Hezhe women]. Harbin: Heilongjiang renmin chubanshe, 2012.

Lu, Zhong-jun. "Rice Cultivation Changes and Its Relationships with Geographical Factors in Heilongjiang Province, China." *Journal of Integrative Agriculture* 16, no. 10 (October 2017): 2274–82. https://doi.org/10.1016/S2095-3119(17)61705-2.

Luebke, Frederick C. *Bonds of Loyalty: German-Americans and World War I*. DeKalb: Northern Illinois University Press, 1974.

Lynteris, Christos. *Ethnographic Plague: Configuring Disease on the Chinese-Russian Frontier*. New York: Springer, 2016.

———. "Plague Masks: The Visual Emergence of Anti-Epidemic Personal Protection Equipment." *Medical Anthropology* 37, no. 6 (August 2018): 442–57. https://doi.org/10.1080/01459740.2017.1423072.

Ma Baojian. "Qingdai zhi minguo shiqi Heilongjiang Sanjiangpingyuan senlin bianqian yanjiu." *Beijing linye daxue xuebao* 7, no. 4 (2008): 11–16.

Ma Menglong. "Mudekeng chabian yu Huangyu quanlantu bianhui—liandui Mukedeng 'shenshipai' chuli weizhi de kaobian [Mukedeng's survey of the border and the compiling of the "Overview Map of the Imperial Realm"—along with a consideration of the location of Mukedeng's border-marking stele]. *Zhongguo bianjiang shidi yanjiu* 19, no. 3 (2009): 85–99.

Marcon, Federico. "The Critical Promises of the History of Knowledge: Perspectives from East Asian Studies." *History and Theory* 59, no. 4 (2020): 19–47. https://doi.org/10.1111/hith.12180.

———. *The Knowledge of Nature and the Nature of Knowledge in Early Modern Japan*. Chicago: University of Chicago Press, 2015.

Mason, David. *Spirit of the Mountains*. Elizabeth, NJ: Hollym International, 1999.

Matsuda Kamezō. *Mantetsu chishitsu chōsajo shiki* [A personal record of the SMRC geological survey institute]. Tokyo: Hakueisha, 1990.

Matsumura Takao. *"Ronsō" 731 Butai* [The controversy of Unit 731]. Shohan. Tokyo: Banseisha, 1994.

———. "Shinkyō, No-an pesuto ryūkō (1940-nen) to 731 butai" [The 1940 outbreak of plague in Changchun and Nong'an and Unit 731]. *Mita gakkai zasshi* 95, no. 4 (January 2003): 647–68.

Matsusaka, Yoshihisa Tak. *The Making of Japanese Manchuria, 1904–1932*. Cambridge, MA: Harvard University Asia Center, 2001.

Matsuura, Shigeru. *Shinchō no Amūru seisaku to shōsū minzoku* [Qing policy in the Amur and ethnic minorities]. Kyoto: Kyōto Daigaku gakujutsu shuppankai, 2006.

———. "The Qing Surveys of the Left Bank of the Amur after the Conclusion of the Treaty of Nerchinsk." *Memoirs of the Toyo Bunko* 68 (2010).

Matthiessen, Peter. *Tigers in the Snow*. New York: North Point Press, 2000.

Maximowicz, Karl Johann. *Primitiae florae amurensis: Versuch einer Flora des Amur-Landes*. St. Petersburg: Kaiserlichen Akademie der Wissenschaften, 1859.

Mayor, Adrienne. *The First Fossil Hunters: Dinosaurs, Mammoths, and Myth in Greek and Roman Times*. Princeton, NJ: Princeton University Press, 2011.

Mazur, Allan. *A Romance in Natural History: The Lives and Works of Amadeus Grabau and Mary Antin*. Syracuse, NY: Garret, 2004.

McDonald, Kate. *Placing Empire: Travel and the Social Imagination in Imperial Japan*. Berkeley: University of California Press, 2017.

"Mechanical Farming." *Contemporary Manchuria* 3, no. 4 (October 1939): 42–62.

Mercier, Laurie. *Anaconda: Labor, Community, and Culture in Montana's Smelter City*. Urbana: University of Illinois Press, 2001.

Métailié, Georges. "Plantes et noms, plantes sans nom dans le 'Zhiwu Mingshi Tukao.'" *Extrême-Orient Extrême-Occident*, no. 15 (1993): 138–48.

———. *Science and Civilisation in China*. Vol. 6, *Biology and Biological Technology*, part 4, *Traditional Botany: An Ethnobotanical Approach*. Cambridge: Cambridge University Press, 2015.

Middendorff, Aleksandr Fëdorovič. *Middendorff's Reise in den äusserten Norden und Osten Sibiriens*. St. Petersburg: Kaiserlichen Akademie der Wissenschaften, 1848.

Miller, Ian Jared, Julia Adeney Thomas, and Brett L. Walker, eds. *Japan at Nature's Edge: The Environmental Context of a Global Power*. Honolulu: University of Hawai'i Press, 2013.

Miller, Ian Jared, and Paul Warde. "Energy Transitions as Environmental Events." *Environmental History* 24, no. 3 (July 2019): 464–69.

Millward, James A. "'Coming onto the Map': 'Western Regions,' Geography and Cartographic Nomenclature in the Making of Chinese Empire in Xinjiang." *Late Imperial China* 20, no. 2 (1999): 61–98.

Min, Jung. "The Shadow of Anonymity: The Depiction of Northerners in Eighteenth-Century 'Hearsay Accounts' (Kimun)." In *The Northern Region of Korea: History, Identity & Culture*, edited by Sun Joo Kim, 93–115. Seattle: University of Washington Press, 2010.

Minami Manshū tetsudō kabushiki kaisha chishitsu chōsajo. *Manshū nanseibu no chishitsu oyobi chishi* [Geology of southwestern Manchukuo, along with a geological gazetteer]. Dairen: Minami Manshū tetsudō kabushiki kaisha chishitsu chōsajo, 1937

Minami Manshū tetsudō kabushiki kaisha eiseika. *Kōtoku ni-nendo pesuto bōeki gaiyō* [Overview of anti-plague measures in the second year of Kangde]. Dairen: Mantetsu chihōbu eiseika, 1936.

Minami Manshū tetsudō kabushiki kaisha eiseika. *Manshū fūdo eisei kenkyū gaiyō* [Overview of local hygiene research in Manchukuo]. Dairen: Mantetsu chihōbu eiseika, 1936.

Minzu wenti wuzhong congshu Heilongjiang sheng bianji zu. *Hezhezu shehui lishi diaocha* [Investigation into the history of Hezhe society]. Beijing: Minzu chubanshe, 2009.

Mitman, Gregg. "Living in a Material World." *Journal of American History* 100, no. 1 (June 2013): 128–30. https://doi.org/10.1093/jahist/jat081.

Mitsuhashi, Hiroshi. "Medicinal Plants of the Ainu." *Economic Botany* 30 (September 1976): 209–17.

Mitter, Rana. *The Manchurian Myth: Nationalism, Resistance, and Collaboration in Modern China.* Berkeley: University of California Press, 2000.

Mizuno, Hiromi. *Science for the Empire: Scientific Nationalism in Modern Japan.* Stanford, CA: Stanford University Press, 2008.

Mizuno, Hiromi, Aaron S. Moore, and John DiMoia, eds. *Engineering Asia: Technology, Colonial Development, and the Cold War Order.* London: Bloomsbury Publishing, 2018.

Moore, Aaron S. *Constructing East Asia: Technology, Ideology, and Empire in Japan's Wartime Era, 1931–1945.* Stanford, CA: Stanford University Press, 2013.

Moore, Gregory. *Defining and Defending the Open Door Policy: Theodore Roosevelt and China, 1901–1909.* New York: Lexington Books, 2015.

Morris-Suzuki, Tessa. *To the Diamond Mountains: A Hundred-Year Journey through China and Korea.* New York: Rowman & Littlefield Publishers, 2010.

Morse, Kathryn Taylor. *The Nature of Gold: An Environmental History of the Klondike Gold Rush.* Seattle: University of Washington Press, 2003.

Mosca, Matthew. "Empire and the Circulation of Frontier Intelligence: Qing Conceptions of the Ottomans." *Harvard Journal of Asiatic Studies* 70, no. 1 (2010): 147–207.

———. *From Frontier Policy to Foreign Policy: The Question of India and the Transformation of Geopolitics in Qing China.* Stanford, CA: Stanford University Press, 2013.

Mueggler, Erik. *The Paper Road: Archive and Experience in the Botanical Exploration of West China and Tibet.* Berkeley: University of California Press, 2011.

Mullaney, Thomas S. *Coming to Terms with the Nation.* Berkeley: University of California Press, 2010.

Muscolino, Micah S. *The Ecology of War in China: Henan Province, the Yellow River, and Beyond, 1938–1950.* Cambridge: Cambridge University Press, 2014.

———. "Refugees, Land Reclamation, and Militarized Landscapes in Wartime China: Huanglongshan, Shaanxi, 1937–45." *Journal of Asian Studies* 69, no. 2 (May 2010): 453–78.

Nagoya hakubutsukan. *Shin hakubutsukan taisei* [The state of the modern museum]. Nagoya: Nagoya-shi Hakubutsukan, 1995.

Nappi, Carla. "Surface Tension: Objectifying Ginseng in Chinese Early Modernity." In *Early Modern Things: Objects and Their Histories, 1500–1800*, edited by Paula Findlen, 31–52. New York: Routledge, 2012.

———. *The Monkey and the Inkpot: Natural History and Its Transformations in Early Modern China.* Cambridge, MA: Harvard University Press, 2009.

———. "Winter Worm, Summer Grass: Cordyceps, Colonial Chinese Medicine, and the Formation of Historical Objects." In *Crossing Colonial Historiographies: Histories of Colonial and Indigenous Medicines in Transnational Perspective*, edited by Anne Digby, Waltraud Ernst, and Projit B. Muhkarji, 21–35. Newcastle-upon-Tyne: Cambridge Scholars, 2010.

Naquin, Susan. "The Material Manifestations of Regional Culture." *Journal of Chinese History* 3, no. 2 (July 2019): 363–79. https://doi.org/10.1017/jch.2018.35.

Nash, Linda. *Inescapable Ecologies: A History of Environment, Disease and Knowledge.* Berkeley: University of California Press, 2006.

———. "The Agency of Nature or the Nature of Agency?" *Environmental History* 10, no. 1 (2005): 67–69.

Nathan, Carl F. *Plague Prevention and Politics in Manchuria, 1910–1931.* Cambridge, MA: East Asian Research Center, Harvard University, 1967.

Needham, Joseph. *Science and Civilisation in China: Volume 3, Mathematics and the Sciences of the Heavens and the Earth.* Cambridge: Cambridge University Press, 1959.

Needham, Joseph, and Gwei-Djen Lu. *Science and Civilisation in China.* Volume 6, *Biology and Biological Technology*, part 1, *Botany.* Cambridge: Cambridge University Press, 1986.

Nie, Jing Bao, Nanyan Guo, Mark Selden, and Arthur Kleinman. *Japan's Wartime Medical Atrocities: Comparative Inquiries in Science, History, and Ethics.* Abingdon, UK: Routledge, 2013.

Nongye dianyingshe. *Guoying Youyi nongchang* [The state-run Friendship farm]. Shanghai: Shanghai renmin meishu chubanshe, 1957.

O'Dwyer, Emer. *Significant Soil: Settler Colonialism and Japan's Urban Empire in Manchuria.* Cambridge, MA: Harvard University Asia Center, 2015.

Oh, Youngchan, and Mark E. Byington. "Scholarly Studies on the Han Commanderies in Korea." In *The Han Commanderies in Early Korean History*, edited by Mark E. Byington, 18–26. Cambridge, MA: Korea Institute, Harvard University, 2013.

Olšáková, Doubravka. "Pugwash in Eastern Europe: The Limits of International Cooperation under Soviet Control in the 1950s and 1960s." *Journal of Cold War Studies* 20, no. 1 (April 2018): 210–40.

Olwig, Kenneth. *Landscape, Nature, and the Body Politic: From Britain's Renaissance to America's New World.* Madison: University of Wisconsin Press, 2002.

Osterhammel, Jürgen. *The Transformation of the World: A Global History of the Nineteenth Century.* Translated by Patrick Camiller. Reprint; Princeton, NJ: Princeton University Press, 2015.

Otis, Laura. *Müller's Lab: The Story of Jakob Henle, Theodor Schwann, Emil Du Bois-Reymond, Hermann von Helmholtz, Rudolf Virchow, Robert Remak, Ernst Haeckel, and Their Brilliant, Tormented Advisor.* Oxford: Oxford University Press, 2007.

Outram, Dorinda. "New Spaces in Natural History." In *Cultures of Natural History*, edited by Nicolas Jardine, James Secord, and Emma Spary, 249–65. Cambridge: Cambridge University Press, 1996.

Overmyer, Daniel. *Local Religion in North China in the Twentieth Century: The Structure and Organization of Community Rituals and Beliefs.* Leiden: Brill, 2009.

Owen, Stephen. *Remembrances: The Experience of the Past in Classical Chinese Literature.* Cambridge, MA: Harvard University Press, 1986.

Pai, Hyung Il. *Constructing "Korean" Origins: A Critical Review of Archaeology, Historiography, and Racial Myth in Korean State-Formation Theories.* Cambridge, MA: Harvard University Asia Center, 2000.

Paine, S. C. M. *Imperial Rivals: China, Russia, and Their Disputed Frontier.* Armonk, NY: M. E. Sharpe, 1996.

———. *The Sino-Japanese War of 1894–1895: Perceptions, Power, and Primacy.* Cambridge: Cambridge University Press, 2002.

Palmer, Morgan. "Colonising in Manchuria." *The China Weekly Review*, November 1, 1924, 270–72.

Pankenier, David W. *Astrology and Cosmology in Early China: Conforming Earth to Heaven.* Cambridge: Cambridge University Press, 2013.

Park, Hyun Ok. *Two Dreams in One Bed: Empire, Social Life, and the Origins of the North Korean Revolution in Manchuria.* Durham, NC: Duke University Press, 2005.

Peattie, Mark R. *Ishiwara Kanji and Japan's Confrontation with the West.* Princeton, NJ: Princeton University Press, 1975.

Pechous, Roger D., Vijay Sivaraman, Nikolas M. Stasulli, and William E. Goldman. "Pneumonic Plague: The Darker Side of Yersinia Pestis." *Trends in Microbiology* 24, no. 3 (March 2016): 190–97. https://doi.org/10.1016/j.tim.2015.11.008.

Peckham, Robert. "Matshed Laboratory: Colonies, Cultures, and Bacteriology." In *Imperial Contagions: Medicine, Hygiene, and Cultures of Planning in Asia*, edited by Robert Peckham and David Pomfret, 123–47. Hong Kong: Hong Kong University Press, 2013.

Perdue, Peter. "Boundaries, Maps, and Movement: Chinese, Russian, and Mongolian Empires in Early Modern Central Eurasia." *International History Review* 20, no. 2 (June 1998): 263–86.

———. *China Marches West: The Qing Conquest of Central Eurasia.* Cambridge, MA: Belknap Press of Harvard University Press, 2005.

———. "Chinese Exploration." In *The Oxford Companion to World Exploration*, edited by David Buisseret, 187–91. Oxford: Oxford University Press, 2007.

———. "Crossing Borders in Imperial China." In *Asia Inside Out: Connected Places*, edited by Eric Tagliacozzo, Helen F. Siu, and Peter Perdue, 195–218. Cambridge, MA: Harvard University Press, 2015.

Perrin, William, et al. *Encyclopedia of Marine Mammals.* New York: Elsevier, 2008.

Perrins, Robert. "Doctors, Disease and Development: Engineering Colonial Public Health in Southern Manchuria, 1905–1931." In *Building a Modern Japan: Science, Technology, and Medicine in the Meiji Era and Beyond*, edited by Morris Low, 103–32. New York: Palgrave Macmillan, 2005.

Peterson, Willard. *Bitter Gourd: Fang I-Chih and the Impetus for Intellectual Change.* New Haven, CT: Yale University Press, 1979.

Pomeranz, Kenneth. "Water to Iron, Widows to Warlords: The Handan Rain Shrine in Modern Chinese History." *Late Imperial China* 12, no. 1 (1991): 62–99.

Pratt, Mary Louise. *Imperial Eyes: Travel Writing and Transculturation.* 2nd ed. Abingdon, UK: Routledge, 2007.

Prieto, Andrés I. "Classification, Memory, and Subjectivity in Gonzalo Fernández de Oviedo's Sumario de La Natural Historia (1526)." *MLN* 124, no. 2 (April 2009): 329–49. https://doi.org/10.1353/mln.0.0111.

Prince, Hugh C. *Wetlands of the American Midwest: A Historical Geography of Changing Attitudes.* Chicago: University of Chicago Press, 1997.

Pritchard, Sara B. *Confluence: The Nature of Technology and the Remaking of the Rhône.* Cambridge, MA: Harvard University Press, 2011.

Pulford, Ed. "Material States: China, Russia, and the Incorporation of a Cross-Border Indigenous People." *Modern Asian Studies* 55, no. 1 (March, 2020): 292–333.

———. *Mirrorlands: Russia, China, and Journeys In Between.* London: Hurst & Co., 2019.

Qiang, Ji, Philip J. Currie, Mark A. Norell, and Ji Shu-An. "Two Feathered Dinosaurs from Northeastern China." *Nature* 393, no. 6687 (June 25, 1998): 753–61.

Qin, Boqiang. *Lake Taihu, China: Dynamics and Environmental Change.* New York: Springer, 2008.

Qinhuangdao difangzhi bangongshi. *Qinhuangdao shi zhi* [Gazetteer of Qinhuangdao city]. Beijing: Fangzhi chubanshe, 2009.

Quan Yuedong. "Qiantan gujin Qing Yongling zhi cha yi" [Discussion of the differences between the past and present Qing Yongling tomb]. *Liaoning sheng bowuguan qikan* (2014): 264–73.

Raj, Kapil. *Relocating Modern Science: Circulation and the Construction of Knowledge in South Asia and Europe, 1650–1900.* New York: Palgrave Macmillan, 2007.

Rauchway, Eric. "Willard Straight and the Paradox of Liberal Imperialism." *Pacific Historical Review* 66, no. 3 (1997): 363–97. https://doi.org/10.2307/3640202.

Rawski, Evelyn S. *Early Modern China and Northeast Asia: Cross-Border Perspectives.* New York: Cambridge University Press, 2015.

———. *The Last Emperors: A Social History of Qing Imperial Institutions.* Berkeley: University of California Press, 2001.

Reardon Anderson, James. *Reluctant Pioneers: China's Expansion Northward, 1644–1937.* Stanford, CA: Stanford University Press, 2005.

Renne, P. R. "40Ar/39Ar Dating into the Historical Realm: Calibration against Pliny the Younger." *Science* 277, no. 5330 (August 29, 1997): 1279–80. https://doi.org/10.1126/science.277.5330.1279.

Richthofen, Ferdinand von. *China: Ergebnisse eigener reisen und darauf gegründeter studien.* Vol. 2. 5 vols. Berlin: D. Reimer, 1882.

Rieppel, Lukas. *Assembling the Dinosaur: Fossil Hunters, Tycoons, and the Making of a Spectacle.* Cambridge, MA: Harvard University Press, 2019.

Ripa, Matteo. *Giornale (1705–1724).* Naples: Instituto Universitario Orientale, 1990.

Ritvo, Harriet. "Zoological Nomenclature and the Empire of Victorian Science." In *Victorian Science in Context,* edited by Bernard Lightman, 334–53. Chicago: University of Chicago Press, 1997.

Roberts, Priscilla. "Willard Straight, The First World War, and 'Internationalism of All Sorts': The Inconsistencies of An American Liberal Interventionist." *Austra-*

lian Journal of Politics & History 44, no. 4 (1998): 493–511. https://doi.org/10 .1111/1467–8497.00033.

Robinet, Isabelle. *The World Upside Down: Essays on Taoist Internal Alchemy*. Mountain View, CA: Golden Elixir Press, 2011.

Robinson, David. "Chinese Border Garrisons in an International Context: Liaodong under the Early Ming Dynasty." In *Chinese and Indian Warfare—From the Classical Age to 1870*, edited by Kaushik Roy and Peter Lorge, 57–73. Abingdon, UK: Routledge, 2014.

Robinson, Kenneth R. "From Raiders to Traders: Border Security and Border Control in Early Chosŏn, 1392–1450." *Korean Studies* 16 (1992): 94–115.

Robson, James. *Power of Place: The Religious Landscape of the Southern Sacred Peak (Nanyue) in Medieval China*. Cambridge, MA: Harvard University Asia Center, 2009.

Rogaski, Ruth. *Hygienic Modernity: Meanings of Health and Disease in Treaty-Port China*. Berkeley: University of California Press, 2004.

———. "Knowing a Sentient Mountain: Space, Science, and the Sacred in Ascents of Mount Paektu/Changbai." *Modern Asian Studies* 52, no. 2 (March 2018): 716–52. https://doi.org/10.1017/S0026749X17001081.

———. "The Manchurian Plague and COVID-19: China, the United States, and the 'Sick Man,' Then and Now." *American Journal of Public Health* 111, no. 3 (March 2021): 423–29. https://doi.org/10.2105/AJPH.2020.305960.

———. "Nature, Annihilation, and Modernity: China's Korean War Germ-Warfare Experience Reconsidered." *Journal of Asian Studies* 61, no. 02 (May 2002): 381–415.

Rogers, Thomas D. *Deepest Wounds: A Labor and Environmental History of Sugar in Northeast Brazil*. Chapel Hill: University of North Carolina Press, 2010.

Rohlf, Gregory. "The Soviet Model and China's State Farms." In *China Learns from the Soviet Union, 1949–Present*, edited by Thomas P. Bernstein and Hua-Yu Li, 197–228. Lanham, MD: Rowman & Littlefield, 2010.

Rojas, Carlos. *The Great Wall*. Cambridge, MA: Harvard University Press, 2011.

Ru Yan. *Guoying Youyi nongchang* [State-run Friendship farm]. Beijing: Zhongguo qingnian chubanshe, 1956.

Rudwick, Martin. *Georges Cuvier, Fossil Bones and Geological Catastrophes: New Translations and Interpretations of the Primary Texts*. Chicago: University of Chicago Press, 1997.

Ruprecht, Franz Joseph. "Die ersten botanischen Nachrichten uber das Amurland: Beobachtungen von C. Maximowitsch, redigirt von Akademiker Ruprecht." *Archiv für wissenschaftliche Kunde von Russland* 17 (1858): 104–44.

Sahlins, Peter. *Boundaries: The Making of France and Spain in the Pyrenees*. Berkeley: University of California Press, 1989.

Samuelson, L. *Tankograd: The Formation of a Soviet Company Town: Cheliabinsk, 1900s–1950s*. New York: Palgrave Macmillan, 2011.

Sárközi, Ildikó Gyöngyvér. *From the Mists of Martyrdom: Sibe Ancestors and Heroes on the Altar of Chinese Nation-Building*. Münster: LIT Verlag, 2018.

Sauvage, M. H. E. "Sur un Prolebias (*Prolebias davidi*) des terrains tertiaries du nord de la Chine." *Bulletin de la Société géologique de France* Series 3 (1880): 452–54.

Scarth, Alwyn. *Vesuvius: A Biography*. Princeton, NJ: Princeton University Press, 2009.

Schäfer, Dagmar. *The Crafting of the 10,000 Things: Knowledge and Technology in Seventeenth-Century China*. Chicago: University of Chicago Press, 2015.

Schafer, Edward. *The Vermilion Bird: T'ang Images of the South*. Berkeley: University of California Press, 1967.

Schama, Simon. *Landscape and Memory*. New York: A. A. Knopf, 1995.

Scherer, Anke. "Japanese Emigration to Manchuria: Local Activists and the Making of the Village-Division Campaign." PhD diss., Ruhr Universität, 2006. http://www-brs.ub.ruhr-uni-bochum.de/netahtml/HSS/Diss/SchererAnke/diss.pdf.

Schiebinger, Londa L. *Plants and Empire*. Cambridge, MA: Harvard University Press, 2009.

Schlesinger, Jonathan. *A World Trimmed with Fur: Wild Things, Pristine Places, and the Natural Fringes of Qing Rule*. Stanford, CA: Stanford University Press, 2017.

Schmalzer, Sigrid. "Layer upon Layer: Mao-Era History and the Construction of China's Agricultural Heritage," 2019. https://doi.org/10.1215/18752160-7498416.

———. *The People's Peking Man: Popular Science and Human Identity in Twentieth-Century China*. Chicago: University of Chicago Press, 2008.

———. *Red Revolution, Green Revolution: Scientific Farming in Socialist China*. Chicago: University of Chicago Press, 2016.

Schmid, Andre. *Korea between Empires, 1895–1919*. Studies of the East Asian Institute. New York: Columbia University Press, 2002.

———. "Tributary Relations and the Qing-Chosŏn Frontier on Mount Paektu." In *The Chinese State at the Borders*, edited by Diana Lary, 126–50. Vancouver: University of British Columbia Press, 2007.

Schmidt, Benjamin. *Inventing Exoticism: Geography, Globalism, and Europe's Early Modern World*. Philadelphia: University of Pennsylvania Press, 2015.

Seow, Victor. *Carbon Technocracy: Energy Regimes in Modern East Asia*. Chicago: University of Chicago Press, 2022.

———. "Sites of Extraction: Perspectives from a Japanese Coal Mine in Northeast China." *Environmental History* 24, no. 3 (July 2019): 504–11. https://doi.org/10.1093/envhis/emz006.

Sepe, Agostino. "Back to the Roots: The Imperial City of Shenyang as a Symbol of the Manchu Ethnic Identity of the Qing Dynasty." *Ming Qing Yanjiu* 16, no. 1 (February 2011): 129–76.

Seth, Suman. "Putting Knowledge in Its Place: Science, Colonialism, and the Postcolonial." *Postcolonial Studies* 12, no. 4 (December 2009): 373–88.

Setton, Mark. *Chŏng Yagyong: Korea's Challenge to Orthodox Neo-Confucianism*. Albany: State University of New York Press, 1997.

Sewell, William Shaw. *Constructing Empire: The Japanese in Changchun, 1905–45*. Vancouver: University of British Columbia Press, 2018.

Shan, Patrick Fuliang. *Taming China's Wilderness: Immigration, Settlement and the*

Shaping of the Heilongjiang Frontier, 1900–1931. Farnham, UK: Ashgate Publishing, 2014.

Shapiro, Judith. *Mao's War against Nature: Politics and the Environment in Revolutionary China*. Cambridge: Cambridge University Press, 2001.

Shen, Grace Yen. *Unearthing the Nation: Modern Geology and Nationalism in Republican China*. Chicago: University of Chicago Press, 2014.

Shen, Yubin. "Pneumonic Plagues, Environmental Changes, and the International Fur Trade: The Retreat of Tarbagan Marmots from Northwest Manchuria, 1900s–30s." *Frontiers of History in China* 14, no. 3 (November 2019): 291–322. https://doi.org/10.3868/s020–008–019–0016–1.

Shen Zhongyi (Sin Ch'ung-il). *Jianzhou jicheng tulu (Kŏnju kijŏng torok)* [Illustrated journey to Jianzhou]. Taipei, 1971.

Shepherdson-Scott, Kari. "Utopia/Dystopia: Japan's Image of the Manchurian Ideal." PhD diss., Duke University, 2012.

Shetler, Stanwyn G. *The Komarov Botanical Institute: 250 Years of Russian Research*. Washington, DC: Smithsonian Institution Press, 1967.

Shi Guangwei, and Liu Housheng. *Manzu saman tiaoshen yanjiu* [Research on Manchu shaman ritual]. Changchun: Jilin wenshi chubanshe, 1992.

Shih, James C. *Chinese Rural Society in Transition: A Case Study of the Lake Tai Area, 1368–1800*. Berkeley: Institute of East Asian Studies, University of California, Berkeley, 1992.

Shimazu, Naoko. *Japanese Society at War: Death, Memory and the Russo-Japanese War*. Cambridge: Cambridge University Press, 2011.

Shin, Ik-cheol. "Travel to Baekdusan and Its Significance during the Joseon Period." *Review of Korean Studies* 13, no. 4 (December 2010): 13–31.

Simonov, Eugene, and Thomas Dahmer, eds. *Amur-Heilong River Basin Reader*. Hong Kong: Ecosystems, 2008.

Simoons, Frederick. *Plants of Life, Plants of Death*. Madison: University of Wisconsin Press, 1998.

Slezkine, Yuri. *Arctic Mirrors: Russia and the Small Peoples of the North*. Ithaca, NY: Cornell University Press, 1994.

Sloan, Christopher P. "Feathers for T. Rex? New Birdlike Fossils Are Missing Links in Dinosaur Evolution." *National Geographic* 196, no. 5 (November 1999): 98–107.

Smil, Vaclav. *Energy Transitions: History, Requirements, Prospects*. Santa Barbara, CA: Praeger, 2010.

Smith, Aminda. *Thought Reform and China's Dangerous Classes: Reeducation, Resistance, and the People*. Lanham, MD: Rowman & Littlefield Publishers, 2012.

Smith, Andrew. *Guide to the Mammals of China*. Princeton, NJ: Princeton University Press, 2008. http://site.ebrary.com/lib/alltitles/docDetail.action?docID=10394779.

Smith, Mike. "Introduction and Bibliography." In *Arthur Smith Woodward: His Life and Influence on Modern Vertebrate Palaeontology*, 1–30. Geological Society, London, Special Publications, 2015.

Smith, Norman, ed. *Empire and Environment in the Making of Manchuria*. Vancouver: University of British Columbia Press, 2017.

Smith, Pamela. *The Body of the Artisan: Art and Experience in the Scientific Revolution*. Chicago: University of Chicago Press, 2004.

Smith, Richard Joseph. *Mapping China and Managing the World: Culture, Cartography and Cosmology in Late Imperial Times*. London: Routledge, 2013.

"Snapshots of Salmonella Serotypes | Salmonella Atlas | Reports and Publications | Salmonella | CDC." http://www.cdc.gov/salmonella/reportspubs/salmonella-atlas/serotype-snapshots.html.

Snyder-Reinke, Jeffrey. *Dry Spells: State Rainmaking and Local Governance in Late Imperial China*. Cambridge, MA: Harvard University Asia Center, 2009.

Sŏ Myŏng-ŭng. "Yu Paektusan gi" [Record of a journey to Paektusan]. In *Pomanje chip*, 7:17a–28b. Database of Korean Classics, 1838. http://db.itkc.or.kr/inLink?DCI=ITKC_MO_0540A_0090_010_0140_2005_A233_XML.

Song Heping. *Manzu saman shenge yizhu* [Manchu shaman spirit-songs, translated and annotated]. Beijing: Shehui kexue wenxian chubanshe, 1993.

Song Heping and Meng Huiying. *Manzu saman wenben yanjiu* [Research on Manchu shaman texts]. Taipei: Wunan tushu chuban youxian gongsi, 1997.

Song, Nianshen. "Imagined Territory: Paektusan in Late Chosŏn Maps and Writings." *Studies in the History of Gardens and Designed Landscapes* 37, no. 2 (2017): 157–73.

———. *Making Borders in Modern East Asia: The Tumen River Demarcation, 1881–1919*. Cambridge: Cambridge University Press, 2018.

Spence, Jonathan D. *Emperor of China: Self-Portrait of K'ang-Hsi*. New York: Vintage Books, 1974.

———. *To Change China: Western Advisors in China*. New York: Penguin, 2002.

Stary, Giovanni. "Il 'vero' esploratore del Changbaishan e il valore delle relative fonti: Un'analisi critica." In *Selected Manchu Studies: Contributions to History, Literature, and Shamanism of the Manchus*, edited by Giovanni Stary and Hartmut Walravens, 266–70. Berlin: Klaus Schwarz Verlag, 2013.

———. "Zue evolution der mandschurischen Übersetzungstechnik anhand einiger Beispiele aus dem 'Buch der Lieder.'" In *Ultra paludes Maeoticas: Zentralasienwissenschaftliche und linguistische Studien für Michael Weiers*, 157–76. Wiesbaden: Harrassowitz, 2006.

Stearn, William T. *Botanical Latin: History, Grammar, Syntax, Terminology, and Vocabulary*. Newton Abbot, UK: David & Charles, 1992.

Steinberg, John, ed. *The Russo-Japanese War in Global Perspective: World War Zero*. 2 vols. Leiden: Brill, 2005.

Steinberg, Ted. "Down to Earth: Nature, Agency, and Power in History." *American Historical Review* 107, no. 3 (June 2002): 798–820. https://doi.org/10.1086/532497.

Sterckx, Roel. *The Animal and the Daemon in Early China*. Albany: State University of New York Press, 2002.

Stevens, P. F. "Haüy and A.-P. Candolle: Crystallography, Botanical Systematics, and Comparative Morphology, 1780–1840." *Journal of the History of Biology* 17, no. 1 (Spring 1984): 49–82.

Stewart, Gordon T. "The Exploration of Central Asia." In *Reinterpreting Exploration: The West in the World*, edited by Dane Keith Kennedy, 195–213. Oxford: Oxford University Press, 2014.

Stilwell, Jeffrey D. "Trilobites and Linnaeus: The First Fossil Reconstruction from 1759." *Archives of Natural History* 33 (2006): 101–8.

Strang, Cameron B. *Frontiers of Science: Imperialism and Natural Knowledge in the Gulf South Borderlands, 1500–1850.* Williamsburg, VA: Omohundro Institute; Chapel Hill: University of North Carolina Press, 2018.

Strassberg, Richard E. *Inscribed Landscapes: Travel Writing from Imperial China.* Berkeley: University of California Press, 1994.

Suleski, Ronald Stanley. *Civil Government in Warlord China: Tradition, Modernization and Manchuria.* New York: P. Lang, 2002.

Summers, William C. *The Great Manchurian Plague of 1910–1911: The Geopolitics of an Epidemic Disease.* New Haven, CT: Yale University Press, 2012.

Sun, Xiaoping. "'War against the Earth': Military Farming in Communist Manchuria, 1949–75." In *Empire and Environment in the Making of Manchuria*, edited by Norman Smith, 263–90. Vancouver: University of British Columbia Press, 2017.

Sun Zhanwen. *Heilongjiang sheng shi tansuo* [Explorations in the history of Heilongjiang province]. Harbin: Heilongjiang renmin chubanshe, 1983.

Suzuki, Makoto. "Tracts of Japanese Geology: Geological Inspection during Russo-Japanese War." *GSJ Chishitsu News* 4, no. 5 (2015): 153–57.

Sweeten, Richard. "The Early Qing Imperial Tombs: From Hetu Ala to Beijing." In *Proceedings of the First North American Conference on Manchu Studies*, 1 (Studies in Manchu Literature and History), 63–104. Wiesbaden: Harrassowitz Verlag, 2007.

Swope, Kenneth M. *The Military Collapse of China's Ming Dynasty, 1618–44.* New York: Routledge, 2014.

Takashi Tsuchiya. "Imperial Japanese Experiments in China." In *The Oxford Textbook of Clinical Research Ethics*, edited by Ezekiel J. Emanuel, 31–45. Oxford: Oxford University Press, 2008.

Tamanoi, Mariko. *Memory Maps: The State and Manchuria in Postwar Japan.* Honolulu: University of Hawai'i Press, 2009.

Tammiksaar, Erki, and Ian R. Stone. "Alexander von Middendorff and His Expedition to Siberia (1842–1845)." *Polar Record* 43, no. 03 (July 2007): 193. https://doi.org/10.1017/S0032247407006407.

Tan Xichou. "Beipiao meitian dizhi" [The geology of Beipiao coal]. *Dizhi huibao (Bulletin of the Geological Survey of China)* 8 (1926): 20–29, 45–57.

———. "Liaoning Rehe jian ji Chao-Chi tiedao yanxian dizhi kuangchan" [The geological mining products along the Chaoyang-Chifeng rail line in Rehe, Liaoning] *Dizhi huibao* [Bulletin of the Geological Survey of China] 16 (1931): 39–82.

———. "On the Existence of the Cretaceous Coal Series in North China." *Bulletin of the Geological Society of China* 6, no. 1 (1927): 53–59.

Tang Gengsheng. "Zhu Yuanzhang san fan jiaxiang Fengyang kaoshu" [A study of Zhu Yuanzhang's three trips back to his hometown of Fengyang]. *Journal of Anhui Science and Technology University* 31, no. 2 (2017): 124–28.

Tang, Youcai. "Changbaishan Volcanism in Northeast China Linked to Subduction-

Induced Mantle Upwelling." *Nature Geoscience* 7, no. 6 (June 2014): 470–75. https://doi.org/10.1038/ngeo2166.

Tanner, Harold M. *Where Chiang Kai-Shek Lost China: The Liao-Shen Campaign, 1948.* Bloomington: Indiana University Press, 2015.

Teng, Emma Jinhua. *Taiwan's Imagined Geography: Chinese Colonial Travel Writing and Pictures, 1683–1895.* Cambridge, MA: Harvard University Asia Center, 2006.

Tie Yuxin, and Wang Peihuan. *Qing di dong xun* [The eastern tours of the Qing emperors]. Shenyang: Liaoning daxue chubanshe, 1991.

Tilley, Helen. *Africa as a Living Laboratory: Empire, Development, and the Problem of Scientific Knowledge, 1870–1950.* Chicago: University of Chicago Press, 2011.

———. "Ecologies of Complexity: Tropical Environments, African Trypanosomiasis, and the Science of Disease Control in British Colonial Africa, 1900–1940." *Osiris* 19 (2004): 21–38.

Tokunaga, Shigeyasu. *Report of the First Scientific Expedition to Manchoukuo under the Leadership of Shigeyasu Tokunaga, June-October 1933.* 25 vols. Tokyo: Waseda University, 1934.

Tollmachoff, I. "The Coal of Soochan and Its Importance in Pacific Trade." *Coal Age* 24, no. 14 (October 4, 1923): 509–14.

Toyoda Hidezō. *Toman to eisei* [Settling in Manchuria and (issues of) hygiene]. Tokyo: Sanseidō, 1933.

Toyoda Taro. *Manshū no iji eisei kotoni densenbyō* [Medical matters in Manchukuo, particularly regarding infectious diseases]. Fukuoka: Kyūshū Teikoku Daigaku igakubu gakuyūkai shuppanbu, 1935.

Tsuneishi Keiichi. *Hyōteki, Ishii: 731 Butai to Beigun chōhō katsudō* [Target, Ishii: Unit 731 and the activities of US military intelligence]. Tokyo: Ōtsuki shoten, 1984.

———. *Senjō no ekigaku* [Epidemiology of the battlefield]. Tokyo: Kaimeisha, 2005.

Tuan, Yi-Fu. *Space and Place: The Perspective of Experience.* Minneapolis: University of Minnesota Press, 2001.

Turczaninow, Nikolai Stepanovich. *Flora Baicalensi-Dahurica, seu descriptio plantarum in regionibus cis- et transbaicalensibus atque in Dahuria sponte nascentium.* Amsterdam: A. Asher, 1969.

Turnbull, David. "Travelling Knowledge: Narratives, Assemblages, Encounters." In *Instruments, Travel and Science: Itineraries of Precision from the Seventeenth to the Twentieth Century*, edited by Marie Noëlle Bourguet, Christian Licoppe, and H. Otto Sibum, 273–94. London: Routledge, 2002.

Twitchett, Denis, John King Fairbank, and Michael Loewe. *The Cambridge History of China.* Vol. 1, *The Ch'in and Han Empires, 221 BC–AD 220.* Cambridge: Cambridge University Press, 1986.

Unschuld, Paul, and Jinsheng Zheng. *Chinese Traditional Healing: The Berlin Collections of Manuscript Volumes from the 16th through the Early 20th Century.* Leiden: Brill, 2012.

Uruno Katsuya. *Manshū no chihōbyō to densenbyō.* Tokyo: Kainan shobō, 1943.

US Geological Survey. "Geologic Time: Index Fossils." https://pubs.usgs.gov/gip/geotime/fossils.html.

Verbiest, Ferdinand. "Journey into Tartary." In *History of the Two Tartar Conquerors of China, Including the Two Journeys into Tartary of Father Ferdinand Verbiest, in the Suite of the Emperor Kang-Hi: From the French of Père Pierre Joseph d' Orleans,* translated by the Earl of Ellesmere, 68–80, 103–20. London: Hakluyt Society, 1854.

Verhaeren, H. *Catalogue de la bibliotheque du Pei-t'ang.* Peking: Impr. des Lazaristes, 1949.

Vinkovetsky, Ilya. *Russian America: An Overseas Colony of a Continental Empire, 1804–1867.* Oxford: Oxford University Press, 2014.

Vollmer, John. *Dressed to Rule: 18th Century Court Attire in the Mactaggart Art Collection.* Edmonton: University of Alberta Press, 2007.

Vucinich, Alexander. *Science in Russian Culture.* Stanford, CA: Stanford University Press, 1963.

Wada Sei, ed. *Seikyō Kitsurin Kokuryūkō tōsho hyōchū senseki yozu ni tsuite (Shengjing Jilin Heilongjiang deng chu biao zhan ji yu tu)* [Annotated Map of Battle Sites in the Region of Shengjing, Jilin, and Heilongjiang]. Dairen: Hakkōnin Satō Shirō, 1935.

Wakeman, Frederic E. *The Great Enterprise: The Manchu Reconstruction of Imperial Order in Seventeenth-Century China.* Berkeley: University of California Press, 1985.

Waldron, Arthur. *The Great Wall of China: From History to Myth.* Cambridge: Cambridge University Press, 1992.

Wang Jiping. *Changbaishan zhi* [Gazetteer of Mount Changbai]. Changchun: Jilin wenshi chuban she, 1989.

Wang Liquan. *Zhonghua long wenhua de qiyuan yu yanbian* [The origins and evolution of China's dragon culture]. Beijing: Qixiang chubanshe, 2010.

Wang Mingqi, and Wang Qingxian. "Shenyang gugong Dazhengdian yu Chongzhengdian jianzhu ji chenshe kao" [Research on the buildings and furnishings of the Dazhengdian and Chongzhengdian palaces in Shenyang]. *Zhongguo bowuguan,* 1984.

Wang Mingxia. "Huifa bu de lishi yu saman wenhua" [The history of the Huifa tribe and its shamanistic culture]. *Jilin shifan daxue xuebao* 5 (October 2005): 92–96.

Wang Peihuan. "Kangxi dong xun shi shi gou bu" [The historical facts of Kangxi's eastern tour] *Lishi dang'an,* no. 1 (1987): 89–93.

Wang Shizhen. *Chibei outan* [Random chats from north of the pond]. Vol. 1. 2 vols. Beijing: Zhonghua shuju, 1982.

Wang Zhitian and Wen Zhixin. *Youyi nongchang shi: 1954–1984* [History of the Friendship farm, 1954–1984]. Youyi xian: Youyixian yinshuachang, 1985.

Ward, Rowena. "Delaying Repatriation: Japanese Technicians in Early Postwar China." *Japan Forum* 23, no. 4 (December 2011): 471–83.

Wei, Haiquan, Guoming Liu, and James Gill. "Review of Eruptive Activity at Tianchi Volcano, Changbaishan, Northeast China: Implications for Possible Future Eruptions." *Bulletin of Volcanology* 75, no. 4 (March 2013): 1–14. https://doi.org/10.1007/s00445-013-0706-5.

Wheatley, Paul. "The Ancient Chinese City as a Cosmological Symbol." *Ekistics* 39, no. 232 (March 1975): 147–58.

White, Richard. "Are You an Environmentalist, or Do You Work for a Living?" In *Uncommon Ground: Rethinking the Human Place in Nature*. New York: W. W. Norton, 1996.

———. *The Middle Ground: Indians, Empires, and Republics in the Great Lakes Region, 1650–1815*. Cambridge: Cambridge University Press, 1991.

———. *The Organic Machine: The Remaking of the Columbia River*. New York: Macmillan, 1995.

———. "The Spatial Turn in History: Spatial Humanities." http://spatial.scholarslab .org/spatial-turn/the-spatial-turn-in-history/index.html#_ftn25.

Whiteman, Stephen. "Kangxi's Auspicious Empire: Rhetorics of Geographic Integration in the Early Qing." In *Chinese History in Geographical Perspective*, edited by Yongtao Du and Jeff Kyong-McClain, 33–54. Lanham, MD: Lexington Books, 2013.

———. *Where Dragon Veins Meet: The Kangxi Emperor and His Estate at Rehe*. Seattle: University of Washington Press, 2020.

Wilford, John Noble. "Feathered Dinosaur Fossils Are Unearthed in China." *New York Times*, April 25, 2001, sec. Science.

Wilkinson, Endymion Porter. *Chinese History: A Manual*. Cambridge, MA: Harvard University Asia Center, 2000.

Willcox, Merlin, Gerard Bodeker, and Elisabeth Hsu. "Artemisia Annua as a Traditional Herbal Antimalarial." In *Traditional Medicinal Plants and Malaria*, 48–68. Boca Raton, FL: CRC Press, 2004.

Williams, Nicholas Morrow. "The Pity of Spring: A Southern Topos Reimagined by Wang Bo and Li Bai." In *Southern Identity and Southern Estrangement in Medieval Chinese Poetry*, edited by Ping Wang and Nicholas Morrow Williams, 137–64. Hong Kong University Press, 2015.

Williams, Peter, and David Wallace. *Unit 731: Japan's Secret Biological Warfare in World War II*. New York: Free Press, 1989.

Williams, Raymond. *The Country and the City*. Oxford: Oxford University Press, 1973.

Witek, John W., ed. *Ferdinand Verbiest*. London: Routledge, 1994.

Won, Changman, and Kimberly G. Smith. "History and Current Status of Mammals of the Korean Peninsula." *Mammal Review* 29, no. 1 (March 1999): 3–36. https:// doi.org/10.1046/j.1365–2907.1999.00034.x.

Woodward, Arthur Smith. *Catalogue of the Fossil Fishes in the British Museum (Natural History)*. London: Printed by Order of the Trustees, 1889.

World Wildlife Federation. "Mongolian-Manchurian Grassland." http://www.eoearth .org/view/article/154670/.

Worster, Donald. *Dust Bowl: The Southern Plains in the 1930s*. 25th anniversary ed. New York: Oxford University Press, 2004.

———. "Transformations of the Earth: Toward an Agroecological Perspective in History." *Journal of American History* 76, no. 4 (March 1990): 1087. https://doi.org/ 10.2307/2936586.

Wright, Colin W. *Artemisia*. London: Taylor & Francis, 2001.

Wu, Hung. *A Story of Ruins: Presence and Absence in Chinese Art and Visual Culture*. London: Reaktion Books, 2013.

Wu, Lien-teh. *Plague Fighter; the Autobiography of a Modern Chinese Physician*. Cambridge: W. Heffer, 1959.

———. "The Second Pneumonic Plague Epidemic in Manchuria, 1920–21: A General Survey of the Outbreak and Its Course (with Map)." *Journal of Hygiene* 21, no. 3 (May 1923): 262–88.

Wu, Lien-teh, J. W. H. Chun, and R. Pollitzer. "Clinical Observations upon the Manchurian Plague Epidemic, 1920–21." *Journal of Hygiene* 21, no. 3 (May 1923): 298–306.

Wu Lien-teh and Dongbei fang yi chu. *North Manchurian Plague Prevention Service Reports (1911–1913)*. Cambridge: Cambridge University Press, 1914.

Wu Lien-teh, Dongbei fang yi chu, and China National Quarantine Service. *Manchurian Plague Prevention Service Memorial Volume: 1912–1932*. Edited by Wu Lien-teh. Shanghai: National Quarantine Service, 1934.

Wu, Lien-teh, C. W. Han, and R. Pollitzer. "Plague in Manchuria: I. Observations Made during and after the Second Manchurian Plague Epidemic of 1920–21. II. The Rôle of the Tarabagan in the Epidemiology of Plague." *Journal of Hygiene* 21, no. 3 (May 1923): 307–58.

Wu, Lien-teh, R. Pollitzer, and Chia-swee Lin. "Studies Upon the Plague Situation in North China." *National Medical Journal of China* 15, no. 3 (June 1929): 274–410.

Wu, Shellen Xiao. *Empires of Coal: Fueling China's Entry into the Modern World Order, 1860–1920*. Stanford, CA: Stanford University Press, 2015.

Wu Weiye. "Beige zeng Wu Jizi" [A lament penned for Wu Jizi]. In *Wu Meicun quanji*, edited by Li Xueyin, 257. Shanghai: Shanghai guji chubanshe, 1990.

Wu Zhaoqian. *Qiu jia ji* [Collected works of the barbarian flute in autumn]. Vol. 8, *Yueyatang congshu*. Annotated by Tan Ying. Taipei: Hualian chubanshe, 1965.

———. *Qiu jia ji* [Collected works of the barbarian flute in autumn]. Annotated by Ma Shouzong. Shanghai: Shanghai guji chubanshe, 1993.

———. *Qiu jia ji, Guilai caotang chidu* [Collected works of the barbarian flute in autumn, and letters from the thatched cottage of returning]. Harbin: Heilongjiang daxue chubanshe, 2010.

Xie Xueshi and Matsumura Takao. "Xinjing shuyi moulüe—1940" [The Xinjing plague strategy—1940]. In *Zhanzheng yu eyi: 731 budui zuixing kao*, 58–122. Beijing: Renmin chubanshe, 1998.

———, eds. *Zhanzheng yu eyi: 731 budui zuixing kao* [War and the evil epidemic: Research into the crimes of Unit 731]. Beijing: Renmin chubanshe, 1998.

Xin Peilin, Zhang Fengming, and Gao Xiaoyan, eds. *Heilongjiang kaifa shi* [History of the opening of Heilongjiang]. Harbin: Heilongjiang renmin chubanshe, 1999.

Xiong Yi. *Zhongguo heliu* [The rivers of China]. Beijing: Renmin jiaoyu chubanshe, 1991.

Xiqing. *Heilongjiang wai ji* [Unofficial record of Heilongjiang]. Harbin: Heilongjiang renmin chubanshe, 1984.

Xu, Bin. *Chairman Mao's Children: Generation and the Politics of Memory in China.* Cambridge: Cambridge University Press, 2021.

Xu, Haigen, GuangQing Zhu, Lianlong Wang, and Haoshen Bao. "Design of Nature Reserve System for Red-Crowned Crane in China." *Biodiversity & Conservation* 14, no. 10 (September 2005): 2275–89. https://doi.org/10.1007/s10531-004-1663-2.

Xu, Jiandong. "Climatic Impact of the Millennium Eruption of Changbaishan Volcano in China: New Insights from High-Precision Radiocarbon Wiggle-Match Dating." *Geophysical Research Letters* 40, no. 1 (January 2013): 54–59. https://doi.org/10.1029/2012GL054246.

Xu Shuming. "Qingdai dongbei diqu tudi kaiken shulüe" [Sketch of land reclamation in the northeast during the Qing era]. *Qingdai bianjiang kaifa yanjiu* (1990): 87–122.

Xu Xin. *Nu'erhachi ling ji Qing zu ling lishi zhi mi* [Secrets of the history of the tombs of Nurhaci and his ancestors]. Shenyang: Liaoning renmin chubanshe, 2016.

Yajima, Michiko. "Japanese Wartime Geology: A Case Study in Northeast China." *Historia Scientiarum* 15 (2006): 222–32.

Yamada Otozō (defendant). *Materials on the Trial of Former Servicemen of the Japanese Army Charged with Manufacturing and Employing Bacteriological Weapons.* Moscow: Foreign Languges Publishing House, 1950.

Yamagata Miyuki. *Manshū no yaseichō* [The wild birds of Manchuria]. Tokyo: Shito kayō no sha, 1942.

Yan Wu. "Long zai zijincheng" [Dragons in the forbidden city]. *Zijincheng*, no. 1 (January 1988): 8–9, 43.

Yang, Daqing. *Technology of Empire: Telecommunications and Japanese Expansion in Asia, 1883–1945.* Cambridge, MA: Harvard University Asia Center, 2010.

Yang, Lihui, Deming An, and Jessica Anderson Turner. *Handbook of Chinese Mythology.* Oxford: Oxford University Press, 2008.

Yang Ruisong. *Bingfu, huanghuo yu shuishi: "Xifang" shiye de Zhongguo xingxiang yu jindai Zhongguo guozu lunshu xiangxiang* [Sick man, yellow peril, and sleeping lion: The "West's" image of China and modern China's discursive imaginary of the Chinese people]. Taipei: Chengchi University Press, 2010.

Yang Yanjun. "Guanyu 731 budui shuyi baogaoshu de chubu jiedu" [Preliminary interpretation of the plague reports of Unit 731]. *Yixue yu zhexue* 34, no. 11 (June 2013): 87–89, 96.

Yi Sŏ-haeng and Chŏng Ch'i-yŏng, eds. *Ko chido wa sajin ŭro pon Paektusan* [Paektusan as seen in old maps and photos]. Sŏngnam: Han'gukhak Chungang Yŏn'guwŏn Ch'ulp'anbu, 2011.

Yi T'ae-jin, and Yi Sang-t'ae, eds. *Kwanbuk ŭpchi—Kapsanpu ŭpchi* [Gazetteers of the Kwanbuk region—Kapsan prefecture gazetteer]. Vol. Hamgyŏng-do 6. Seoul: Han'guk Inmun Kwahagwŏn, 1990.

Yim, Lawrence C. H. *The Poet-Historian Qian Qianyi.* London: Routledge, 2009.

———. "Traumatic Memory, Literature and Religion in Wu Zhaoqian's Early Exile." *Zhongguo wenzhe yanjiu jikan* 27 (September 2005): 123–65.

Yinghe. *Bukui ji lüe* [A record of Bukui (Qiqihar)]. Vol. 2. *Xiaofanghuzhai yudi cong-chao.* Taipei: Guangwen shuju, 1962.

Yosano, Akiko. *Travels in Manchuria and Mongolia: A Feminist Poet from Japan Encounters Prewar China.* Translated by Joshua Fogel. New York: Columbia University Press, 2001.

Young, John. *The Research Activities of the South Manchurian Railway Company, 1907–1945: A History and Bibliography.* New York: East Asian Institute, Columbia University, 1966.

Young, Louise. *Japan's Total Empire: Manchuria and the Culture of Wartime Imperialism.* Berkeley: University of California Press, 1999.

Yu Changqin, and Han Zhifei. *Guoying Youyi nongchang fangwen ji* [A visit to the Friendship farm]. Beijing: Tongsu duwu chubanshe, 1956.

Zatsepine, Victor. *Beyond the Amur: Frontier Encounters between China and Russia, 1850–1930.* Vancouver: University of British Columbia Press, 2017.

Zhang, Baichun. "The Introduction of European Astronomical Instruments and the Related Technology into China during the Seventeenth Century." *East Asian Science, Technology, and Medicine,* no. 20 (2003): 99–131.

Zhang Boying. *Heilongjiang zhi gao* [A draft gazetteer of Heilongjiang]. Harbin: Heilongjiang renmin chubanshe, 1992.

Zhang Fuyou. *Xunfang Eheneyin: Manjiang wenshi kaochaji* [In search of Eheneyin: Research into the history and culture of the Man River]. Changchun: Jilin wenshi chubanshe, 2015.

Zhang Jinyan. *Ningguta shanshui ji, yu wai ji* [A record of the landscape of Ningguta (and) A record of the land beyond the pale]. Edited by Li Xingsheng. Harbin: Heilongjiang renmin chubanshe, 1984.

Zhang, Lijuan. "Spatially Precise Reconstruction of Cropland Areas in Heilongjiang Province, Northeast China during 1900–1910." *Journal of Geographical Sciences* 25, no. 5 (June 2015): 592–602.

Zhang, Qiong. "From 'Dragonology' to 'Meteorology': Aristotelian Natural Philosophy and the Beginning of the Decline of the Dragon in China." *Early Science and Medicine* 14, no. 1/3 (2009): 340–68.

Zhang Wenli. "Riwei zai dongbei diqu chuangjian de bowuguan ji qi pingjia" [An evaluation of museums established by the Japanese puppet government in the northeastern region]. *Zhongguo bowuguan,* no. 4 (1992): 89–94.

Zhang, Xuezhen. "Vegetation of Northeast China during the Late Seventeenth to Early Twentieth Century as Revealed by Historical Documents." *Regional Environmental Change* 11, no. 4 (December 2011): 869–82. https://doi.org/10.1007/s10113-011-0224-y.

Zhang, Yong. *Shenyang gugong jianzhu zhuangshi yanjiu* [Studies in the decoration of the Shenyang palace buildings] Nanjing: Dongnan daxue chubanshe, 2010.

Zhang Yuliang. *Guoying Youyi nongchang de zhiwu diaocha* [Survey of the plants of the Friendship farm]. Beijing: Kexue chubanshe, 1956.

Zhang Yuxing, ed. *Qingdai dongbei liuren shi xuanzhu* [A selection of poems by exiles to the northeast during the Qing period]. Shenyang: Liaoshen shushe, 1988.

Zheng Jiazhen. *Zhandou zai Beidahuang: Mudanjiang qingnian kenhuangdui de gushi* [Struggle in the Great Northern Wasteland: Stories of the Mudanjiang youth land reclamation corps]. Shanghai: Shaonian ertong chubanshe, 1964.

———. *Zhongguo dongbei jiao: Beidahuang liushinian (1947–2007)* [The northeast corner of China: Sixty years in the Great Northern Wasteland (1947–2007)]. Harbin: Heilongjiang renmin chubanshe, 2007.

Zhongguo renmin zhengzhi xieshang huiyi, Heilongjiang sheng wenshi he xuexi weiyuanhui. *Zhishi qingnian zai Heilongjiang* [Educated youth in Heilongjiang]. 2 vols. Harbin: Heilongjiang renmin chubanshe, 2005.

Zhou Jingnan. "Zijincheng li de long wenzhuang shi" [History of dragon patterns in the forbidden city]. *Jiaju*, no. 2 (March 2012): 50–56.

Zhou Yuling, *Heitushang de Zhongguo* [Black-earth China]. Harbin: Heilongjiang renmin chubanshe, 2014.

Zhu, Zhenhua. "Using Toponyms to Analyze the Endangered Manchu Language in Northeast China." *Sustainability* 10, no. 2 (February 2018): 563. https://doi.org/10.3390/su10020563.

Zizzamia, Daniel. "Restoring the Paleo-West: Fossils, Coal, and Climate in Late Nineteenth-Century America." *Environmental History* 24, no. 1 (January 2019): 130–56. https://doi.org/10.1093/envhis/emy092.

INDEX